沼气工程与技术
Biogas Engineering and Application

董仁杰　[奥]伯恩哈特·蓝宁阁　主编
Edited by: Prof. Dr. DONG Renjie
Prof. Dr. Bernhard Raninger

第2卷
Volume 2

- **Papers of the Sino-German GIZ Biomass Utilization Project of MOA in China**
  德国国际合作机构与中国农业部合作的中德生物质能优化利用项目论文集
- **Proceedings of 2011 Forum of Renewable Energy Promotion in Developing Countries**
  发展中国家可再生能源发展论文集

中国农业大学出版社
·北京·

图书在版编目(CIP)数据

沼气工程与技术:第2卷/董仁杰,[奥]伯恩哈特·蓝宁阁主编.—北京:中国农业大学出版社,2012.3
ISBN 978-7-5655-0497-6

Ⅰ.①沼… Ⅱ.①董… ②伯… Ⅲ.①沼气池-工程技术-国际学术会议-文集 Ⅳ.①S216.4-53

中国版本图书馆 CIP 数据核字(2012)第 029445 号

书　　名 沼气工程与技术(第2卷)
作　　者 董仁杰 [奥]伯恩哈特·蓝宁阁 主编

责任编辑 季 方 邓 旭 张秀环　　**责任校对** 陈 莹 王晓凤
封面设计 季 方 秦怡思 郑 川
出版发行 中国农业大学出版社
社　　址 北京市海淀区圆明园西路2号　　**邮政编码** 100193
电　　话 发行部 010-62818525,8625　　**读者服务部** 010-62732336
编辑部 010-62732617,2618　　**出 版 部** 010-62733440
网　　址 http://www.cau.edu.cn/caup　　**E-mail** cbsszs@cau.edu.cn
经　　销 新华书店
印　　刷 涿州市星河印刷有限公司
版　　次 2012年3月第1版　2012年3月第1次印刷
规　　格 889×1 194　16开本　21.5印张　680千字
定　　价 150.00元

PREFACE

After the so-called 19th century of 'coal' and the 20th century of 'oil', the 21st century would be probably a century of 'gas'. Carbon neutral gases from renewable sources will have priority to play an important role in energy supply, in order to tackle fossil carbon emissions, which is another challenge and threat to our planet's ecological condition to survive.

The power sector is one of the main carbon emitters and therefore Germany intends to provide in the year 2050 about 60 percent of its energy from renewable sources, from which again 60% will derive from biomass. The natural gas demand in Germany should be covered in 2030 by at least 10 billion m^3 or by 11.5% through biogas and the electric power supply will derive in 2050 by 100 percent from renewable sources with a potential of at least 12% up to 20% deriving from biogas. The transport shall be fuelled in the future by electricity for small cars for mainly local traffic and trains and with biogas for heavy duty vehicles and far distance ground transportation. Bio-methane from biogas can easily be stored in the national gas grids and storage systems and it will turn out, as one of the most likely options, to solve the issue of electric power storage, derived from surplus wind and solar power, at times when the production exceeds the demand, through producing hydrogen through electrolysis and to convert this hydrogen into bio-methane. By this way bio-methane can be produced at biogas plants even without additional purification steps.

Biogas is basically a base-load energy supply regime and available when wind or solar is not available or not sufficient. In the future biogas can be produced more predictable on demand for peak load coverage, to be available when the power market prices are high, without compromising on specific performances. Biogas for peak load coverage can be achieved on one hand by using the before mentioned natural gas storage options, but as well by producing biogas or bio-methane on demand through advanced biogas technologies.

In Germany the potential to extend biogas is somewhere between 2-3 times of the current capacities and only in combination with a strict energy saving target, which anticipates to reduce energy consumption by 50% till 2050 will help to keep biogas playing a prominent role in the renewable energy mix. In China, as the development of grid connected biogas is still at the beginning, the potential to develop biogas is still much higher. Chinas biogas potential is at least 20 times that of Germany and as the energy produced from Chinese biogas plants is currently only about 10% of that in Germany, the extension factor in China will about at least

200 times. But reliable and matured technologies and suitable business and market conditions are required to make biogas a competitive business.

In the future the agricultural sector will take more and more over an important role in energy provision, with an expected overall agricultural market share of 11%. China can use this opportunity as well to strengthen agribusinesses and to further support the socio-economy in rural areas.

Highly required new waste management solutions in China growing urban areas to catch up the demand of waste disposal and low carbon society, where biogas from the bioorganic waste fraction will and must play a key function, and the growing understanding that bio-ethanol as liquid transport bio-fuel is also not the best possible economic and sustainable solution and the ecological threats from nuclear power, experienced once more in Fukushima this year, will propel further the biogas sector. The Chinese Government is taking action to support the biogas sector framework conditions, inter alia by settling further close cooperation with the German government to jointly develop biogas from domestic to international performance.

The 2nd volume of Biogas Engineering and Application is reflecting this development and we wish all readers the insights they expected.

The Chief Editors

Prof. Dr. DONG Renjie
China Agricultural University (CAU)
Bio Energy Engineering and Low Carbon Technology Laboratory
Biomass Engineering Center

Prof. Dr. phil.-habil. Bernhard Raninger
Shenyang Aeronautical University (SAU)
Institute for Clean Energy and Environmental Engineering
GIZ, Sino-German Biomass Utilization Project

Beijing, December 2011

前　言

如果说十九世纪是“煤炭”时代，20 世纪是“燃油”时代，那么 21 世纪或将是“燃气”时代。21 世纪以来，化石燃料燃烧导致的碳排放严重威胁着地球的生态环境，由可再生资源生产的碳中性燃气将得到优先重视。

能源领域是主要的碳排放源之一。以德国为例，计划到 2050 年，可再生能源将占国内能源产量的 60%，且其中 60%源于生物质。到 2030 年，天然气需求的 11.5%（即 1×10^{10} m^3）将由沼气代替。到 2050 年，可再生资源将提供德国全部的电能，其中沼气发电将达到电力生产总量的 12%～20%。未来小型运输工具如小轿车，当地主要交通工具和火车的燃料将由电能所代替；大型交通工具和远距离地面运输将使用沼气为燃料。沼气所产生的生物甲烷并入国家天然气管网或储存于燃气贮存系统中，这也为电力的存储提供了另一种解决方法。当电力产量大于需求量时，风能与太阳能所产电能可以用于电解产氢，再将氢气转化为生物甲烷；使用这种方法，沼气厂甚至可以不需要额外的提纯工艺便可生产生物甲烷。

沼气主要是一种基本负荷型能源，在风能或太阳能不足时可以供给能源。未来沼气能够根据用能需求定量生产，而且能够在能源市场价格较高时生产。沼气能够通过上述天然气储存方法实现用能高峰期补充能源，也能够通过先进的沼气技术生产沼气或生物甲烷来定时补充能源。

德国沼气潜力可以是目前沼气产量的 2～3 倍；只有在严格遵循节能目标，即在 2050 年减少 50%能耗的前提下，才能使沼气在可再生能源中扮演重要的角色。在中国，沼气并网尚处于初级阶段，因此有着巨大的发展潜力。中国沼气潜力约为德国的 20 倍，但是目前沼气产量只有德国的 10%，这就意味着中国沼气的增长空间可以达到目前的 200 倍。为了使沼气工程成为一种具有竞争力的产业，中国需要发展成熟可靠的沼气技术并完善合理的商业与市场环境。

农业在未来能源供给中所扮演的地位将越来越重要，预计占整个农业市场份额的 11%。中国可以借此契机加强农村商业发展并进一步巩固农村经济。

中国急需能够满足废弃物处理需求的新型废弃物管理方案以实现向低碳社会转型，而利用生活垃圾生产沼气则是实现该目标的最佳选择。人们已经逐渐了解到燃料乙醇作为液体生物燃料或许不是最经济可行的可持续方案，而今年日本所经历的福岛核泄漏事故，也使人们认识到核能对生态环境有着极大的威胁，进一步发展沼气技术成为时代的需要。中国政府制定了各种政策来支持沼气工程的发展，并通过与德国政府的合作，共同推进沼气技术，并使之逐步走向国际市场。

沼气工程与技术系列丛书第二卷将展示上述成果，希望能满足各位读者的需求。

董仁杰 博士　教授
中国农业大学生物质能工程与低碳技术研究室
中国农业大学生物质工程中心常务副主任

伯恩哈特·蓝宁阁 博士　教授
沈阳航空航天大学清洁能源和环境工程研究所
德国 GIZ 中德生物质能优化利用项目主任

2011 年 12 月于北京

TABLE OF CONTENT

目　录

BIOGAS TO GRID IN CHINA-CHALLENGES AND OPPORTUNITIES OF A NEW MARKET FOR INDUSTRIAL LARGE SCALE BIOGAS PLANTS

B. Raninger[1], DONG Renjie[2], QIAN Mingyu[1]

1. Deutsche Gesellschaft für Internationale Zusammenarbeit (GIZ) GmbH, SINO-German Project of Optimization Biomass Utilization; 2. Energy Engineering and Low Carbon Technology Laboratory of China Agricultural University (EELC)

Contact: Prof. Dr.-habil Bernhard Raninger, GIZ, Project Director, FECC of Ministry of Agriculture No. 55 Nongzhan Beilu, Chaoyang, Beijing 100125, PRC, Tel.: +86-10-6508-7080-200, E-Mail: Bernhard.Raninger@giz.de, www.biogas-china.org;

Prof. Dr. DONG Renjie, Biomass Engineering Centre of China Agricultural University (BECAU), Energy Engineering and Low Carbon Technology Laboratory of China Agricultural University (EELC), Mail Box 184, No. 17 Qinghuadonglu, Haidian District, Beijing 100083, P. R. China, E-Mail: rjdong@cau.edu.cn

EXECUTIVE SUMMARY

By setting the target to reduce the intensity of CO_2 emissions per GDP by 40% to 45%, compared to 2005, till 2020 China has committed itself to a stringent climate policy. At the same time fossil fuels are getting more expensive. This drives the development to generate 15% renewable energy in 2020 by inter alia using the energetically potential of China's abundant biomass resources. They derive mainly from large-scale livestock breeding, agro-industries, forests, residential areas as bioorganic municipal waste, from straw and even potentially from energy crops grown on marginal land (considering 'no food for energy'!). Biomass has the potential to replace one third of China's coal consumption in 2050, which is equivalent to one billion tons of standard coal. Biogas from large-scale biogas plants shall contribute at least three Giga Watts grid-connected energy in 2020 (NDRC, 2006). The actual contribution of middle and large scale biogas plants to energy generation is very limited. Therefore 'international best practice' technology shall support the change from environmental protection driven to rather power producing industrial scale biogas plants, wherever condition permits. The SINO-German Project of Optimization of Biomass Utilization, supported by the German Ministry for International Cooperation (BMZ) and implemented by the 'Deutsche Gesellschaft fuer Internationale Zusammenarbeit (GIZ)' together with the Foreign Economic Cooperation Center (FECC) of the Chinese Ministry of Agriculture (MOA) aims to provide technical support to design and operate large scale biogas plants financed through an ADB loan, according to best practice and to create demonstration cases which landmark the future development in this sector. International equipment suppliers will have, in the course of an improved technical standard, increasingly more opportunities to get involved.

Key words: China biogas sector; biogas technology; biogas plant performance; biogas utilization; Sino-German biogas cooperation

1 Biogas development in China

China is well known for its rural biogas utilization program. Since the 50ies more than 35 million household scale biogas digesters with an estimated annual biogas output of 12 billion m^3 were installed. The development target is 80 million units by 2020, including the climatically colder regions in the northern parts of China.

The small, middle and large scale biogas plants construction has developed since the early 1990s, mainly using Up-flow Sludge (USR), Up-flow Anaerobic Sludge Blanket (UASB) and simple High Concentration Flow (HCF) reactors. This environmentally friendly treatment is suitable to treat the liquid effluent from animal husbandries and agro industries, such as breweries, alcohol production factories, and slaughterhouses. Biogas was at that time more a by-product, which was used for self supply or local biogas supply to neighboring households. In 2009 about 34 000 small-scale biogas plants and 22 900 medium and large scale biogas plants (MLBGPs, Fermenter > 50 m^3), with 3 717 large scale installations (Fermenter >300 m^3) included (*Figure 1*). A biogas-driven power generation capacity of 42 MW_{el} was installed on biogas plants sites by 2009.

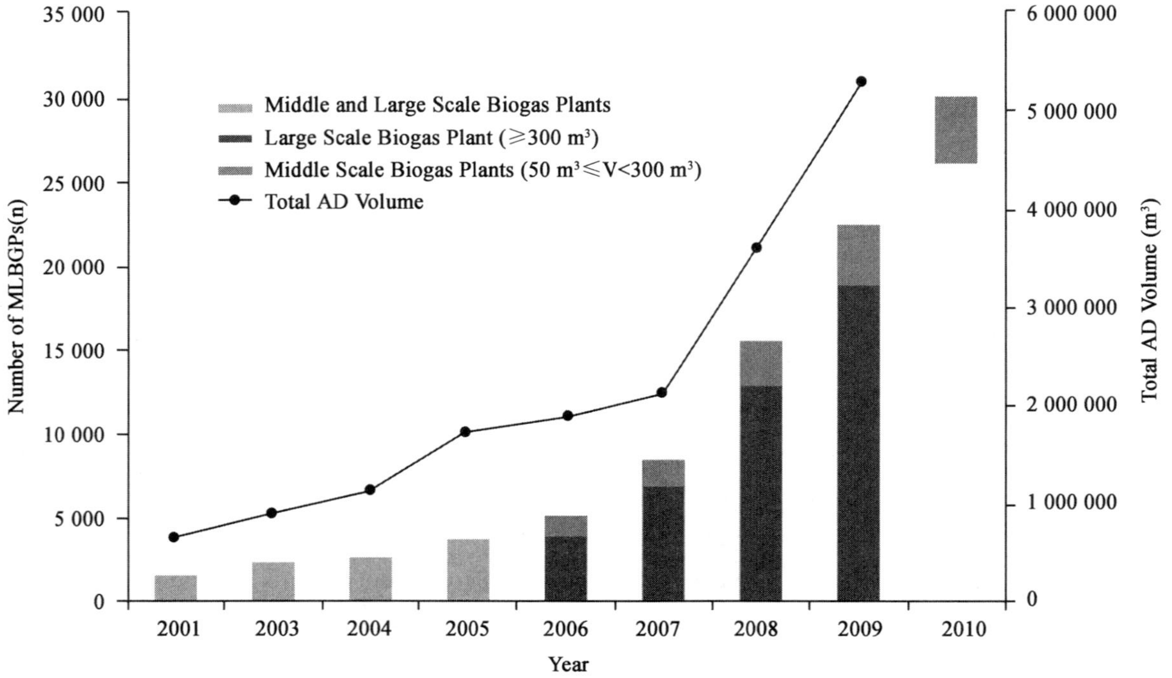

Figure 1 Middle and large scale biogas plants developed in China, number of plants and fermenter volume 2001—2010 (China Biogas Association, BIOMA and NDRC, 2006—2020). NDRC-Chinese National Development and Reform Committee, ministerial authority in the government.

Biogas plants built under the aspect of the overall energy generation are fairly new. At waste water treatment plant, biogas as a by-product was usually not collected and used. So far the self-use or the supply to local biogas grids was realized, but due to lack of temperature control very commonly the biogas plants are just effectively run in the warm seasons. The Chinese Ministry of Agriculture Asian Development Bank 'Integrated Renewable Biomass Energy Development Sector Project', should be supportive to the improvement of biogas plants management and to a much proper and reasonable development of new power generation oriented MLBGPs in China.

2 Political and financial supports

The policy drivers to support biogas generation in China are quite similar to Germany (see *I*). The awareness, that biogas plants are not only contributing to environmental protection, but seriously can supply energy, was promoted after the China Renewable Energy Law (NDRC, 2006) was enacted.

Figure 2 Qionglai Jinli pig farm near Chengdu, Sichuan Province, a traditional USR Biogas plant for the liquid farm effluent operates a small 30 kW_{el} gas engine, the effluent is treated as waste water, built in 2009.

Renewable energies derived from wind turbines, hydropower, solar, geothermic, and biomass (solid and liquid bio-fuels and biogas) shall reach 15% in the overall primary energy mix in 2020.

Ⅰ. Main Biogas Policy Drivers in China

- Environmental Protection (waste water pollution; ground- and surface waters, and soil protection);
- Renewable Energy (substitution of fossil fuels, sustainable energy resources development);
- Circular Economy (Organic fertilizer);
- Greenhouse Gases Mitigation (Kyoto Protocol 2002, Copenhagen 2009, low carbon economy);
- Rural Development (agro-industrialization, farmers income, diminishing urbanization).

During the last 10 years, climate change mitigation emerged as an additional driver to biogas plant construction. In the course of China's Greenhouse Gases emission reduction policy, China signed the Kyoto Protocol in 2002. During the 11th Five Year Plan, the Chinese government had reduced CO_2 emission per unit GDP by 20%. A total investment of about CNY 2 trillion (240 billion€) had been made to achieve this goal (NDRC, 2010). In the course of the Copenhagen UN Climate Change Conference 2009, the 2020 voluntary target, to reduce the intensity of CO_2 emissions per GDP by 40% to 45%, compared to 2005, was announced by China. China has recognized already a negative impact from climate change on agriculture and infrastructure, and has recognized the importance of international efforts for preventing global climate change.

To support the biogas sector and the development of biogas plants in China, there are various systems and mechanisms. Till 2010 MOA invested more than CNY 24 billion (2.8 billion€) into biogas construction. During the last 3 years, the average annual investment support was CNY 5-6 billion. Per plant the central government subsidy varied between 25% and 45% of the total project investment, similar investment comes from the provincial and municipal governments, based on an application and evaluation process; the optimistic situation for a biogas plant owner is he just needed to cover about 15% of the total cost. Some important supports for large scale biogas plants constructions came in types of international loans, donations, and businesses. International loans used to be from ADB and World Bank. Donations were from the Global Environmental Facility, UNDP and bilateral cooperation partner countries such as the Netherlands, Sweden, Denmark, Finland, and Germany (BMZ, GIZ, CIM, BMBF, Robert Bosch Stiftung). Business cooperation was conducted with the United States, Canada, Japan and so on. The specific Investment costs per metric ton of feedstock are between 30 € and 60 € Per kW installed power capacity the Investment costs of CSTR biogas plants in the megawatts range of size, are between, depending on plant size and type of feedstock between 1 600 €/kW and 3 300 €/kW.

Figure 3 Biogas plant Deqingyuan in Yanqing Beijing, UNDP, GEF, USAID demonstration CSTR technology, the biogas from 2. 2 million layers drives 2 GE Jenbacher gas engines with 2. 4 MW_{el}, grid connected, operation since 2007, CDM since 2009.

Biogas electricity generation in China is not popular because of year-round biogas availability and power installation size requirement. Only when electricity generation capacity exceeds 500 kW (there are exceptions locally) will be an obligation to have access to the grid and then can benefit from the financial supporting policy. There however is one successful case in China. The biogas plant in Deqingyuan is to produce electricity for grid connection, receiving feed-in tariff support according to the Renewable Energy Law, with a top up of at least 0. 25 CNY /kWh (0. 03 €/kWh) to the local coal power generation price (*Table 1*).

Carbon trading revenues from the UNFCC Clean Development Mechanism (CDM) would ensure the financial viability of large scale biogas plants (PPTA, 2009). Large scale biogas plants can reach an amount of between 40 000 and 100 000 t CO_{2eqv}. Certified Emission Reduction (CERs) (UNFCCC, PDDs). Since 2008 NDRC requests for at least 7 €/t CO_{2eq}. and depending on the quality of the projects and the market situation 14 € and more can be achieved. But so far, due to the complexity and costs of registration and monitoring and due to the low performance of most current plants, Biogas CDM projects developers have not make sufficient use of this possibility in China.

Table 1 Grid feed-in tariff and tax privilege for biogas and-mass power (status 2010)

<table>
<tr><th></th><th>Power benchmark tariff</th><th>Power subsidy (obligatory for grid companies >500 kW)</th><th>Grid connection subsidy (to grid comp.)</th><th>Tax concession</th></tr>
<tr><td>Livestock- & poultry waste</td><td>Provincial price of desulfurized coal power in 2005 (app. 0. 45 CNY/ kWh*)</td><td>0. 25 CNY/kWh, Since 2010, new projects get 2% decrease than the last year projects. Duration 15years.</td><td rowspan="2">0. 01 CNY/kWh (<50 km)
0. 02 CNY/kWh (50-100 km)
0. 03 CNY/kWh (>100 km)</td><td>No income tax (first three years), 50% income tax (second three years).</td></tr>
<tr><td>Agro- & forestry-biomass waste</td><td colspan="2">a. 0. 75 CNY/kWh (incl. tax)
b. For the approved project or tendering project, the feed-in tariff required approval</td><td>If 70% of feedstock is crop straw, husk and /or corn crop, 10% income is tax free.</td></tr>
</table>

* This tariff was increased by the Chinese Government by 0.025 CNY/kWh in December 2011.

3 Potential of large-scale biogas plant development

According to MOA in 2010 there were 4.2 million commercial livestock and poultry farms holding livestock equivalent to 610 million pigs, which produced about 3 billion tons of manure that year. Amongst about 1.1 billion tons was available as feedstock for biogas plants with a potential biogas production of 67 billion m^3. In addition, 90 million of pigs, 16 million of cattle, 85 million of chickens and 26 million of sheep were raised by 140 million small farms in the suburbs of the metropolis cities. It is estimated to have an additional annual biogas production potential of 54 billion m^3. Those distributed small farms can either hold their own household digesters or put the manure treated in centralized co-digestion biogas plants (see GEF Component of the ADB Project).

NDRC anticipates in the 'Medium and Long Term Program of Renewable Energy Development Plan 2006-2020' that 10 000 MLBGPs, at mainly animal husbandries, and 6 000 MLBGPs processing industrial organic wastewater, with an overall annual biogas yield of 14 billion m^3 and an installed electricity capacity of 3 GW by 2020 will be in place. As in 2010 about only 15% of the biogas electricity generation capacity (3 GW) was installed (but not fully operated), the biogas power generation has a big step to go. For the total energy potential from biomass in 2050, see *Table 2*.

Table 2 Potential of Bio-energy till 2050 (1 billion standard coal equivalent-SCe, per Year)

Biomass source	standard coal equivalent/a
Waste from livestock production	90*
Agro-industry waste	40*
Straw (corn, rice, grain) 60% of potential	340**
Forest residues and firewood	100
Bioorganic municipal waste and restaurants waste (50% of potential)	90*
Industrial organic wastewater	20*
Energy crops, including bioethanol, biodiesel, biogas (200 million ha marginal, reclaimed and unused land)	320**
Total	1 000

Note: 1. * mainly biogas, ** including biogas
2. 120 billion Nm^3 biogas from about 1.1 billion tons(1 Nm^3 biogas = 0.75 kg SCe)

4 Current biogas plant performance

No systematic assessment of biogas plant operation conditions has been accomplished. China Agriculture University (CAU) has started to develop an assessment in four pilot provinces, which should be further developed to a national cadastre. But as a matter of fact the biogas plants performance is far behind international standard, showing a low availability and productivity, insufficient specific biogas production, operating at low dry matter contents, low organic loading rates, lacking of process control and monitoring. Only a few plants are connected to the electricity grid, only seven biogas plants are registered CDM projects, from which so far only one can generate some CDM carbon credits based on certified emission reduction (CER). On environmental aspect such as the flaring of unused biogas and the minimization of biogas leakage is not observed. The mainly subsidy oriented support policy had boosted the construction activities without putting attention on operation and use of outputs, gas and digestates.

Figure 4 Traditional UASB biogas plant in Suichuan, Jiangxi, to treat the liquid effluent from a pig farm after separation of the solid phase (left picture), which leads besides of a low level technology and poor construction to such low biogas production that the CHP cannot be operated, constructed in 2005.

5 Technical services for biogas development

The subsidy policy of the Chinese government has led to a remarkable development of the biogas industry. In total there are about 4 000 biogas companies, service and equipment suppliers and the gross output value of biogas sector was about CNY 26 billion in 2010. Design and planning is done by design and research institutes on different design levels. Some have developed as market leaders, acting across the country. Design companies are often involved in delivery of equipment which makes a clear definition of their responsibility and liability difficult. To separate planning and delivery completely, but to support the one-shop approach in project engineering and implementation supervision according to International Federation of Consulting Engineers (FIDIC), would at the end of the day be helpful to achieve a better performance of the biogas plants.

In 1980s, some research groups have grown up in the agricultural universities through the country; some continue to be strengthened into majors for undergraduate and postgraduate education. In the same time, some national organizations and research institutions like China Biogas Association (CBA) and Biogas Institute of the Ministry of Agriculture (BIOMA) were also founded to promote the technology development and extension. The middle scale biogas plants, such as the Hangzhou Fushan Biogas Plant which was built in 1988 using chicken and pig manure as feedstock and Beijing Liuminying Biogas Plant which was built in 1992 using chicken manure, as two examples supplying biogas for about 200 households, were development in the same period. In the new century, biogas plants are scaled up to large size and advanced technologies become more critical for proper design and proper management. Two alliances of biogas technological organizations have been built and accepted/accredited by Ministry of Science and Technology of the state (MOST), one is Biogas Innovation Alliance, the other is Technology Innovation Alliance for Municipal Centralized Biomass Gas. Meanwhile, almost all important universities and research institutions have reorganized to include biogas research and educational bodies. Biomass Engineering Center (BECAU) and Energy Engineering and Low Carbon Technology Laboratory (EELC) have been set up in China Agricultural University in 2004 and 2009, respectively. The structure of a huge national research center, Biomass Engineering Center, has also been initialized under MOA.

About 10 technical standards regarding MLBGPs have been developed. However, these technical standards do not correspond to the international state of the art. And all these standards are recommended industrial standards of MOA and not compulsory national standard and they are often not considered in biogas planning. For the future a national biogas sector standardization administration is being organized and compulsory national standards for MLBGPs will be drawn up.

6 GIZ 'Biomass Utilization' via biogas project

With the overall objective 'to improve the technical standard and operational performance of MLBGPs, and to attract private and governmental organizations to invest into the biogas sector', the Sino-German Project on Optimization of Biomass Utilization, implemented by Deutsche Gesellschaft für Internationale Zusammenarbeit (GIZ) GmbH, in cooperation with the Foreign Economic Cooperation Center of Ministry of Agriculture (FECC/MOA), was launched in January 2009 (see *Ⅱ* and *Ⅲ*). The BMZ project is linked to the ADB loan to finance the 'Integrated Renewable Biomass Energy Development Project', where in total U$ 130 million will be invested to set up 109 MLBGPs in four provinces in China.

Ⅱ. Expected results of the GIZ Biomass Utilization Project in China

- Demonstration of "International best practice biogas projects", through technical assistance to planning, implementation, operation and management;
- Capacity development in the biogas sector through trainings, workshops and information events to disseminate the biogas knowledge to a wider group of stakeholders (governmental decision makers, project owners, designers, operators, laboratory personnel) Circular Economy (Organic fertilizer);
- Policy development by supporting the policy dialogue, setting up draft technical standard to make policy makers aware how to improve the biogas industry sector through policies.

Ⅲ. GIZ/FECC PPP Sino-German Demo projects: Innovation and business concepts

- Large scale biogas plants (up to 5 MW_{el});
- Technology (CSTR, dry-fermentation, fixed bed technology,);
- Co-digestion of alternative feedstock (BMW, straw, energy crops from marginal & reclaimed land);
- Business concepts revenue oriented at centralized biogas plants (energy sales, gate fees);
- Concept of centralized biogas plant;
- Design and operation fitting extreme cold climatic conditions in North & West China;
- Biogas purification to bio-methane (micro gas grid, NG grid injection, CBG);
- Recommending new CDM methodologies (PoA CDM);
- Cooperation with German designers, investors (DEG), CDM developers, equipment suppliers, research and strategic partners.

7 Opportunities for international cooperation

During the last ten years, a strongly supported biogas plant development has lead to huge number of biogas plants and however so far to insufficient results. From now on the public investment subsidies may gradually decrease and Chinese investors may care more about the IRR of their investment. Under the aspect of 'lessons learned' the biogas industry is getting more and more aware about the necessity to employ 'state of the art' technology to achieve process performance and finally revenues (e. g. biogas energy and carbon emissions trading). Increasing energy costs (the retail price for petrol has reached about half of the price in Germany, but in China only 30% taxes are included) makes biogas energy more competitive.

These are reasons that biogas investors are seeking reliable technologies and international cooperation, and especially with Germany which is well known for its successful development of a worldwide leading biogas industry sector. The trend to apply more quality equipment, such as pumps, process control systems, gas engines, gas purification devices and so on clearly exists. German equipment suppliers are welcomed, especially if they are in cooperation with Chinese partners. Further strategic partners such as investors and companies with operational experiences are welcomed on the Chinese biogas market.

In China biogas is going to be produced more and more following market mechanisms. Public investment subsidies are going gradually to be reduced and the fossil fuel prices are increasing.

8 Conclusions

The number of biogas plants, the installed fermenter volume and power generation capacity in China is ahead of the policy targets. However due to the weak process performance the biogas and biogas energy generation is far below expectations.

China was not really in the position to develop their own reliable large-scale biogas technology in the past two decades, and investors are lacking 'best practice' examples which are financially viable. After years of trials, it is recognized that based on the existing performance of Chinese biogas plant sustainable business models cannot be operated and that large scale biogas plants may disappear from the market in China. A redirection of this development only can be achieved if the 'state of the art' technology is applied and the domestic biogas industrial sector needs to cooperate with international suppliers to meet the requirements of the anticipated biogas energy generation. Therefore the investors are getting increasingly aware that international technology, especially for key components and international cooperation for operation and education of biogas technicians is required.

References

ASAE, (2003): *Manure Production Characteristics*. American Society of Agricultural Engineers, Standard, USA.

Asian Development Bank, (2009): *ADB Technical Assistance Consultant's Report*. Project Number: 40682 January 2009, PRC: Preparing the Integrated Renewable Biomass Energy Development Project, SINOC & HASSAL Associates International, PRC, http://www.adb.org/Documents/Reports/Consultant/40682-PRC/40682-PRC-TACR.pdf.

China Ministry of Agriculture MOA and ADB, (2008): *Proceedings of International Seminar on Rural Biomass Energy & ASEAN plus Three (China, Japan and Korea)*. Forum on Biomass Energy, China Agricultural Press, ISBN 978-7-109-1270-2, pp. 414.

China Ministry of Agriculture, (2007): *National Rural Biogas Construction Plan*. China.

China Ministry of Agriculture, (2007): *National Agricultural Biomass Energy Industry Development Plan* 2007-2015.

China National Bureau of Standardization, (2007): *NY/T1222-2006 Design Standards for Biogas Projects in large livestock breeding farms*.

China National Bureau of Standardization, (2007): *NY/T1221-2006 Technical Standards for Operation, Maintenance and Safety of Biogas Projects in Large Livestock Animal Breeding and Poultry Farm*.

China National Bureau of Standardization, *NY/T 388-1999*, *SEPA*, *Environmental Quality Standards for Livestock Farms*.

China National Bureau of Standardization, (2003): *NY/T 667-2003 Classification of scale for biogas engineering*.

China National Development and Reform Commission, NDRC, (2006): *The Renewable Energy Promotion Law*. http://ndrcredp.com/english/download/The%20Renewable%20Energy%20Law.doc.

China National Development and Reform Commission, NDRC, (2007): *Medium and Long Term Program of the Renewable Energy, Development Plan 2006-2020*.

China National Development and Reform Commission, NDRC, (2010): *China's Policies and Actions for Addressing Climate Change, the Progress Report*.

Gao Ding, et al., (2006): *Release of pollutants from livestock manure in China and recommended strategies for pollution prevention*. Geographical Research, 25, 2, pp. 311-319.

Hansen T. L., et al., (2004): *Method for determination of methane potentials of solid organic waste*. Waste Management, pp. 393-401.

Qu Pingmei, Cheng Chiying, Long Chunlin, (2008): *Comparison of biogas plants between China and Germany*. Journal of Renewable Energy Resources, 26, 1, pp. 110-115.

Raninger, B., Li Rundong, X. Chen, Xu Haiyun, (2010): Biomass *Wastes-to-Energy in China, Biogas from Landfills or Anaerobic Digestion*? Depotech, Abfallwirtschaft, Abfalltechnik, Deponietechnik & Altlasten, ISBN 987-3-200-02018-4, pp. 105-111.

Raninger B., Mang H. P., Heidelmann N., (2007): *CDM Projects in the Waste Management and Agroindustrial Sector in China - Status, cases, lessons learned and recommendations*. GIC-AHK, Germany, CDM conference, Beijing.

沼气与天然气并网——中国规模化沼气工程的挑战与机遇

B. Raninger[1]，董仁杰[2]，钱名宇[1]

(1. 德国国际合作机构，中德生物质能优化利用项目，北京 100125；
2. 中国农业大学能源工程与低碳技术研究室，北京 100083)

【摘要】中国在制定单位国内生产总值二氧化碳排放量比 2005 年减少 40%到 50%的目标后，到 2020 年中国将建立起严格的气候政策。同时由于化石能源价格的不断攀升，以多种生物质资源为原料的可再生能源得以快速发展，预计到 2020 年生物质能源将占可再生能源的 15%。生物质能资源主要来自大型畜禽养殖场、农业企业、森林、城市可降解废弃物、秸秆以及潜在的种植于边际土地上的能源作物(基于“不争粮”的生物能源政策)。到 2050 年，生物质能具备取代中国煤炭消耗总量 1/3 的潜力，相当于 1×10^9 t 标准煤。根据可再生能源中长期发展规划，截止到 2020 年大型沼气工程沼气发电并网将达 3×10^6 kW。但由于目前大中型沼气工程实际电力生产十分有限，有必要利用“国际最佳实践”技术将国内某些环保型沼气工程向产能工业型沼气工程进行转变。中德生物质能优化利用项目由德国联邦经济合作与发展部资助，并通过德国国际合作机构与中国农业部对外经济合作中心共同执行，对从亚洲开发银行获得贷款的大型沼气工程的设计和运营提供技术援助，目标是根据沼气工程的最佳实践创造本行业未来发展的示范效果。在这个技术升级期间，国际设备供应商将会得到更多在华合作的机会。

【关键词】中国沼气行业；沼气技术；沼气工程运行绩效；沼气使用；中德沼气合作

1 中国沼气的发展

中国农村户用沼气项目闻名于世。20 世纪 50 年代以来，中国建设了超过 3.5×10^7 座户用沼气池，年

沼气产量达到 1.2×10^{10} m^3。其发展目标为到 2020 年农村户用沼气总数达到 8×10^7 户，包含中国北部极端寒冷地区。

小型、中型和大型沼气工程的建设自 20 世纪 90 年代开始就得到发展。这种环境友好型处理方法适用于畜禽养殖厂和农业企业如啤酒厂、酒精生产厂和屠宰厂的液体排放物的处理。沼气此时只作为副产品自用或供当地临近居民使用。截止到 2009 年，已建有 34 000 座小型沼气工程和 22 900 座大中型沼气工程（MLBGPs，容积 >50 m^3），其中大型沼气工程（容积 >300 m^3）有 3 717 座（图 1）。截止到 2009 年，沼气发电装机容量达 42 MW。

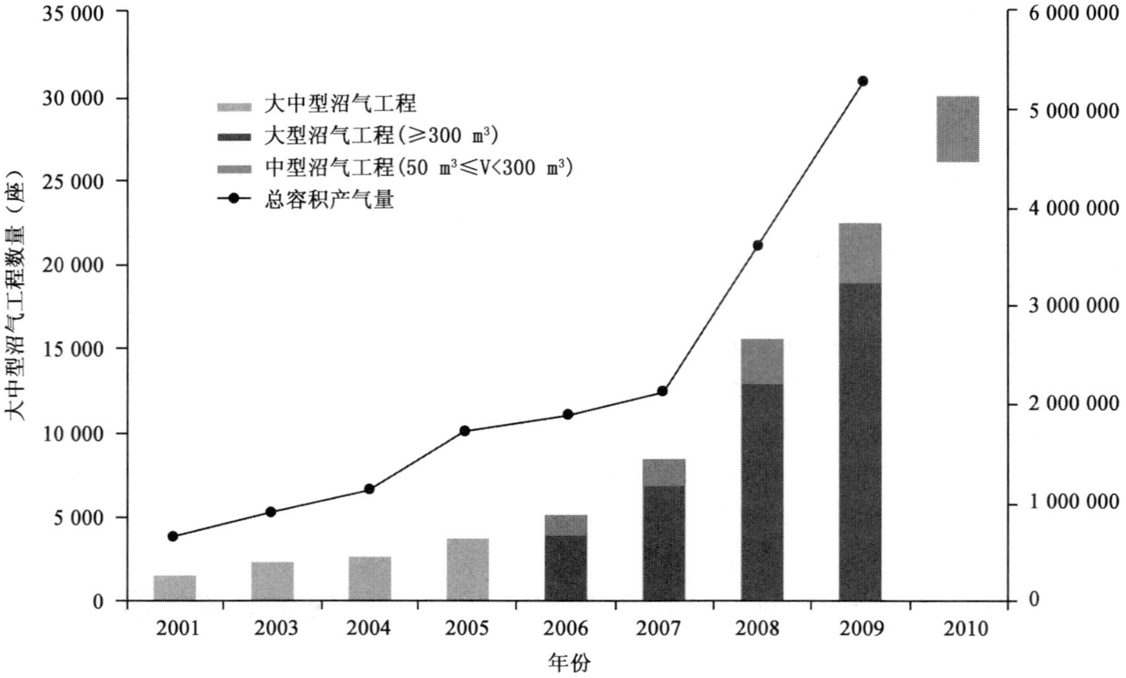

图 1　2001—2010 年中国大中型沼气工程数量和总容积产气量（中国沼气协会．中国农业部沼气科学研究所和中国发改委．2006—2020）

图 2　四川省邛崃市金利养猪厂，2009 年建立一座利用 USR 技术处理养殖厂液体排放物的沼气工程（排放物处理方法与污水处理类似），所产生的沼气用来运行一台 30 kW 的小型沼气发电机

建立以生产能源为目的的沼气工程的概念较为新颖。从目前来看，沼气自用或供当地沼气管网已逐步实现。但由于缺乏对发酵罐温度的控制，大部分沼气工程通常只能在暖季才能有效运行。中国农业部亚洲开发银行(ADB)“中国农村能源生态建设二期项目”应当致力于提升沼气工程的管理水平并大力支持中国建设以生产新能源为目的的大中型沼气工程的发展。

2 政策与财政支持

中国推动沼气发展的政策与德国的相关政策非常相似(见Ⅰ)。沼气工程的建造不仅用于环境保护,同时还能大量提供能源这一概念在中国可再生能源法(中国发改委,2006)制订后得到促进。计划到2020年,风能、水电、太阳能、地热能和生物质能(固体和液体生物燃料、沼气)等可再生能源将占据能源消费总量的15%。

Ⅰ.中国沼气主要的政策驱动力

- 环境保护(废水污染;地下、地表水和土壤的保护);
- 可再生能源(替代化石能源、可持续能源发展);
- 循环经济(有机肥料);
- 温室气体减排(2002年京都议定书,2009年哥本哈根会议,低碳经济);
- 农村地区发展(农业工业化、农民增收、减缓城市化进程)。

在过去的10年中,缓解气候变化成为沼气工程建设的又一驱动力量。中国实施了温室气体减排政策,并于2002年签署了京都议定书。在十一五期间,中国政府投资约2×10^{12}元(合2.4×10^{11}€)用于实现单位GDP二氧化碳排放量减少20%的目标(中国发改委,2010)。在2009年哥本哈根联合国气候变化大会上,中国还提出了自愿减排目标,即到2020年单位GDP二氧化碳排放量比2005减少40%~45%。中国已经看到气候变化给农业和基础设施建设带来的负面影响,同时也认识到全球共同努力阻止气候变化的重要性。

为了支持沼气工程的建设发展,中国出台了许多政策。截止到2010年,中国农业部已投入2.4×10^{10}元(合2.8×10^{9}€)用于沼气工程建设。在过去的三年里,国家平均每年的投资额均在$5\times10^{9}\sim6\times10^{9}$元之间。对于每座沼气工程的建设,在申请和评审的基础上,国家会给予总投资额25%~45%的补贴,当地政府也会给予类似的补贴。这样,项目业主实际最少只需要支付沼气工程建设总投资额的15%。对于大规模沼气工程还会有一些来自国际贷款、捐赠和商业贷款等机会。国际贷款一般来自于亚洲开发银行和世界银行,捐赠主要来自全球环境基金、联合国开发计划署和双边合作成员国如荷兰、瑞典、丹麦、芬兰、日本等国。其单位原料投资额为每吨30~60€之间。对于兆瓦级全混式沼气工程(completely stirred tank reactor, CSTR)的单位装机容量投资额在每千瓦1 600~3 300€之间。

图3 北京延庆德青源沼气工程,联合国开发计划署,全球环境基金,美国国际开发总署示范CSTR技术,以2.2×10^{6}只蛋鸡粪便为材料进行厌氧发酵,所产沼气用于驱动2台颜巴赫发电机,共计2.4 MW,于2007年开始并网发电,2009年获得清洁发展机制(CDM)认证

由于中国沼气全年产量不稳定和对发电机装机容量的高要求，沼气发电并网在中国并没有得到长足的发展。只有当发电能力超过 500 kW(地方省县除外)才可能实现并网发电，并从国家财政补贴政策中获益。德青源沼气工程是目前中国成功实现并网发电，且根据可再生能源法的相关政策得到国家入网补贴(在当地脱硫燃煤机组标杆上网电价的基础上每度电多加 0.25 元(合 0.03 €)的一个成功案例(表 1)。

表 1　沼气发电并网价格和税收待遇(2010 年)

原料（沼气工程类型）	发电基准单价	发电补贴（功率＞500 kW）	并网补贴（补贴给沼气工程）	税收优惠
畜禽粪便	2005 年地方脱硫燃煤机组标杆上网电价（约 0.45 元/kWh*）	0.25 元/kWh，从 2010 年起，每年新批准建设项目补贴电价比上一年批准建设项目的补贴电价递减 2%，持续时间 15 年	0.01 元/kWh（＜50 km） 0.02 元/kWh（50～100 km） 0.03 元/kWh（＞100 km）	免收企业所得税（头三年）；交企业所得税的 50%（后三年）
农林废弃物	a. 0.75 元/kWh(含税) b. 对已批准的项目或正在建设的项目，其税收还需要核实			如果原料的 70%是作物秸秆等农林废弃物，其中 10%的收入不用交纳企业所得税

* 该价格在 2011 年 12 月由中国政府上调了 0.025 元/kWh。

来自联合国气候变化框架协议清洁发展机制(CDM)的碳交易收益能够确保大型沼气工程的经济可行性(PPTA，2009)。大规模沼气工程的核准减排量(CERs)能够达到 $4\times10^4\sim1\times10^5$ t CO_2 当量(UNFCCC，PDDs)。从 2008 年起国家发改委规定每吨二氧化碳当量转让价格不低于 7 €，从该项目质量和市场行情看，价格甚至可以超过 14 €。但由于 CDM 项目注册和监控的复杂性和费用问题，且国内大部分沼气工程运行状况不佳，到目前为止，沼气工程 CDM 项目开发商还没有充分利用中国沼气市场。

3　大规模沼气工程的发展潜力

中国农业部的调查结果显示，截止到 2010 年已建成 4.2×10^6 座商业化畜禽养殖厂，存栏量为 6.1×10^8 头猪当量，所排粪便达到 3×10^9 t。其中 1.1×10^9 t 粪便可作为沼气发酵原料，沼气生产潜力达 6.7×10^{10} m^3。同时，在郊区的 1.4×10^8 座小型养殖厂中，还饲养有 9×10^7 头猪、1.6×10^7 头牛、8.5×10^7 只鸡和 2.6×10^7 只羊。如果有效的利用这些动物产生的粪便，每年可增加沼气产量 5.4×10^{10} m^3。这些小型养殖厂可以独立或联合建立沼气工程处理畜禽粪便沼气工程。

在《2006—2020 年可再生能源中长期发展规划》中，国家发改委计划截止到 2020 年，建立 1×10^4 座大中型畜禽养殖厂沼气工程和 6 000 座工业有机废水沼气工程，沼气年产量约 1.4×10^{10} m^3，发电装机容量达到 3 GW。但在 2010 年，只实现了计划发电装机容量(3 GW)的 15%(没有全负荷运行)，因此沼气并网发电仍有很长的路要走。2050 年生物质能总潜力情况见表 2。

表 2　2050 年生物能潜力情况表(1×10^9 t 标煤/a)

生物质资源	$\times10^6$ t 标煤/a
畜禽生产的废弃物	90*
农业废弃物	40*
60%秸秆废弃物(玉米、大米、谷物)	340**
林业废弃物和薪柴	100
城市有机废弃物和餐厨垃圾	90*
工业有机废水	20*
能源作物，包括生物乙醇、生物柴油和沼气(2×10^8 hm^2 边际土地)	320**
总量	1 000

注：① * 主要为沼气，** 包含沼气；②$1.2\times10^{11}$标方沼气来自 1.1×10^9 t 标煤(1 标方沼气 = 0.75 kg 标煤)。

4 目前沼气工程的运行情况

中国目前还没有实现对沼气工程运行状况的系统评估。中国农业大学已经着手对四个试验省份的沼气工程进行系统评估，从而进一步实现全国范围内的评估体系。但由于中国沼气工程运行状况远远落后于国际标准，如：产气率低、有机负荷低，单位容积产气量不足，原料干物质含量低，缺乏系统控制检测等。所以只有很少的沼气工程能够实现并网发电，且在目前成功注册 CDM 的 7 座沼气工程中，只有一家获得项目产生的经核准减排量(CER)。而在环境方面，有些沼气工程将未使用的沼气进行燃烧，并未对沼气泄漏等进行检查。由此可以看出，虽然目前政府提供的财政补贴加快了沼气工程的建设速度，但忽视了沼气工程的运行水平及后续产品的利用，如沼气和沼液沼渣的利用。

图 4 建设于 2005 年的江西遂川传统 UASB 沼气工程，处理养猪厂废弃物固液分离(左图)后的液体排放物，技术含量低、沼气产量低，无法进行发电

5 沼气工程发展的技术服务

中国政府的补贴政策促进了沼气工程的蓬勃发展。目前中国已有 4 000 家提供沼气相关服务与设备的供应商，2010 年，沼气工程产值高达 2.6×10^{10} 元人民币。沼气工程的设计与规划由不同的设计研究单位承担，它们中的一些已经发展成为国内市场的领军企业。目前一些设计公司还经常进行设备引进的业务，因此很难明确定义它们的责任与义务。但若能将设计与设备交付完全分开，根据国际咨询工程师联合会的体系来监督工程建设，将会对沼气工程的良好运行提供帮助。

从 20 世纪 80 年代起，一些研究小组已经在国内的大学里不断成长，有一些高校还设立了相关的本科与研究生专业。同时，一些国家机构和研究院所，如中国沼气协会(CBA)、中国农业部沼气科学研究所(BIOMA)也得到资助用于促进沼气技术的发展。中型沼气工程，如建于 1988 年的以鸡粪和猪粪为原料的杭州浮山沼气工程；建于 1992 年的以鸡粪为原料的北京留民营沼气工程，它们均为 200 户居民提供沼气并于同一时期得到发展。进入新世纪后，随着沼气工程规模的不断扩大，先进技术也成为正确设计与管理的重要因素。中国科学技术部(MOST)建立并认可了两个沼气技术组织，一个是沼气产业联盟，另一个是城市集中式生物质燃气产业技术创新联盟。同时，几乎所有重点大学和研究机构都已将沼气工程纳入教育体系。中国农业大学分别在 2004 年与 2009 年建立了生物质工程中心(BECAU)和能源工程与低碳技术实验室(EELC)。在农业部的支持下，大型国家研究中心框架下的生物能源工程中心也已经开始建立。

用于鉴定大中型沼气工程的 10 项技术标准已经出台，但这些技术标准并没有涉及到国际领先的技术水平，而且这些标准都是中国农业部的行业标准，并不是国家标准，因此许多沼气工程的规划并没有参考这些标准。相信在不久的将来，中国将建立国家级的沼气标准管理部门，同时起草国家大中型沼气工程的行业标准。

6 GIZ“生物质利用项目”沼气工程部分

在“提升大中型沼气工程技术标准并优化大中型沼气工程绩效，同时吸引政府机构与个人投资沼气行业”的目标下，通过德国国际合作机构(GIZ)与中国农业部对外经济合作中心(FECC/MOA)的合作，于2009年1月实施了中德生物质能优化利用项目(见*II*和*III*)。该德国经合部项目与亚行(ADB)贷款援助的“中国农村能源生态建设二期项目”(即亚行贷款1.3×10^8美元在中国4个省份投资建立109座大中型沼气工程)结合执行。

Ⅱ. GIZ 生物质能优化利用项目预期在中国取得的成果

- 通过对项目规划、执行、运营和管理提供技术援助，建立“国际最佳实践”沼气示范项目；
- 通过培训、研讨会和其他活动提高沼气工程的建设能力，将沼气知识传播给更广大的群体(政府决策者、项目业主、设计人员、运营人员和实验室人员)；
- 通过支持政策对话和建立技术标准草案使政策制定者认识到如何利用政策法规促进沼气产业的发展。

Ⅲ. GIZ/FECC PPP 中德示范项目：创新与商业理念

- 大规模沼气工程(大于5 MW)；
- 技术(CSTR、干发酵、固定床技术)；
- 多原料混合发酵(城市生物垃圾、秸秆、边际土地上种植的能源作物)；
- 面向集中式沼气工程的商业理念收入(能源销售、垃圾处理费)；
- 集中式沼气工程理念；
- 在中国北部与西部设计和运营适应极端寒冷地区的沼气工程；
- 将沼气提纯为生物甲烷(微型燃气管网，并入天然气管网，压缩沼气)；
- 引进新的CDM方法学(规划类CDM)；
- 与德国设计公司、投资方(DEG)、CDM开发商、设备供应商、研究和战略伙伴合作。

7 国际合作机遇

在过去的10年里，中国对沼气工程发展的大力支持促进了大量沼气工程点的建设，但这带来的效果并不理想。目前政府投资补贴逐年减少，而中国投资者考虑更多的是投资内部收益率。在经历众多的教训后，沼气产业越来越注重引进先进工艺技术以保证沼气工程的良好运行并从中获益(如沼气能源和碳排放交易)，增加能源成本(目前中国汽油零售价已经达到德国价格的一半，但是税收比例只有30%)以使沼气能源更具竞争力。这些都是国内沼气投资商寻求可靠技术和国际合作的原因，尤其是与德国这种在沼气领域技术处于世界领先地位的国家合作。引进并应用高质量设备的趋势(如泵、工艺控制系统、沼气发电机、沼气提纯设备等)明显。因此德国设备供应商，特别是有与中国合作经历的供应商很受中国市场欢迎，相信在不久的将来，更多投资商和拥有沼气工程运营经验的公司也将同样受到中国市场的欢迎。

随着石油价格不断攀升，在市场机制的动作下，中国政府对沼气工程的投资补贴将逐年减少，但中国的沼气产量仍将不断增加。

8 总结

目前中国的沼气工程的数量、发酵罐容积和发电能力都领先于政策设定的目标。但由于沼气工程运行效果不佳，沼气产量和实际沼气发电产量仍远远落后于预定的目标。

在过去的 20 年里，中国并没有把重点放在大规模沼气工程技术的发展上，且投资者缺乏经济可行的最佳实践案例进行参考。经过多年探索，按照中国目前建成的沼气工程的运行水平来看，如果可持续商业模式仍无法建立起来，大规模沼气工程也许会从中国市场中逐渐消失。因此，只有应用先进工艺技术，这种情况才能够得到改变。国内沼气产业需要与国际供应商合作才能达到沼气产能的预期目标。因此，越来越多的投资者意识到国际先进技术，特别是关键设备和技术的重要性，以及沼气工程运营和技术员工培训方面实现国际合作也是十分必要的。

德国沼气产业持续高速发展的启示

陈晓夫[1]，邓良伟[2]，周红军[3]，钱名宇[4]

（1. 中国农村能源行业标准化技术委员会，北京 100125；2. 农业部沼气科学研究所，成都 610041；3. 中国石油大学新能源研究中心，北京 102249；4. 德国国际合作机构，北京 100125）

【摘要】德国历来非常重视可再生能源的发展，把沼气提高到能源安全战略、环境、资源和可持续发展的重要目标来大力发展。早在 2000 年 4 月德国联邦政府就颁布了可再生能源法，为可再生能源的优先生产、并网、传输与使用提供了法律基础，并制定了切实可行的补贴细则。目前，该法案已被全世界 40 多个国家借鉴。截止到 2010 年，德国共建有沼气工程 6 000 座，发电装机容量 2 500 MW，规模主要集中在 70～500 kW 之间，年发电量 2.2×10^{10} kWh，相当于德国全年电量消费的 3%。2010 年沼气行业年产值达 4.7×10^{9} €，就业人数超过 4×10^{4} 人。在德国现有的沼气工程中，采用玉米青贮为原料的占 80%，投资回收年限为 5～7 a。通过建立沼气工程，不仅增加了可再生能源生产，增加了农民收入，促进了就业，同时对农村发展，种植结构的改变等都起到了深远的影响。沼气提纯替代天然气是德国沼气发展的一个新方向，新修订的可再生能源法中也明确了对沼气提纯的补贴政策。目前，德国有沼气提纯工厂 57 座，每小时产气量可达 37 470 标准 m^3，规模主要集中在 350～700 标准 m^3/h。德国规划或在建的沼气提纯工厂超过 80 座，绝大多数的沼气经提纯后进入天然气管网。德国沼气提纯的发展目标为，到 2020 年和 2030 年，分别有 6%和 10%的天然气消耗将由沼气替代，这要求每年再建 100～120 座每小时产气 700 m^3 的沼气工程。

【关键词】德国沼气领域；沼气工程；沼气发电；沼气提纯；可再生能源法

1 德国的可再生能源与沼气政策目标

德国可再生能源的发展目标为，到 2050 年，一次能源消费总量比 2008 年降低 50%；电耗下降 25%；未来 15 年内彻底结束核能使用。可再生能源年生产量比 2008 年提高 3 倍，生物质能提高 2.5 倍。温室气体排放比 1990 年减少 80%。到 2050 年，可再生能源发电量占总发电量的比例从现在的 16%提高到 80%，可再生能源占能源总消耗的比例从现在的 11%提高到 60%，生物质能占可再生能源的比例从现在的 17%提高到 60%[1]。

2 德国可再生能源法的进展

德国自 2000 年颁布可再生能源法后，于 2004 年和 2009 年分别进行了两次修订，第三次的修订法案将在 2012 年开始实施。该法案的要点是：可再生能源具有进入能源供应系统的优先权；持续的长期补贴（20 年），保障投资者的利益；设置了为促进新技术应用的专门补贴，且新建的沼气工程每年基本补贴降低 1%[2]；补贴计算简单，公众易于接受；所有补贴费用来自于用户电价的提高（约 5%）[3]，相较传统电力的价格，可再生能源电价每度电较 2010 年提高 2.047 €-ct。该法案多次修订的主要原因是为了更好地指导沼气工程的发展方向，促进行业健康快速的发展。2000 年可再生能源法案的出台为沼气工程并网、补贴和长期的支持提供了法律保障，2004 年的法案新加入了小型（≤150 kW）沼气工程的补贴政策，并丰富了额外补贴的种类，开始促进能源作物的使用。2009 年对小于 150 kW 的工程给予了更多的补贴，降低了大于 150 kW 沼气工程的补贴额，进一步促进了沼气工程的分散布局，并新增加了粪污补贴和减排补贴，整体补贴额度较 2004 年有了一定的提高（对于小于 150 kW 的沼气工程 2004 年版补贴最大额为 0.180 7 €，2009 年版为 0.257 7 €）[4]。但 2009 年版的修订也起到了一些负面效应，例如青贮原料的大量使用造成了能源作物与粮食作物争地，土地租金上涨。法案中规定，只要原料中含至少 30%粪污就可以获得补贴，使得 81%的沼气工程获得了粪便原料补贴，而实际处理的粪污量却只有 20%[5]。

2012 年实施的新可再生能源法修订案，将提高运行效率，简化补贴方案，缩减核查成本，促进污染类原料应用，鼓励沼气工程分散布局，以及将利用沼气工程调整农业种植结构与增加农户收入等放在了首位。限制青贮玉米的利用量不得超过 60%，对于粪便、园林修剪废弃物给予了更高的补贴额度；减少对 500 kW 以上沼气工程的补贴额度，增加对 75 kW 以下且以粪便为主的小型沼气工程的补贴额度；为保证沼气提纯工程的经济性，提纯补贴只提供给大的项目，平均降低补贴额度 10%～15%，放缓了沼气工程的建设速度[4]。根据德国农业部的预测，新的法案将有效提高沼气工程的效率，促进沼气工程分散布局，优化农业种植结构，未来农民的收入中将有 1/4 来自沼气工程[1]。

可再生能源法为德国沼气工程的发展提供了重要的政策支持，优先生产、并网、传输、使用等保障了沼气的出口问题；长达 20 年不变的补贴政策使得农户可以向银行贷款并使用好的设计与设备，且保证了大中小型沼气工程的盈利；不断调整的政策导向促使更多类型沼气工程的出现，更多新技术的应用，以及更加合理的农业种植结构；吸引了更多不同领域的人才和公司进入沼气行业，保障了德国沼气工程快速健康的发展。

3 沼气工业发展的经济基础

德国在沼气发电补贴政策的激励下，许多农场主纷纷建造沼气工程。农业沼气工程从 20 世纪 90 年代初开始建设，到 1999 年，全国仅有 850 个沼气工程；2000 年可再生能源法颁布以后，到 2003 年，德国已建设了 900 个沼气工程；2004 年修订后，2004—2008 年建设的沼气工程数量达到 2 141 个；2009 年的再次修订，使得 2009 年和 2010 年沼气工程建设数量分别达到 1 093 个和 1 016 个[3]。不同原料和规模的沼气工程，其发电上网电价补贴从 0.104 €增加到 0.257 7 €(常规上网电价为 0.075 €)[4]。银行为沼气工程投资业主提供了长期低息贷款，可以分 15～20 a 归还，贷款额可达工程建设总投资的 80%。目前，一个装机 250 kW 的沼气发电工程，总投资为 1.1×10^6 €，每年获得利润约 6×10^4 €，5～7 a 可收回投资[3]。可以说，“经济可行”成了德国沼气发展的核心要素，沼气工程的发展速度主要取决于上网电价的补贴以及工程的赢利能力。

尽管美国、法国均为高度发达的工业化国家，科学、技术与设备先进，但由于对沼气工程没有有力的激励政策，农场沼气工程发展缓慢。美国正常运行的沼气工程只有 111 个，法国目前只有 8 个农场沼气工程(且多数沼气厂为市政垃圾沼气工程)。这也从另一个角度充分说明经济可行是德国沼气工程大发展的主要因素。欧盟认为沼气能源非常重要，在 2020 年交通运输所用的化石能源中，将有 20%由可再生能源代替，而其中三分之一来源于沼气(参考自欧盟-Biomax 项目)。

4 成熟的沼气技术与设备

德国沼气工程普遍采用产气率高的专用青贮玉米作为主要发酵原料(产气率是鸡粪的 2.5 倍，猪粪的 3.4 倍，牛粪的 4.5 倍)。为了保证沼气工程较高的池容产气率以及冬季的稳定运行，大部分沼气工程采用热电联产的沼气利用模式，沼气发电上网，发电余热为沼气发酵装置增温、保温，从而保证了沼气发酵可以在中高温条件下稳定进行，如果不是热电联产项目，热源可来自沼气锅炉或者额外的热源。中温沼气发酵装置产气率达 1.2～1.8 $m^3/(m^3\cdot d)$，高温沼气发酵装置可超过 3.0 $m^3/(m^3\cdot d)$，最大达到 5.0 $m^3/(m^3\cdot d)$。即使在冬季气温低至－20℃时，沼气工程仍然能正常运行。为了保证沼气工程较高的池容产气率，德国沼气工程普遍采用干物质浓度 8%～12%的高浓度发酵，干发酵技术干物质浓度可超过 25%。发酵工艺主要采用适合高浓度、高固体的全混合厌氧反应器。德国沼气工程无论规模大小均采用自动控制，基本只需一人，或一个兼职管理者即可稳定运行，大大降低了运行成本。相关的发电设备、机械搅拌装置、自动控制系统、余热利用技术等已进入专业化设计和市场化制造阶段。工程质量可靠，运行安全稳定，技术服务专业组织完善，质量标准检测体系健全。先进的沼气发酵与净化技术以及性能优良的配套设备支撑了德国沼气工程的快速发展，使德国处于世界沼气行业的领先地位。

5 沼气工业的良性发展

德国沼气行业已达到高度工业化水平，较小的沼气工程规模也达到 70 kW，发酵罐规模在 2 000 m^3 以上，最大的沼气工程可达到每天 1.3×10^5 m^3 沼气的产量。在过去的十多年中，德国沼气工程依靠自己的建立和发展，已成功地成为一个新兴产业。目前，德国大约有 470 个沼气公司从事沼气工程的设计、建造和设备生产供应。近三年来，每年用于沼气工程建设的投资达 4.5×10^9 €左右，其中，约 20%是通过出口获得[3]。德国沼气产业之所以能得到迅速发展，主要是找到了适合德国沼气发展的产业链以及可行的商业模式：(1)可大量获得的青贮及畜禽粪便，2010 年玉米青贮的种植面积高达全国可耕地面积的 5%，以及各种食品工业和市政产生的废弃物；(2)经过厌氧发酵的沼气发电既可优先上网，也可净化纯化后无障碍进入当地天然气管网；(3)沼液在一定区域内可完全无成本或低成本消纳(是政府准入条件，也是商业计划的一部分)，沼渣加工后的有机肥市场认知度很高。根据德国最新研究表明，沼肥中含有的氮可以直接替代 60%的化肥用量，而磷和钾可以替代更多的化肥用量[6]；(4)沼气工程建设带来的效益，吸引了沼气生产商、电(管)网运营商、能源供应商投资的积极性，带动了沼气发酵装置、沼气发电系统及自动化控制系统等相关技术的发展和相关设备、产品的工业化制造水平，从而逐步推动沼气产业迅速发展壮大。

参考文献

[1] Germany Ministry of Agriculture (BMELV), (2011): *EEG 2012, (Erneuerbare-Energien-Gesetz) Die Gesetzesnovelle*. Presentation at BMELV in Germany.

[2] German Biomass Research Centre (DBFZ), (2011): *Development of bioenergy/biogas in German and legal regulations*. Presentation at DBFZ in Germany.

[3] Fachverband Biogas e. V., (2011): *Biogas market in Germany*. Presentation at Fachverband Biogas in Germany.

[4] Daniel-Gromke J., Gierdorf J., (2011): *Comparative overview of EEG Version 2000-2012: Example Biogas*. Biogas Engineering and Application, 2, Beijing.

[5] DBFZ, (2011): *Evaluation of the effects of the EEG*. Presentation at DBFZ in Germany.

[6] Bayerische Landesanstalt für Landwirtschaft (LfL), (2011): *Eco-fertilizer Research*. Presentation at LfL in Germany.

A RAPID BUT SUSTAINABLE DEVELOPING GERMAN BIOGAS INDUSTRY

CHEN Xiaofu[1], DENG Liangwei[2], ZHOU Hongjun[3], QIAN Mingyu[4]

1. Committee on Rural Energy Standardization of China; 2. Biogas Research Institute of Ministry of Agriculture; 3. New Energy Research Center, China University of Petroleum; 4. Deutsche Gesellschaft für Internationale Zusammenarbeit (GIZ) GmbH, SINO-German Project for Optimization of Biomass Utilization

Contact: Mr. CHEN Xiaofu, Committee on Rural Energy Standardization of China, Tel.: +86-10-65031567, E-Mail: chxiaofu@126.com;

Prof. Dr. DENG Liangwei, Biogas Research Institute of Ministry of Agriculture, Tel.: +86-28-85320701, E-Mail: dlwbrtc@yahoo.com.cn;

Prof. Dr. ZHOU Hongjun, New Energy Research Center, China University of Petroleum, Tel.: +86-10-89731979, E-Mail: zhouhongjun@cup.edu.cn;

Mr. QIAN Mingyu, GIZ, SINO-German Project for Optimization of Biomass Utilization, Tel.: +86-10-65087080, E-Mail: mingyu.qian@giz.de

EXECUTIVE SUMMARY

Germany attaches great importance to the development of renewable energy, and the development of biogas is an important goal in regards to the energy security strategy, environment, resource, and sustainable development. In April 2000, the German government promulgated the Renewable Energy Law to provide the legal basis for the priority of renewable energy production, grid combination, transmission and utilization, and to constitute practical subsidy rules. Currently, these rules have been drawn by more than 40 countries around the world. By 2010, Germany has built 6 000 biogas projects, the power generation installed capacity has reached 2 500 MW which are mainly in the size of 70 to 500 kW, and annual generation capacity has reached 22 billion KWh that is equivalent to 3% of annual electricity consumption in Germany. The annual turnover of biogas industry was 4. 7 billion euros and the employee numbers exceeded 4 million in 2010. In the existing biogas projects in Germany, using corn silage as feedstock accounts for 80% and the investment recovery period is 5 to 7 years. The establishment of biogas plants can not only help to increase the renewable energy production, farmers' income and promote employment, but also play a far-reaching impact on the changing of rural cultivation structure and rural development. Using biogas purification to substitute natural gas is a new direction of German biogas development; the revised new renewable energy law also defines the subsidies on the biogas purification. At present, Germany has 57 biogas purification plants, biomethane production can be 37 470 Nm^3/h, which is mainly in the scale of 350 to 700 Nm^3/h. In Germany there are more than 80 biogas purification plants which are planned or already in construction, and the vast majority of purified biogas is combined with the natural gas gird. The development goal of German biogas purification is, in 2020 and 2030, 6% and 10% of natural gas consumption will be replaced by biogas respectively. A further goal is to build 100 to 120 biogas projects every year with a biogas production of 700 Nm^3/h.

Key words: German biogas sector; biogas plant; biogas power generation; biogas purification; renewable energy law

1 Biogas and Germany's renewable energy policy

The goal of German renewable energy development is that the total primary energy consumption in 2050 will decrease by 50%, with a power consumption decrease of 25% compared with 2008. In the next 15 years, Germany will completely stop using nuclear energy. The production of renewable energy will increase by 3 times and biomass energy by 2. 5 times, based on the production in 2008. The GHG emission will reduce 80% based on 1990. By 2050, the proportion of the renewable energy production in the total generating capacity will increase from the current 16% to 80%, the proportion of renewable energy in the total energy consumption will increase from the current 11% to 60%, and the proportion of biomass energy in the renewable energy will increase from the current 17% to 60%.

2 Progress of the German EEG

After Germany's enacting of the renewable energy law in 2000, it has been revised twice in 2004 and 2009 respectively, and the third amendment will be implemented in 2012. This law's key points are: the renewable energy has a priority to be integrated into the energy supply system, it includes continuous long-term subsidies (20 years) and protects the benefits of investors; a bonus to promote the application of new technologies is paid; subsidies of new biogas projects (reduce by 1% per year after establishment of the law) are implemented; simple subsidy calculation; promotion of public acceptance. All subsidies come from the increase of the consumer electricity price (about 5%). The main reason of the repeated amendment of the law is to better guide the development of biogas projects and to promote a healthy and rapid development of the German biogas industry.

The purpose of the 2004 edition is to promote the development of energy crops; the purpose of the law of 2009 edition is to promote the development of new technologies, to give more subsidies to projects with less than 500 kW, and to increase subsidies for energy crops and subsidies of emission reduction. However, the amendment of 2009 edition resulted in a number of negative effects: For example, the extensive use of silage resulted in the competition for land between energy crops and food crops, and land rents increased. According to the 2009 law, a project can receive subsidies as long as the raw material contains at least 30% manure. As a result, in the year 2010, 81% of biogas projects received subsidies for the manure feedstock, but the handling amount of the total available manure is only 20%.

In 2012, the implementation of the third amendment of the renewable energy law will put the improvement of the operational efficiency of biogas plants, the promotion of pollution-based feedstock application, the encouragement of biogas plants following a scattered layout, and the use of biogas projects to adjust agricultural cultivation structure and to increase farmers' income on the top. The use of corn silage must not exceed 60%, but higher amounts of subsidies are given for manure and wastes (landscaping, biowaste). The amount of subsidies for projects more than 500 kW will be reduced, whereby the amount of subsidies paid for small manure-based biogas projects of less than 75 kW will be increased. To ensure that biogas purification projects work economically, subsidies are only available for large scale biogas projects. According to the prediction from German Ministry of Agriculture, the new law will increase the efficiency of biogas projects, promote the scattered layout of biogas plants, optimize agricultural cultivation structure, and a quarter of farmers' income will come from the biogas projects in the future.

The Renewable Energy Law provides an important policy support for the development of biogas plants in Germany. Priority production, grid connection, transmission and utilization ensures the biogas exportation; a 20-year fixed subsidy makes farmers eligible to receive loans from bank and use good design and equipment, which ensures that both MLBGPs and small scale biogas plants can operate profitable. Constant adjustment of the policy orientation promotes the development of more types of biogas projects, the application of more new technology, and the development of a more reasonable agricultural cultivation structure. It further helps to attract more companies and talented people to enter the field of biogas industry, herewith guaranteeing a rapid and sound development of the German biogas sector.

3 Economic foundation of biogas development

Under the inspiration of subsidies on biogas power generation, many farmers have built biogas plants. Agricultural Biogas plants were built from the early 1990s, and there were only 850 biogas plants till 1999. After the establishment of the renewable energy law in 2000, Germany constructed 900 biogas plants until 2003. After revising the law in 2004, 2 141 biogas projects had been built from 2004 to 2008 in Germany. After revising the law again in 2009, the number of biogas construction has reached 1 093 and 1 016 in 2009 and 2010, respectively. According to the different feedstock and the scale of biogas plants, the subsidy amount for power generation has increased from 0. 104 € to 0. 215 € (conventional grid-electricity price is 0. 075 €). Banks provide long-term low interest loans that may cover 80% of the total investment for the investor of the construction of biogas projects, and they can pay back the loans in 15 to 20 years.

Currently, the total investment of a biogas power generation project with installed capacity of 250 kW is 1.1 million €, its profit is about 6 mil €/a, and it needs 5 to 7 years to take back the investment. Hence, it can be said, "financial viability" has become the core element for the development of the German biogas, and the speed of biogas development mainly depends on subsidies on electricity price and the project profitability.

Although the United States and France are highly developed industrialized countries and they have advanced science, technology and equipments, the farm biogas projects develop slowly because they have no strong incentive policies on biogas engineering. There are only 111 biogas plants under normal operation in United States, and France only has 8 farms biogas projects so far, but biogas plants in the municipal waste sector. It also fully explains from another point of view that financial viability is the core element of the development of German biogas projects. The EU policy considers biogas energy very serious, so f. e. in 2020 20% of fossil fuel in transport should be replaced by renewable energy, from which 1/3 should come from biogas (see EU-Biomax Project).

4 Matured status of biogas technology and equipment

German biogas projects commonly use corn silage as main feedstock, as it provides a high rate of biogas production (2.5 times of the rate of biogas production of chicken manure, 3.4 times of the rate of biogas production of pig manure and 4.5 times of the rate of biogas production of cow manure). In order to ensure a high rate of specific biogas production and a stable operation during the winter, most biogas plants adopt the CHP biogas utilization method, a combination of electricity grid and use of waste heat to keep fermentation devices warm, hereby ensuring that the fermentation process is carried out in constant middle and high temperature conditions. If no CHP is installed at the biogas plant, a gas boiler or heat from external sources is used. The specific biogas production of mesophilic fermentation is 1.2-1.8 $m^3/(m^3 \cdot d)$, and for thermophilic fermentation is more than $>3.0\ m^3/(m^3 \cdot d)$ up to 5 $m^3/(m^3 \cdot d)$. The biogasplant can still operate at temperatures as low as −20℃ during the winter. In order to ensure a high rate of specific biogas production, biogas projects in Germany widely use high concentration fermentation during which dry matter concentrations reach 8% to 12% in liquid technologies and more than 25% in dry-fermentation technologies, and choose CSTR for the fermentation process. Regardless of the scale of the biogas plants in Germany, all adopt automatic control systems. Basically one person alone, even a part-time manager, can achieve stable operation of the biogas plant, which can reduce the operating costs greatly. Related to power generation equipments, mechanical stirring devices, automatic control systems and the technology of waste heat utilization has entered the phase of professional design and is manufactured for the market. German biogas projects provide reliable quality, and can operate safely and stably. Professional technical services are available and the organization is very professional. The German inspection system in regards to quality standards is sound. The advanced fermentation & purification technologies and equipments with excellent performance support a rapid biogas development in Germany, and make German gas industry stand at the world leading position.

5 Sound development of the biogas industry sector

The German gas industry has reached a high degree of industrialization: small scale biogas plants have reached 70 kW, with a fermenter volume of more than 2 000 m^3, and the biogas production of the largest biogas plant can reach 130 000 cbm per day so far. In the past decade, the German biogas was established and developed successfully by its own and became an emerging industry. Currently, there are about 470 German biogas companies engaged in the design, construction and equipment supply of biogas plants. In the last three years, the construction investments in biogas plants are about 4.5 billion , of which about 20% are obtained through export. Reasons of the fast development of the German biogas industry are mainly because they have found a suitable industry chain and viable business model of German biogas development: (1)Silage and manure and widely available. The planting area of corn silage tooks up 5% of arable land in Germany in 2010, but various wastes from food industries, and municipal sources are used; (2)The biogas through anaerobic fermentation can be connected to the electricity grid and also be purified for natural grid injection; (3)Liquid residues can be processed for free or low-cost in a certain area (as a governmental permitting condition its part of the business concept), and the organic fertilizer produced from solid residue has a high market awareness. The nitrogen in the effluent can directly replace 60% of mineral fertilizer consumption; potassium and phosphorous can even be replaced by more than 60%, acc. to the latest research in Germany; (4) Benefits from biogas plants attract investment from biogas producers, electricity- and gas grid operators and energy suppliers. It also supports the development of biogas fermentation devices, biogas power generation systems, automated control systems and other relevant technologies. Meanwhile it can improve the industrial manufacturing standards of related equipments and products, so as to promote the development of the biogas industry gradually.

THE STATUS QUO OF COMPREHENSIVE UTILIZATION OF LIQUID PHASE OF BIOGAS DIGESTATE IN CHINA AND ABROAD

YUAN Yi

Qingdao Tianren Eco-Mansion Environment Co., Ltd.

Contact: YUAN Yi, Qingdao Tianren Eco-Mansion Environment Co., Ltd. No. 30, Ke Yuan Wei Si Road, Laoshan District, Qingdao 266101, China, Tel.: +86-532-8099-6035, E-Mail: yuanyi@tianren.com

EXECUTIVE SUMMARY

This paper describes the three general types of biogas slurry utilization in China and abroad: direct application of the biogas slurry to the field (eco-farm type), development of commodity fertilizer, and algae cultivation. Research and comparative analysis show that the biogas slurry eco-farm type model with its simple processing, matured technology, the low initial investment and low operation cost, can reach a high economic benefit, so its wide utilization is recommended.

Key words: Biogas slurry; comprehensive utilization; biogas status quo

国内外沼液综合利用模式分析

袁怡*

（青岛天人环境股份有限公司，青岛 266101）

【摘要】本文主要介绍国内外沼液综合利用的三种模式：沼液直接还田（生态型利用）、开发成商品肥和养藻，通过调查研究和对比分析，可知沼液生态型利用模式工艺简单，技术成熟，前期投资和运行成本较小，经济效益较高，建议推广应用。

【关键词】沼液；综合利用；利用模式

1　引言

沼液是沼气工程厌氧发酵后的产物，其速效养分含量丰富，能迅速被作物吸收利用[1]，而且还具有防病抗逆作用，是一种优质的液体肥料[2]。研究表明，施用沼液能使水稻和小麦增产5%～10%，卷叶螟、叶蝉等虫害的发生率下降60%以上，对蚜虫、红蜘蛛的杀灭率更是达到了90%以上[3]，因其具有生物肥料和生物农药双重功效以及不会带来抗性问题[4]，在国际上被誉为广谱性的厌氧微生物加工剂[5]。但同时沼液存在量大、连续和集中的特点，且有效成分含量相对较低，深加工和远距离运输成本偏高等问题，如何对沼液进行资源化利用、妥善消纳已成为沼气工程正常运行的瓶颈问题。因此，调研国内外大中型沼气工程沼液的利用途径和技术，探讨适合我国国情的沼液利用模式，对推动沼气产业可持续发展，具有重要意义[6]。

---

* 通信作者：袁怡，男，硕士，青岛天人环境股份有限公司，青岛市崂山区科苑纬四路30号 266101，yuanyi@tianren.com，0532-80996035

2 国外沼液综合利用模式

欧美发达国家已具有二十多年的工程实践，积累了丰富的经验，探索出一条“有机废弃物—沼气生产—沼肥还田”的循环发展之路，沼液利用主要通过采用“以料换肥”和“以粪换肥”的形式，以免费或低价出售的方式返回到农场，实现循环利用。

2.1 沼肥生态型利用

沼肥生态型利用的研究和应用主要集中在一些欧洲国家。他们通常对沼气发酵残留物不进行固液分离，而是直接通过沼液车将其运输并喷洒在农田中回收利用。主要在前期调研、沼肥利用的安全性和施肥设备三个方面上进行了着重考虑。

2.1.1 前期调研

欧洲国家在沼气工厂建设前期，通常要对原料的运输收集、沼渣沼液的消纳进行详细、深入的调查，一些国家还就土地对沼渣沼液消纳的承载力进行了深入的研究，并对相匹配的土地和施肥时间都有明确的规定，如表 1[7] 所示。

Table 1 National regulations on the bearing capacity of earth with nutritient anaerobic digestion residue

表 1 一些国家有关土地对厌氧消化残余物营养物质承载力的规定

国家	最大营养负荷(kg N/hm^2 · a)	需要的储存时间(月)	强制的施用季节
奥地利	100	6	2 月 28 日至 10 月 25 日
丹麦	2003 年前：210～230(牛)140～170(猪) 2003 年后：170(牛)140(猪)	9	2 月 1 日至收获
意大利	170～500	3～6	2 月 1 日至 12 月 1 日
瑞典	基于畜禽数量	6～10	2 月 1 日至 12 月 1 日
英国	250～500	4	
法国	150		
美国	第一年 450，其后 280	12	

2.1.2 沼肥安全利用

在沼肥的安全利用方面，德国最早提出并通过建立《沼肥法》和具体的技术规定对沼肥的安全利用进行监控和管理，如在《循环经济法》中规定了“农业施肥中物质闭合循环管理的要求”，并对沼渣沼液在农业及林业用地上的利用设置了使用条件。如重金属限定值的设置，见表 2[8]。

Table 2 Heavy metals limitations in sludge application

表 2 污泥农用的重金属限定值

金属	欧共体法规限定值(μg/kg)	德国新污泥农用条例限定值(μg/kg)
Ph	750～1 200	900
Cd	40～120	10
Cr	1 000～1 500	900
Cu	1 000～1 750	800
Ni	300～400	200
Hg	16～25	1
Zn	2 500～4 000	2 000～2 500

2.1.3 施肥设备

在一些北欧国家，为了提高沼肥的利用效率，避免 NH_4^+-N 的挥发，已不主张用喷灌的方式进行土地施肥。如丹麦将沼液车进行改装，在其周边加载管道进行触地式的施肥方式(图 1)，并在此基础上增设了机械臂，通过机械臂实现“翻土—灌溉—覆土”三重作用，有效地解决了 NH_4^+-N 养分易挥发和施肥不均、空气污染等问题。

Figure 1　Special vehicle for sludge fertization in Denmark

图 1　丹麦沼肥施用专用车

2.2　商品化开发

欧美等发达国家对沼液的商品化开发主要通过低压或渗透工艺对沼液进行浓缩。如德国 Kaarssen 一个名为 Bioconstruct 的沼气工程每年处理 1×10^5 t 的奶牛粪和 3×10^4 t 的玉米秸秆。使固液分离后的沼液进入三个低压凝水器(见图 2)，收集的冷凝水排入城市污水管网，浓缩的沼液(浓度为原来的 2.5 倍)储存后作为浓缩液肥卖给附近的农场，该方式也是沼液商品化最常见的一种方式。在蒸发环节，多效蒸发器也是一种常见的设备。

Figure 2　Water condenser

图 2　凝水器

2.3　沼液养藻

目前已有大量的试验研究表明，藻类是一种优质的生物质能源，可以用来制取沼气、生物柴油和生物乙醇等多种高附加值的清洁能源[9~11]。用沼液培养藻类的过程中，沼液能够为藻类生长提供必需的生长元素，而藻类则通过消耗沼液中大量的 N、P 等营养元素，使沼液达标排放，解决环境污染和沼液处理难等瓶颈问题，从而形成“沼液养藻—回收能源—水质净化”的循环利用模式。但是当水体中 NH_4^+-N 浓度大于 50 mg/L 时，会对藻类生长产生抑制作用。因此需对沼液进行曝气处理和藻种耐氨驯化，使藻种能够在高浓度的 NH_4^+-N 环境中生长[12,13]。目前荷兰在这方面作了一些探索，如 MARIS 公司通过养藻沟渠和光反应器相结合的方式实现沼液养藻技术的产业化，再用微藻来制取生物柴油。通过该技术，每年可处理约 3 000 m^3 的沼液。

Figure 3　Live-picture of Algae Planting in Slurry by MARIS Company, Holland

图 3　荷兰 MARIS 公司利用沼液养藻实景

3　国内沼液综合利用模式

我国沼液资源化利用的起步相对较晚，由于耕作方式的差异和应用条件的限制，在沼液肥的开发技术、质量保证、施用安全性的管控和监督上，与国外相比尚有不小距离。通过近几年的时间摸索，部分业主已积累了一定的成功经验，并探索出了一些综合利用模式。

3.1　生态型利用

生态型利用是根据当地的土地资源、农作物情况和地形等特征设计一套沼液输配系统，实现沼液的生态化用。先将沼液进行熟化—无害化处理(也可固液分离后直接进行灌溉)，然后再根据当地作物的具体情况和沼液营养元素的含量，有针对性地添加营养成分或稀释，并根据地形用管道或沟渠设计一套灌溉系统。该技术相对简单，投资和运行成本都较小，管理方便。但是该方式需要有足够的土地资源与之相匹配，若以每亩土地消纳 30 t 沼液计算，则一个日产气 3 000 m^3 的沼气工程需要将近 1 000 hm^2 的土地来匹配其产出的沼渣沼液。目前该模式在国内已有部分典型的成功案例。

如北京留民营沼气站每年产生沼液约 3.2×10^4 t，主要通过北京农林科学院的沼液预处理设备，利用三级过滤工程措施和微泡曝气、水汽联合反冲洗等技术使沼液达到滴灌要求。再将沼液与灌溉用水进行自动化配比，通过管道输配，用于该村 21 个大棚和 16 个温室的滴灌，实现水肥一体化，提高肥料和水资源的利用率。又如在重庆开县某沼气工程利用沼液对 1 000 hm^2 山地柑橘进行灌溉，先通过重金属预处理装置除去沼液中的重金属等有害物质，然后以管道的形式对山地柑橘进行浇灌和喷灌。

Figure 4　Drop irrigation to vegetable greenhouse

图 4　沼液大棚滴灌现场图

Figure 5 Drop irrigation to tangerine in hilly area

图 5 山地柑橘滴灌

3.2 商品化开发

3.2.1 生物菌肥

该技术是通过现代生物工程技术，在沼液中接种固氮菌、根瘤菌等功能菌剂，并添加适量的能量物质再次厌氧发酵，从而对功能菌种进行激活和富集培养。通过添加料的催化作用，沼液与功能菌种共惠互生。功能菌种在利用沼液各种营养元素进行自身繁殖的同时，又激活沼液中的各种营养元素和氨基酸等众多的生理活性物质，既可用来做农作物的生物菌肥，也可用来对水产养殖区进行水质净化，还可用作饲料和肥料的发酵剂[14]。如湖北某公司通过该技术，利用沼液生产液体菌肥，年生产规模可达 500 t，产品售价为 8 元/kg。由于主要通过简单的半地下式户用沼气池来对菌原液进行富集培养，所以整套设备成本较低，投资约为 5×10^5 元。

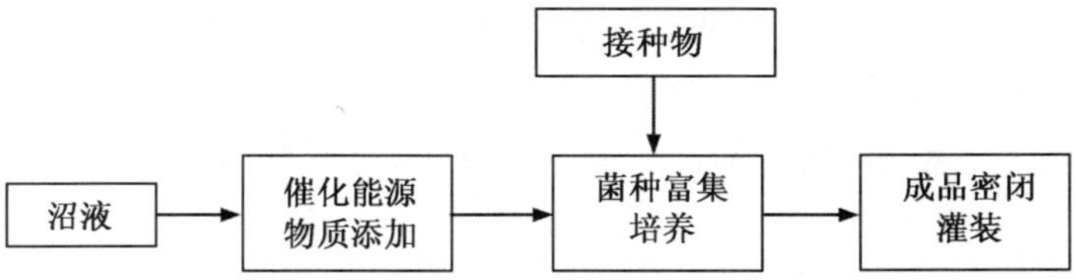

Figure 6 Slurry bacteria fertilizer production process chart

图 6 沼液菌肥生产工艺图

3.2.2 营养液

该技术利用秸秆、棉籽壳等与沼液进行再次厌氧发酵，以增强原液的有机酸和其他肥效成分，然后通过絮凝、沉降塔和超细过滤器等工序去除沼液中的小颗粒物质，再根据不同作物的生长特性调配多种营养元素，并通过螯合技术和乳化剂增强沼液肥的稳定性，从而使沼液营养更加全面，缓急相济，其多种生理活性物质又保证植物健康生长[15]。当前市场上已出现一些沼液肥产品，如北京某公司的花卉营养液和有机复合冲施肥均是通过该类技术研制而成。目前已在新疆和北京安装完毕三套生产设备，但都还没有正式生产运营。建设规模都为 4 000 t/a，总投资在 $1.2\times10^6\sim1.5\times10^6$ 元之间，其中每吨产品的生产成本为 800 元，销售价格在 1 800～2 000 元之间，具有较高的经济效益。沼液肥生产工艺图见图 7。

3.3 优劣势比较分析

对以上几种模式分析比较如下：

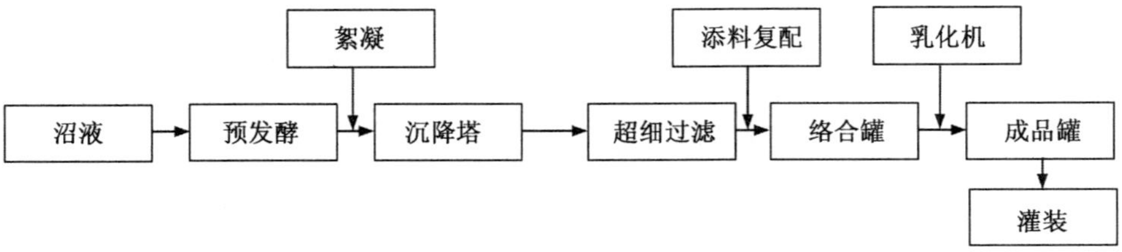

Figure 7　Nutrient solution products

图 7　沼液肥生产工艺图

Table 3　Comparison of various slurry application methods

表 3　沼液利用方式比较

沼液利用方式	生态型利用	商品肥开发	沼液养藻技术
投资	低	较高	一般
运营成本	低	高	一般
占地规模	大	小	较大
技术成熟度	成熟	不成熟	不成熟
技术通用性	好	较差	好
经济效益	一般	高	一般
市场预测	好	较差	一般
工程案例	北京留民营、重庆开县柑橘滴灌等	北京花卉营养液等系列产品、黄石液体菌肥	荷兰 maris 公司

通过比较分析，对以上几种模式的特点归纳如下：

(1)沼液生态化利用是通过设计一套沼液输配系统，利用丰富的土地资源和农作物来对沼液进行安全消纳，实现经济和生态的双重效益，最主要的目的是实现沼液的就地、及时消纳和处置，避免直接排放造成水体污染。该模式简单易行、技术难度小，投资和运行成本均较小，有较好的应用前景，是现阶段沼气工程业主主要采用的一种处理方式。

(2)将沼液开发成高附加值商品，不管是生物农药、菌肥还是营养液，目前在技术上都不够成熟。以菌肥为例，不管是生产过程中对于杂菌率的控制，还是产品的稳定性和保质期都还不能得到很好的保证。其次在经济上，虽然这些产品具有较高的价值，但是目前市场还没有形成规模。根据笔者对多家有机肥厂的调查得知，目前液体肥的市场容量较小，因肥效不突出、使用不便等原因，使得农民对液体肥的认可度不高，由于成本较高、销量不大，生产厂家积极性也不高。在产品外观上，与其他商品液体肥相比，沼液肥不仅色泽浑浊，而且异味较重，竞争优势不大。笔者认为沼液商品肥的开发将会是将来的一个发展方向，但当前各方面条件还不成熟，不宜作为大规模推广的沼液利用模式。

(3)沼液养藻技术投资和运行成本都较低，也有较好的经济效益，同时通过藻类的吸收和净化使沼液得到净化处理，而藻类则可作为厌氧发酵的原料或用来制取生物柴油、乙醇等高附加值产品。但该模式首先对地域和时间的限制较大，北方高寒地区的冬季寒冷漫长，而藻类属于喜温作物，在北方无法正常生长，若通过阳光棚加热等方式来养藻，则会因能耗成本过高而得不偿失。其次在技术上还需进一步完善，如用藻类产沼气极易产生酸化，藻类的收集存在一定困难，需防止微藻收集不完善排入周边水体，导致周边湖泊、水体产生水华等现象。

4　结论与建议

(1)沼液生态型利用模式工艺简单，消纳量大，投资和运行成本较小，经济效益较高，建议作为重点推广应用模式。

(2)我国各个地区的气候、环境及经济条件差别较大，遵循因地制宜原则，在生态型利用模式的基础上，走多元化发展之路。

(3)加强各专业联合攻关，提高沼液深度利用和商品化水平，增加产品附加值，完善沼气工程产业链，推动沼气产业发展。

参考文献

[1] 周孟津. 沼气实用技术[M]. 北京：化学工业出版社，2004.

[2] 刘荣厚，赵玲，武丽娟. 北方农村能源生态模式沼气发酵原料及其产物特性的测试与分析[J]. 中国沼气，2005，23：218-221.

[3] JOHN K. Pear-biogas-pig production simulation model [D]. Zhejiang University，2003.

[4] 农业部环能司. 沼气技术手册[M]. 成都：四川科学技术出版社，1990.

[5] 汪善锋，陈安国，汪海峰. 规模化猪场粪污处理技术研究进展[J]. 家畜生态，2004，25(1)：49-54.

[6] 吴飞龙，叶美锋，林带炎. 沼液综合利用研究进展[J]. 能源与环境，2009，1：94-95.

[7] 邓梁伟，陈子爱，袁心飞，等. 规模化猪场粪污处理工程模式与技术定位[J]. 养猪，2008，6：21-24.

[8] 姚刚. 德国污泥利用与处置[J]. 城市环境与城市生态，2000，13(1)：43-50.

[9] Sheehan J，Dunahay T，Benemann J，et al. A Look Back at the US Department of Energy Aquatic Species Program：Biodiesel from Algae. Close-out report，1998.

[10] International Energy Developing Algae to Oil for Many Applications. Renewable Word，2007，11-7.

[11] Algenol Biofuels Has Big Plans for Algae Ethanol. Green Energy Trends，2008.

[12] V Witt and J A，Borchardt. The removal of nitrogen and phosphorus from sewage effluents through the use of algal culture[J]. J. Biochem. and Microbiol. Technol. and en-gng. 1960，2：187-203.

[13] 张国治，等. 藻类对沼液中氮、磷去除作用的初步研究[J]. 中国沼气，1997，15(4)：11-15.

[14] 熊小京，曹晓婷，李博. 有效微生物复壮过程特性研究[J]. 厦门大学学报，2007，46：107-110.

[15] 范振山，张全国，等. 生态型沼液(渣)加工利用技术和设备研究[J]. 中国沼气，2006，24(1)：24-26.

提高有机废弃物的能源利用效率——多重标准方案

T. Bahr, W. Jacobs, T. Thiel, K. Fricke

(*Braunschweig 工业大学*, *Braunschweig 38106*, *德国*)

【摘要】已有不少描写和评估生物质能源提取最大化的研究和模型。优化潜力主要取决于对一般效率因素的理论评估。本文对有机残渣和废弃物的评估进行了进一步讨论。

大多数情况下，研究者没有充分考虑经济需求和对设备安装相关解决方案的汇总，也没有按照类似工程建造的单个设施进行相关分类。因此不难发现目前尚缺少实际优化潜力和可行的操作方式。

要在现有以及计划建设的设施的基础上实现最优化，就需要确定可行的行动计划，并了解对生物质能源工程的影响。比较不同技术解决方案与程序并推导出提高能源效率的技术，就需要相应的数据矩阵，可比较的背景条件。

在最近的一项研究中第一次为这些研究目标采用了 SWOT 的分析法。SWOT 的分析法(S-优势，W-劣势，O-机会，T-威胁)是一种战略分析法，早已被应用于形成性评价与质量管理方面。SWOT 分析法是一种战略规划方法，用于评估为实现预先设定的结果或目标所需采取的可行性战略解决方案。它也有助于展示这些结果。

其中 SW(优劣势)部分用来处理内部因素，可将其定义为关于工艺或工厂设备的技术可能性和潜在的发展能力。这种方式被用于众多技术方案和工艺上。根据不同的发展阶段对每个过程进行划分，并对技术状况定义一个基准值。OT 部分(机会和威胁)包括各个工艺中整体和特定情况带来的趋势和改变。例如，这些情况会受到法律法规、限值或计划支持的影响。

关于初步数据的获取，被检验的工厂及其组成部分是分开分析的。通过数学模型，可以确定能源效率的核心数据。由该数据产生的基准则直接被用来比较工艺步骤、设备组成部分以及完整的系统。在这个模型基础上推算出工厂的指数，可用于工厂之间的区分比较。利用一个与 LCA(生命周期评估)或者生态指示器类似的系统，也可对不同类型的工程(如有机废弃物发酵、堆肥、热处理)指数进行比较。根据分析的复杂程度，整个工艺系统的能量和物质平衡可以细分到几个工艺单元。由此就可以对每个单元实现不同评估，进而确定某具体单元的优化潜力。

通过 SWOT 分析法得到的综合结论可推出有效的优化方案。这些方案既可以供工厂的制造者和操作者参考，也可以作为政治决策的现实背景。本文选取目前在能源生物质利用上具有代表性的德国工程(好氧、厌氧生物处理以及热处理)来开发上述方法，以满足实际技术所需。

【关键词】生物质能源利用；生命周期评价；能源效率

ENHANCING THE ENERGY EFFICIENCY OF UTILISING BIOGENOUS RESIDUES-A MULTI-CRITERIA APPROACH*

T. Bahr, W. Jacobs, T. Thiel, K. Fricke

Technical University of Braunschweig

Contact: Dr. Tobias Bahr, Wiebke Jacobs, Timo Thiel, Prof. Dr.-Ing. Klaus Fricke, Technical University of Braunschweig, Leichtweiß-Institute, Waste and Resource Management, Beethovenstraße 51a, 38106 Braunschweig, Germany, Tel.: +49-531-391-3970, E-Mail: t.bahr@tu-bs.de

---

* This paper was initially presented at the Orbit 2009 CHINA. 本文已发表于 Orbit 2009 CHINA。

EXECUTIVE SUMMARY

There have been several research projects and models to describe and evaluate the potential for optimising energy recovery within the range of biomass utilisation. The optimisation potential was determined mainly by theoretical assessments of generalised efficiency factors. This paper refers in particular to those estimations that have been made for organic residuals and waste.

In most cases economical needs as well as the compilation of solution factors relevant to facility installations have not been taken into account adequately. A relative classification of single facilities with reference to similar existing plants has not been made. Therefore a lack of realistic optimisation potentials and practical implementation approaches can be identified.

To realise specific optimisation measures on existing and intended facilities it is compelling to identify realisable courses of action and the effect on the stock of bio-energy plants. A data matrix and a comparative background are needed to realise an alignment of different technical solutions and procedures as well as the deduction of the potential techniques for increasing the energy and resource efficiency.

In a recent research project a SWOT analysis is adopted for these aims of research for the first time. The SWOT analysis (S-Strength, W-Weaknesses, O-Opportunities, T-Threats) is an application of the strategic management, which was already used for formative evaluations and quality management. SWOT Analysis is a strategic planning method used to evaluate suitable strategic solutions of a former defined and desired end state or objective. It also assists presenting these results.

The SW-part (strengths and weaknesses) deals with the internal factors, which can be defined as the technical possibilities and the development potentials of the regarded process or plant. The method is to be applied on a multiplicity of technical solutions and processes. To classify the individual procedures by their stage of development, a benchmark is defined referring to state of technology. The OT-part (opportunities and threats) includes trends and changes as a result from the general and specific conditions for the respective process. As an example these conditions can be caused and affected by legal regulations, limit values or support programs.

Referring to a preliminary data acquisition, the examined plants as well as their components are analysed separately. By means of a mathematical model key data of the energy efficiency is determined. The benchmark developed of this data is used to compare process steps, equipment components as well as complete systems directly. Based on this model a plant index is calculated which is used to classify single plants in comparison to each other. Due to a system similar to LCA (Life Cycle Assessment) or Eco-indicators also different types of plants (e. g. bio waste fermentation, composting, thermal treatment) can be compared by this indices. Depending on the complexity of the analysis, energy- and mass-balances of complete process steps can be determined in detail to several units. As a result a so far unequalled itemisation can be achieved which enables to identify optimisation potentials for specific units.

Including the integrated view of the SWOT analysis effective optimisation approaches can be derived. Those can be used both for plant manufacturers and - operators and as a realistic background for political decision-making processes. Currently representative German plants (aerobe and anaerobic biological as well as thermal procedures) for energetic biomass utilisation are selected to develop the described methods meeting the actual requirements of technology.

Key words: Biomass energy utilization; Life Cycle Assessment; energy efficiency

1 Introduction

The effective use of recyclable waste contained residual and waste materials is an important contribution to the conservation of resources and environmental protection. Given a considerable, but still limited amount of biogenic residual materials and acreage for renewable resources the efficient use of biomass is a high priority issue.

German forestry, agriculture, landscaping, industry and waste management produces about 110 million Mg dry matter of biogenic residues each year. The potential contained within must be used for the purpose of a sustainable resource management to gain for energetic and materialistic utilisation. Using these potentials consisting of the energy potential and substantial quantities of organic fertilisers, they will substitute mineral fertilisers and nutrients can be substituted to save energy and raw materials (fossil resources).

1.1 Background

In order to achieve maximum reduction on exhaustible energy resources and on the emission of greenhouse gases, it is necessary to accomplish the highest possible efficiency of the deployed systems and procedures. The considered research methods are fermentation, composting and incineration processes. Publications from research projects and models to describe the optimisation potential of energetic biomass utilisation, especially of organic residues are present on a large scale. Concrete optimization measures based on theoretical assessments are only partially suitable to be realized on specific plants. This results from the economic constraints and because the established procedures are not sufficiently taken into account.

The energy efficiency and therewith the energy demand of treatment facilities is strongly affected by the implemented process. As an example composting plants can be divided in three energetic categories (Witzenhausen, 2008).

- Simple plants, for example open, unventilated rotting with aggregates; middle energy consumption ca. 25 kWh/Mg;
- More complex, partly capsule plants; middle energy consumption ca. 48 kWh/Mg;
- Very complex plants, capsule, extensive technical arrangement against air pollution, control pressure ventilation or draw ventilationwith organic filter; middle energy consumption ca. 77 kWh/Mg.

As a result of a high energy demand Edelmann & Schleiß (Edelmann et al., 2001)conclude that based on the energy potential the capsule plants are inferior against the open rotting. In the fermentation process the energy efficiency is less, known from service experience. This results of:

- low operational availability resulting out of long afresh-starting time, corrosion, abrasion;
- high priority of facilities energy demand;
- low efficiency in the use of waste heat of block heat and power station;
- low degradation rate at long retention time.

In the context of thermal utilisation the focal point is on developing optimised RDF (refuse derived fuels). In a first step a choice of combustible residuals must be made.

Concrete measures on optimisation based on theoretical assessments are only partially suitable to be implemented an specific plants, as economic pressure and the established procedures in practice are not sufficiently taken into account.

To use the potential energy and material resources of biomass residues effectively, the energy efficiency of the processes must be apprehended. At the same time the energy efficiency depends on the electrical and thermal efficiency as well as on the nature and amount of the additional power supply.

There are deficiencies in the availability of data on the technical systems, in particular the efficiency of resource allocation. Thus a data matrix and a comparative approach are missing, which would allow a relative classification of individual plants in the entirety of existing facilities and its potential for optimisation in both thermal processes as well as in biomass utilisation.

It appears that in fermentation systems, where evaluations have been conducted, efficiencies below 30% where achieved. Thus the aim of the project is to examine the scope for increasing energy efficiency in the utilisation of biogenic residues.

1.2 Research objectives

Developing an evaluation matrix to compare the procedures, both for energy and utilisation in terms of energy and resource efficiency is the aim of the research project. As a result a plant index can be created, which also allows a comparison of different procedures and thus makes known solutions transferable.

As part of the research project an actual recording and analysis of existing strengths and weaknesses is created in particular of the energy efficiency influencing factors. Furthermore strategies to optimise the infrastructure, technology, operation and products should be developed to optimise existing plants.

2 Methodology

Recording and evaluating data of existing procedures and technical solutions, a data matrix is developed from which optimisation potential can be derived. The derivation of the potential for optimisation is possible through a comparative study of resource conservation and energy and material utilisation potential. To the authors knowledge this is the first time, the SWOT analysis is applied to such a problem. The SWOT analysis is used for a variety of technical solutions and procedures. An appropriate benchmark has to be defined, which enables the development of individual methods to classify them. A tool is to be developed to complement the SWOT analysis and to collect and weight relevant operational and key performance indicators in high detail. It can create an evaluation index has to be developed, which allows the classification of the plants to each other.

2.1 Data acquisition

Among the biotechnological and thermal process technologies, particularly the efficient allocation of resources are only incomplete, no systematic data available. Similarly, there is a lack of treatment specific product information. First, the present data (scientific papers e. g. Ökoinstitut, 2004; Ifeu, 2007a; Ifeu, 2008) set based on an extended literature review is recorded and the data bases on the utilisation routes have to be cross-checked with each other and tested. It can be assumed that the data collection of previous projects was not done using a common standard and consistent level of detail. The collection of essential data on material flow uses in addition to literature values, the data base of the program GEMIS. This includes information on the availability of energy carriers, heat and electricity, materials, and transport processes. GEMIS calculates for all processes and scenarios so-called life-cycles and has been successfully implemented in other research projects on the creation of scenarios. To create a complete database which is consistent structured and complete for the purposes of processing data, commonly available data and separate surveys on a variety of representative treatment facilities are planned. It is known from previous work that the collection of detailed operating data on existing facilities is hard to gather using questionnaires or similar. Software solutions for process control and data acquisition systems are produced individually, or contain at least plant-specific characteristics. A consistent data collection needs a consistent matrix and the cooperation of operators and / or manufacturers. Through early involvement of the plant operators and manufacturers, it must be within the framework of the SWOT analysis (see Section

2. 2) to assume that a high level of identification and a strong self-interest of stakeholders can be achieved.

2.2 SWOT-Analysis

Taking into account the relevant boundary conditions of a comparative approach, the application of a SWOT analysis is intended for the comparative study of resource conservation and material use and energy potential. The SWOT analysis (an acronym formed of **S**trengths, **W**eaknesses, **O**pportunities and **T**hreats) seeks the intention to determine strategies within organisations and to assess them. It derives from the area of strategic management, but is also applicable for formative evaluations and quality improvement. This instrument is used for situation analysis. It is generally a prerequisite for subsequent development of measures. As part of a strategic planning strengths and weaknesses of current and future opportunities and risks are identified. The SWOT analysis can be divided into an internal and an external part. The internal part covers the Strengths/Weakness-dimension. It lists those factors that are controlled directly by or through the object of study. The presentation of strengths and weaknesses is used to obtain the current state. Whether a criterion is to be classified more strength or a weakness is subject to a closer examination of the overall concept of an organisation or facility. The external part deals with the Opportunities-/Threats-Dimension. The issue of opportunities and risks depends on the future prospects, future opportunities and threats which derives from the considered organisation. This SWOT analysis is used to derive alternative appropriate strategic solutions for achieving previously defined aims and present those in a clearly laid out way.

For the employment of the SWOT analysis to the consideration of technical procedures, it is necessary to adapt the method to the different issues. The strengths and weakness-dimension deals with the technical possibilities and development potentials of a considered proceeding. The Opportunity-/Threats-Dimension implies trends and changes which result from the general and specific conditions for each procedure. When applying the SWOT analysis on technical procedures external factors are those factors, over which a plant operator has no control. These include legal regulations, limits and funding programs. The deduced alternative solutions by the SWOT analysis are therefore development and optimisation potentials in the light of existing and expected conditions.

The application of the SWOT analysis on the subject efficiency of plant systems will be held for the first time and holds a high degree of innovation. A huge advantage of the SWOT analysis is that it can be applied both during the analysis phase and in the design phase. Finding out whether the strengths of the currently applied technologies and procedures for utilisation of available biomass are suited and in the context of current and anticipated future conditions are relevant is the aim of the SWOT analysis.

The SWOT analysis will always be created based on a target of the project and are not kept abstract. A basic objective of the project is to increase energy efficiency. This has to be concretised-based on the investigation and using existing literature and surveys carried out during the course of the project.

Strengths and weaknesses which have been designated in the course of the SWOT analysis are relative to each other. Only through benchmarking with other methods or usage examples they are given significance.

To ensure a consistent approach and comparable results in the analysis of different proceedings (anaerobic digestion, composting, thermal utilisation) within the framework of the SWOT analysis, a list of criteria has to be defined. Such lists will be refined in turn by secondary criteria. Such lists of the SWOT analysis will be used for facility inspections of selected, representative plants in the form of individual workshops. Through early involvement of the plant operators and manufacturers in the framework of the SWOT analysis a high level of identification and a strong self-interest of actors will be achieved.

Within the SWOT analysis no specific measures will be determined, but the relative classification of the plants themselves will be made possible and in a second step strategies are formed. In the course of an action plan, the strategies are transformed into concrete actions.

2.3 Plant index

In order to compare plants with different processes and design in energy efficiency with each other, the creation of a complex index is needed. In the first instance, the specific energy consumption is recorded in high detail and weighted. *Figure 1* shows the flow chart how to determine the plant index. At first the different plants will be considered separately in Part A, based on the different data collection systems.
For this purpose it is necessary to split the process steps within the plant into the main proceeding steps to allow differentiation from other process steps (Part B). It is for example specified where the range of delivery ends and where the sorting area starts.

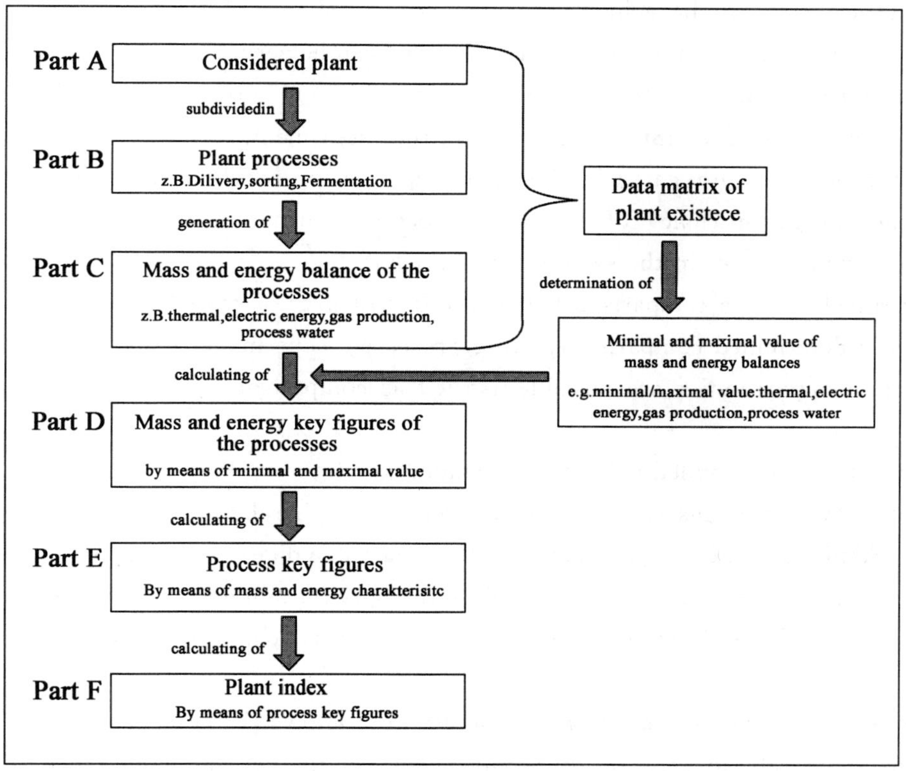

Figure 1 Flow chart for determination of the plant index

For the separate process steps to guarantee comparability has to be guaranteed by normalising to a fixed unit of mass (e. g. 1 Mg dry matter) and specific characteristics derived (e. g. specific thermal and/or electrical energy). Expecting that the specific energy consumption of large plants is lower than those of small plants a classification based on the plant capacity is made to allow comparisons and to avoid distortion of the results. A final classification results after the compilation of the data base.
Plant parts not correlating with the mass flow rate will be considered separately. Including chambers and any buildings, that are not part of the plant and are not correlated directly to the plant, such as administration, central weighing station, canteen, social buildings, etc. Thermal and electrical energy are distinguished for the energy balance. For the mass balance, depending on plant type, the corresponding mass flows are taken into account. This includes for example process water, discharged air, input and output of the processed substrate (Part C). In the case of especially important process-specific materials (special pollutants, potential recyclable waste), reviewing is needed. Produced potential recyclable

material, e. g. compost and other fertiliser substitute are considered in the evaluation, as these products pose a key figure, which can affect the evaluation in a positive way (taking into account the energy/raw material savings by substitution of conventional fertilisers).

Data identified in Parts A-C are collected in a data matrix. From this matrix a minimum and maximum value for each mass and energy balance is calculated. For example the minimum value of the gas formation potential of the gas yield of anaerobic systems is provided by the least efficient plant. The maximum value of gas yield is applied by the most efficient plant. Using these minimum and maximum values an energy/mass ratio is calculated for each process step. Through this key figure it should be possible to compare the results of energy and mass balance of the considered process step with energy and mass balances of the same process measured in other plants (Part D). Depending on the current load or feedstock quality, different power efficiencies can be achieved for the same system and the same kind of process. The processes can have a different "minimum load-capacity" and can be very inefficient while running at reduced load. This is relevant for example with pronounced seasonality in quantity and/or quality of the feedstock. Again, these differences are taken into account during the assessment.

Energy and mass ratios are aggregated in the next step to a process indicator. This figure allows the comparison of the energy efficiency of the same process in different plants as well as the relative comparison with other processes in the same plant (Part E).

With the mathematical model a comparison between different process steps is possible, since the relative specific efficiency is determined comparing existing processes. The model holds the advantage that both systems-optimisation opportunities are possible as well as comparisons of the same process in different plants.

Finally the plant index is determined using the calculated key figures (Part F). The plant index gives the proper context of all the plants analysed relative to each other, whereby it is possible through the model to compare different kinds of plants (e. g. organic waste: anaerobic digestion, composting) with each other. The model is structured in such a fashion that using the specific minimum and maximum values of each energy and mass balance for each process step, a normalisation is achieved, independent of the plant kind and structure.

The rating index's calculation method of the developing tool allows the comparison of single components and process steps separately to each other. Thus, plants with low and high technical equipment are comparable in detail, as each individual process step is compared with the same process step in other plants. *Figure 2* illustrates an exemplary approach for calculating the evaluation index.

The model can be progressively updated and the simulations and forecasts of the optimisation potential maintained at the state of the art. Since the mathematical model is based on the real existing minimum and maximum values, changes in the asset existence and thus the technique directly affect the overall index of each plant. Through this link, it is possible to constantly update the data of the entire asset existence, and to simulate scenarios for further optimisation steps. The maximum value used for the normalisation is a real best-practice value, which is extracted from the data matrix (*Figure 2*). For the purposes of an ideal limiting case of such a consideration the real maximum value for the determination of the energy and mass ratios may be replaced by a hypothetical maximum value, determined from theoretical considerations and the latest research findings. This is not yet implemented in practice, but allows the estimation of the best possible progress in implementation of the current state of research.

This method of simulation can also be used to develop a facility with the current state of the art (best practice) approach. Therefore the maximum value for each energy and mass ratio is adopted (from the data

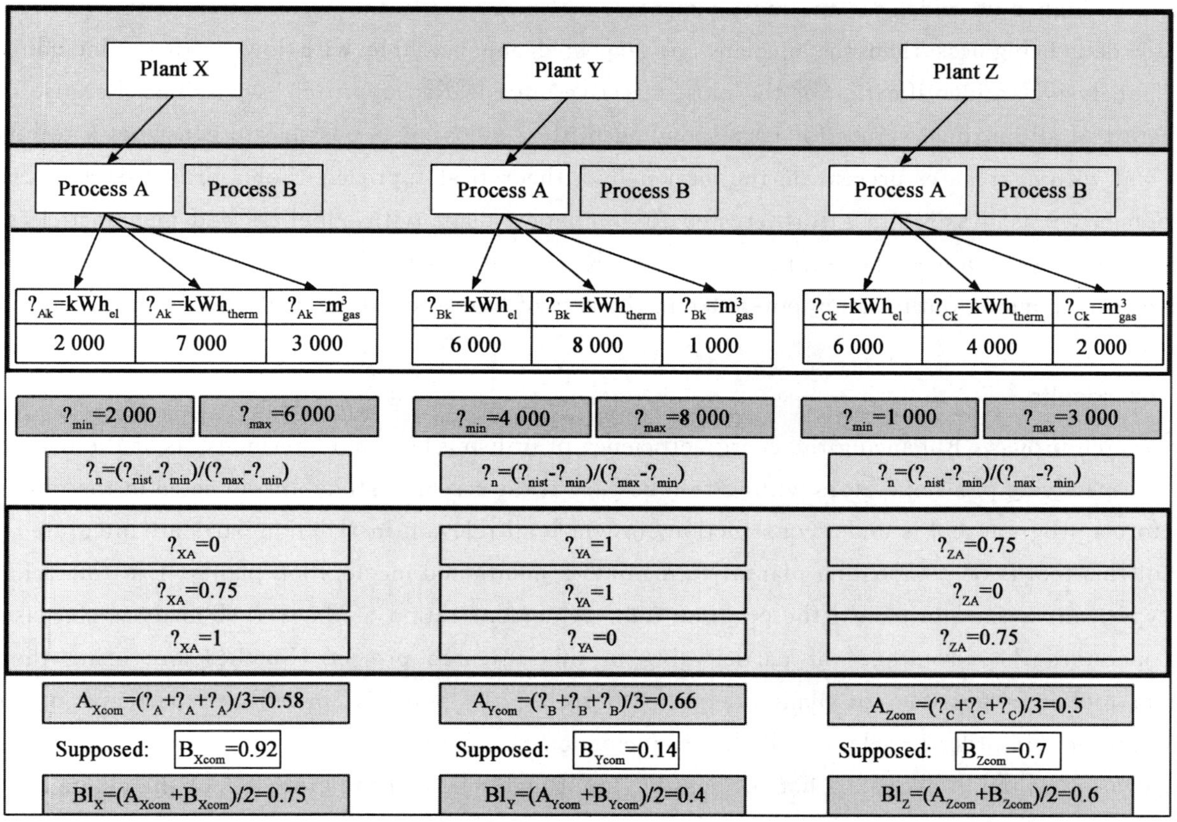

Figure 2 Exemplary extension for the calculation of the estimation index

matrix, see *Figure 2*). Using the maximum process indicator the plant key figure is calculated. Comparing this figure for each facility can help to identify and enhance the potential of optimisation due to real-life values. The focus can be extended under consideration of actual hypothetical, but in the near future realistically achievable energy efficiency of relevant parts of the plant (state of research).

2.4 Valuation for enhancing the energy efficiency

On the basis of the index, which results from the several levels of the plant index, it is possible to implement a four-stage estimation of the individual procedure efficiency.

(1) Comparing respective procedures with the best valued plant at the market (real comparison);

(2) Comparing the respective procedures with one hypothetical best-practice-plant by using a combination of the best valued part processes of several plants present at the market (hypothetical comparison-state of technology);

(3) Comparing the respective procedures with one hypothetical best-practice-plant through idealised grouping of optimised part processes under consideration of the last state of research (hypothetical comparison-state of the research);

(4) Comparing the respective procedures with a hypothetical theoretical extension. Reference is the energy content as a target value (100% effectiveness)-exploratory comparison of basic utilisation variants (thermal utilisation, material utilisation, fermentation, ethanol production).

The optimisation of individual plants is another way of using this model. By means of this model it can be assessed in a simple manner, which installation parts offer the greatest optimisation potential. This is possible because of the standardisation for different processes which then become comparable with each other. The higher the index is, the more expensive and more costly the optimisation of this process will be. Key data being less than the median, are expected enhance able with lower effort. Including the SWOT analysis, an identification of the most effective optimisation approach is possible. Because of the summation of all maximal values for corresponding process steps, it is possible to construct a technically best plant with aggregates present on the market by a theoretical approach. For a prognosis and scenario development, it is possible in a further step to simulate a plant with suppliers and manufacturers that reflects the newest state of the technical research. Regarding this results existing plant can be measured and the corresponding optimisation potentials in the area of biomass and residual utilisation can be shown.

3 First results

The research project "Enhancing the energy efficiency of utilising biogenous residues" is at the beginning. Plants respectively plant operators willing to take part in the research study are collected in the first step. Therefore a program tool is under construction, in which all relevant installation parts are integrated. The focus of this tool is on composting plants, fermentation plants and incineration plants. Creating a kind of archetype-plant is the aim to test the program tool. A demonstration of the SWOT-analysis can visualise plant operators the advantages of participating in this research project. Furthermore using the first program tool it is possible for plant operators to realise the potential benefits of participating in the research project in order to achieve a higher participation rate for the project.

In the context of the project start lists of possible input materials and the processes of the plants has been purposed. As a next step, it is intended to contact specific plants or plant operators who are potentially willing to participate in the research project.

A selection is been made, because it is not to the point to include all existing treatment plants, composting plants and incineration plants. In a first step process concepts and systems are selected that are representative for all remaining facilities in Germany. *Figure 3* shows the representation on the selection of process concepts. To ensure the comparability it is necessary to classify the plants. Plants using online data logging are preferred, as a result of more effective and economical project execution. Special consideration is needed for plants with an additional nutrient utilisation because of sustainable utilisation processes and systems with innovative combinations like fermentation with composting. Considering the possibility of composting plants with a high optimisation potential through special characteristics are favoured at the choices of plants.

4 Conclusions

Known from evaluations, the efficiency of biotechnological processes and with lesser significance thermal processes are considered to be marginal. Regarding the rotting process, the energy efficiency of closed complex plants is much higher than the one of open basic plants (Edelmann, 2001). Fermentation plants exhibit lower energy efficiency as a result of plant availability because of afresh-starting time, corrosion, abrasion. The minimum utilisation of the rejected heat, resulting from block heat and power plants is another challenge. Reliable data is limited for technical systems. Consequently collecting data a SWOT analysis will be made.

Based on the results an energy efficiency optimised plant system will be identified and developed resulting from the analysis. In the context of this research project the energetic and also the material treatment of biological residuals is considered-differing from several existing research approach so far.

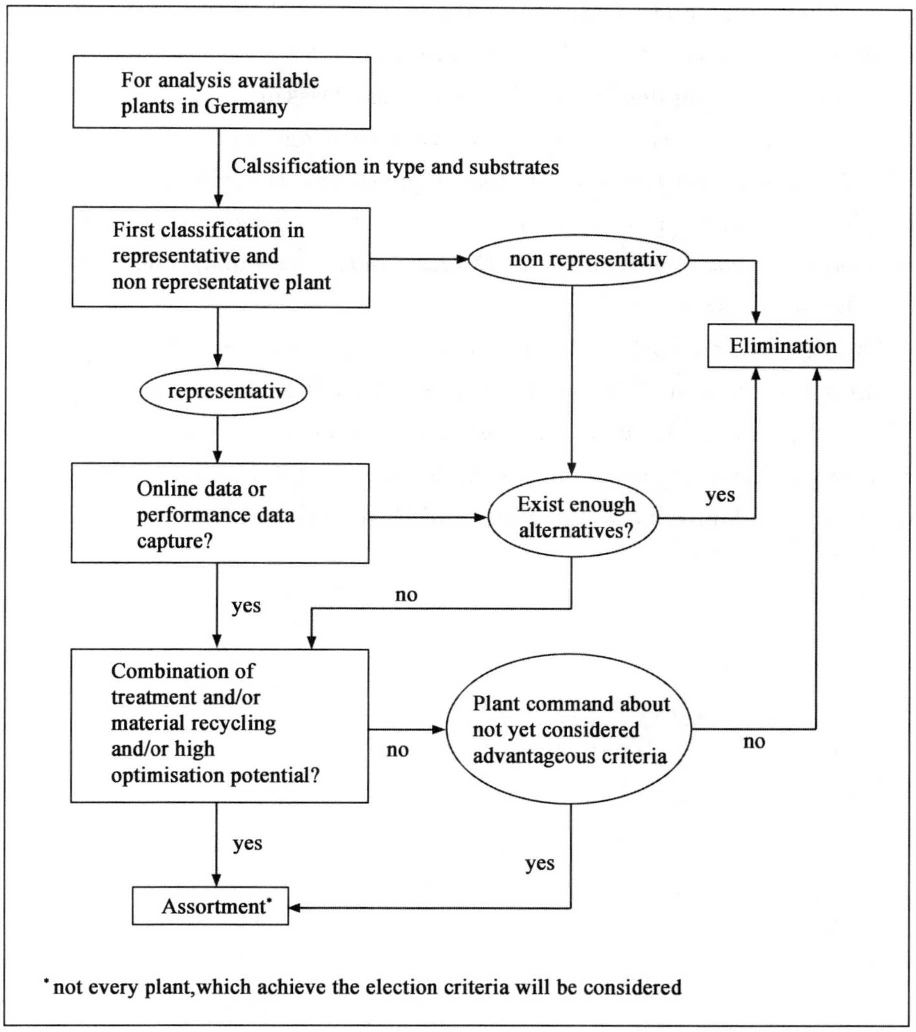

Figure 3 Representation on the selection of process concepts

For the appraisement of the ecological and economic value, the energy generation as well as the potential for energy saving by product substitution (e. g. fertilizers) is taken into account.

As a result of the research project realistic energetic optimisation of treatment facilities for organic residues can be achieved. Gathered data and results will be published consecutively.

Acknowledgements

Funded by Federal Ministry for the Environment, Nature Conservation and Nuclear Safety, a support programme of Germany's Climate Initiative

References

Drews G., Hillebrand N., (2007): *Lexikon der Projektmanagementmethoden*. Rudolf Haufe Verlag, München.

Edelmann W., Schleiß K., (2001): *Ökologischer, energetischer und ökonomischer Vergleich von Vergärung, Kompostierung und Verbrennung fester biogener Abfallstoffe*. Studie im Auftrag des Bundesamtes für Energie und des Bundesamtes für Umwelt, Wald und Landschaft.

IFEU-Institut für Energetik und Umwelt; Öko-Institut, (2007): *Stoffstrommanagement von Biomasseabfällen mit dem Ziel der Optimierung der Verwertung organischer Abfälle.* Forschungsbericht im Auftrag des Umweltbundesamtes; Dessau.

Pehnt M. , Vogt R. , (2007): *Biomasse und Effizienz Vorschläge zur Erhöhung der Energieeffizienz von § 8 und § 7- Anlagen im Erneuerbare-Energien-Gesetz*, Heidelberg.

IFEU-Institut für Energetik und Umwelt, et al. , (2008): *Optimierung für einen nachhaltigen Ausbau der Biogaserzeugung und -nutzung in Deutschland.* Verbundprojekt gefördert vom BMU, Projektträger: Forschungszentrum Jülich.

Öko-Institut. , (2004): *Stoffstromanalyse zur nachhaltigen energetischen Nutzung von Biomasse.* Verbundprojekt gefördert vom BMU im Rahmen des ZIP, Projektträger: FZ Jülich.

Witzenhausen. , (2008): *Aufwand und Nutzen einer optimierten Bioabfallverwertung hinsichtlich Energieeffizienz, Klima- und Ressourcenschutz.* Forschungsbericht im Auftrag des Bundesministeriums für Umwelt, Naturschutz und Reaktorsicherheit, Witzenhausen.

荷兰经验:从市政污泥中回收能源

H. Guo[1], E. Koornneef[1], B. Lue[2]

(1. *DHV BV*,*Amersfoort*,荷兰;2. *DHV BEEC*,*Amersfoort*,荷兰)

【摘要】 随着城市化发展的加快,中国有大量的废弃物资源。而有机废弃物对环境的影响是中国目前面临的一个挑战。高人口密度的荷兰和中国快速城镇化的地区有着许多相似之处,因而荷兰在有机废弃物处理方面的经验和探索对中国目前的情况具有参考借鉴价值。

生物质可以通过厌氧消化转化为沼气,并进一步转化为能源,如电、热和燃料。生物质产能对控制全球二氧化碳排放、节约能源和管理有机废弃物有着重要影响。

在荷兰,市政污泥是生物质来源之一。从19世纪60年代开始荷兰就开始处理市政污泥,如今的重点是从市政污泥中回收能源和资源,从而促进了有机废弃物的管理。现在有超过50%的污泥被厌氧消化处理,几乎所有的污泥最终都被焚烧,焚烧灰用于生产沥青、水泥或者混凝土的骨料。

混合发酵在荷兰有很大的市场潜力。污泥处理厂(STP)现有的厌氧发酵系统通过混合发酵不仅可以用来处理市政污泥,还可以处理其他生物质如畜禽粪便和餐厨垃圾。这种处理方法一方面因有机质的额外增加而使沼气产量增加,另一方面则处理了更多的有机废弃物。

在荷兰,利用沼气的传统途径是用热电联产(CHP)技术发电和产热。因为从有机废弃物里产生的二氧化碳属于二氧化碳短循环,不会增加二氧化碳排放,所以沼气发电对环境有着积极影响。而其生产的热量主要用于提升污泥处理的温度,但在夏天污泥处理厂会有多余的热量。新的方法是联合其他相关单位如工厂或者市政部门,使沼气或者剩余热量能够更好的被利用。

另一个有效利用途径是将沼气提纯为压缩天然气(CNG)。压缩天然气可以注入天然气管网或者用作车用燃气加气站的替代燃料。

将生物质转化为能源的效率取决于很多因素,包括技术工艺、经济因素以及不同机构之间的合作。例如利用沼气提纯与用化石天然气生产压缩天然气的经济性对比。还需要注意污泥与其他生物质资源混合发酵的技术问题。不同机构有机废弃物的收集也会影响生物质管理方式的选择。

本论文描述了荷兰采用的方法,即市政污泥能源回收和污泥与其他生物质资源联合发酵。同时还讨论了这一方式的机遇、问题以及解决途径。

【关键词】 回收能源;污泥;荷兰;有机废弃物处理;混合发酵;回收磷

DUTCH APPROACH: ENERGY RECOVERY FROM SEWAGE SLUDGE*

H. Guo[1], E. Koornneef[1], B. Lue[2]

1. *DHV BV*; 2. *DHV BEEC*

Contact: MSc. Hanning Guo, DHV B. V., P. O. Box 1132, 3800 BC Amersfoort, The Netherlands, Tel.: +31-334-682-481, hanning.guo@dhv.com;
MSc. Eddie Koornneef, DHV B. V., P. O. Box 1132, 3800 BC Amersfoort, The Netherlands, Tel.: +31-334-682-405, E-mail: eddie.koornneef@dhv.com;
Brian Lue, DHV (Beijing) Environmental Engineering Co. Beijing 100028, China, Tel.: +86-10-8440-8442-108, E-mail: brian.lue@dhv.com

* This paper was initially presented at the Orbit 2009 CHINA. 本文已发表于 Orbit 2009 CHINA。

EXECUTIVE SUMMARY

Due to the rapid urbanization, China has a large quantity of waste resources. The impact of organic waste disposal to the environment is a challenge for China. Numerous similarities exist between the densely populated Netherlands and the urbanized developed areas in China. The Dutch experience and considerations about biomass management can be of interest and beneficial for the current situation in China.

Biomass can be converted into biogas by the process of anaerobic digestion, and can further be converted to energy, like electricity, heat and fuels. The conversion of biomass to energy becomes a significant factor in the global management of CO_2 emission, energy savings and organic waste management.

Sewage sludge is one of the biomass sources in the Netherlands. The Dutch sludge treatment started in the sixties of the last century (1960) and is currently focusing on the recovery of energy and resources from sewage sludge which stimulates the improvement of biomass management. Nowadays more than 50 % of the produced sludge is digested. Almost all the sludge is burned in incinerators and the ash is used as filling materials in asphalt, cement or concrete.

Co-digestion is attractive to the market in the Netherlands. With co-digestion, the existing infrastructure for anaerobic digestion at STP's can be used not only for sewage sludge but also for other biomass like livestock manures and food waste from restaurants. Therefore, on the one hand the yield of biogas can be increased due to the additional biomass sources and on the other hand the organic waste can be well controlled well.

The traditional application of biogas in the Netherlands is to produce electricity and heat using a CHP installation. The electricity generated from biogas has a positive effect on the environment because CO_2 out of waste is short cyclic CO_2 and doesn't contribute to the CO_2 emission. The produced heat is used for increasing the process temperature of the sewage sludge. In the summer however there is a surplus of heat at the STP. A new approach nowadays is to look for combinations with other stakeholders, like industries or municipalities where the biogas or surplus heat can be used more economically.

Another new efficient opportunity to make full use of biogas is to upgrade it to CNG (compressed natural gas) quality. It can be injected in a natural gas network or used as a substitute fuel at gas stations for transport purposes.

The efficiency of converting biomass to energy depends on several factors including technical processes, economical factors and cooperation among different organizations. For example, an economy scale is used to produce CNG which should be competitive with fossil natural gas. The technical problems such as mixing of sewage sludge and other biomass resources should be paid attention to. The collection of organic wastes from different organizations also has influence on the selection of approach for biomass management.

This article describes the Dutch approach, namely energy recovery from sewage sludge and co-digestion of sewage sludge with other biomass resources. Opportunities, bottlenecks and the solutions are also mentioned in this article.

Key words: Energy recovery; sewage sludge; the Netherlands; organic waste disposal; co-digestion, phosphorous recovery

1 Introduction

Biomass disposal has become a more important issue in China as a consequence of rapid economic growth and its associated environmental degradation issues. Recently, interests have been growing with regard to sustainable biomass management in China. The technology for biomass disposal should be economic and sustainable, supporting energy savings and decreasing CO_2 emissions.

As a biomass resource, sewage sludge has a large part of renewable resources, including organic compounds from which energy can be extracted, inorganic compounds and nutrients like phosphorous. During the 11th Five Year Plan the Chinese government planned to invest about 40 billion euros for the construction of sewage treatment plants (STP) during the following five years. Therefore a large amount of sewage sludge has to be produced from wastewater treatment. In China, the current national policies are supporting the development of biomass management in compliance with the renewable energy production.

An important reason for energy recovery from sewage sludge is that the price of energy is anticipated to rise (Youssouf Kalogo, Hugh Monteith, 2008). Energy consumption represents about 25% of the O&M costs of STP's (Smith, Clark, 1995). As energy costs continue to rise, energy recovery will become a major goal for STP's. Shizas and Bagley (2004) showed experimentally that sewage contains at least 10 times the energy required for treatment. This means that new technological developments and/or the optimization of the current technologies are necessary to fully take advantage of the maximum energy available in sewage sludge.

Besides the increasing price of energy, global warming is another concern of the increasing energy consumption. Sustainable biomass disposal, with a reduced carbon footprint, is now becoming a goal of major interest. Such an interest will stimulate the development of technology over energy reduction and resources recovery from sewage sludge. Energy produced from sewage sludge is considered as renewable energy that can be directly used for the wastewater treatment, instead of depending on the conventional electricity. This energy production will help to reduce emissions of greenhouse gases (GHG).

In the Netherlands, the potential of energy and resources recovery from sewage sludge have been recognized for many years. Our consideration shifted from disposing wastes to recovering energy and resources. This paper focuses on the Dutch approaches of that. The Dutch sludge treatment started in the late sixties of the last century (1960) and is currently focusing on the recovery of energy and resources from sewage sludge which stimulates the improvement of biomass management. Nowadays more than 50% of the produced sludge is digested whereby the biogas is produced and then generated to electricity and heat or (further) upgraded to be natural gas quality. The Dutch target aims to increase the ratio of the renewable energy to the conventional energy to be 20% and to reduce 30% emission of green house gases till 2020. In addition, there is an agreement with the 24 Dutch Water Boards that 2% reduction of energy consumption for wastewater treatment should be realized every year from 2005 to 2020. In this case about 30% of energy for wastewater treatment should be saved during the 15 years. In order to cope with Dutch situations, onsite generation of energy from sewage sludge has becomes an attractive option.

This paper will describe the development of Dutch approaches towards directly converting sewage sludge (or with other biomass) to biogas on site of the STP and the associated alternatives of efficient utilization of biogas from the economic and environmental point of view, including co-digestion with other biomass is discussed as well as a possibility of reuse recovery. The Dutch experience and considerations about biomass management can be of interest and beneficial for the current situation in China.

2 Energy recovery from sewage sludge

2.1 Traditional method

Due to the prohibition of application of sewage sludge for agricultural purpose, 50% of the produced sludge is nowadays anaerobic digested and then dewatered and incinerated. The rest, mostly surplus sludge from low loaded activated sludge plants, is dewatered directly before incineration. The tendency is that the amount of sludge which will be digested increases. After incineration the ash is used as filling materials in asphalt, cement and concrete, since the Dutch restriction claims that the residues containing municipal sludge can not be used as a fertilizer and dumping on landfills with more than 10% of organic matter is prohibited.

After 1995, there is also a tendency in the Netherlands to concentrate dewatering installations on a few larger STP's in the Netherlands, which means the small-scale of dewatering facilities on site of each individual STP are taken out of operation and the thickened sludge is transported by trucks to the central dewatering site for dewatering (H. Guo, E. Koornneef, B. Lue, 2009).

From an economical point of view, meaning investment costs for dewatering and digestion equipment and exploitation costs (labour, transportation, and maintenance costs), there is a strong tendency for central digestion and dewatering. The Netherlands is a densely populated country and the distance from city to city is relatively short. For example, in Utrecht province the distance between each STP and the central dewatering site is mostly less than 10-20 km. Because of the short distance between the cities, the transportation costs are low.

In the Netherlands, the most traditional method of energy recovery from sewage sludge is to convert the biogas (which mainly consists of CH_4 and CO_2) to electricity and heat via a Combined Heat and Power (CHP) installation. The produced heat is used for increasing the process temperature of the digestion at mesophilic conditions (30-38℃) and for the heating of buildings in the STP. *Figure 1* gives the schematic view of the traditional way of energy recovery from sewage sludge.

A conventional STP can produce 30%-40% of the energy required for wastewater and sludge treatment processes because it only digests itself produced sludge. A STP functioning as a larger central digestion/dewatering plant can produce more energy, even up to more than 100% of the energy required, because it treats external sludge from other STP's.

However, with the traditional method, the production of biogas is limited and the energy potential is only partly applied. Sometimes, especially in summer, there is a surplus of heat production by the CHP which cannot be used efficiently. The conversion process in the CHP installation itself is not efficiently. Also the CHP efficiency is not used efficiently at peak events (for instance peaks in the biogas production).

2.2 Alternatives for energy recovery from sewage sludge

To produce more biogas and recover more energy from sewage sludge, the new approaches can be co-generation and heat treatment. The more efficient applications of the extra biogas and energy are:

- To produce more electricity with CHP generation;
- To sell the biogas to a third party;
- To sell the green gas to a natural gas network or transport company;
- To produce more electricity by using a fuel cell.

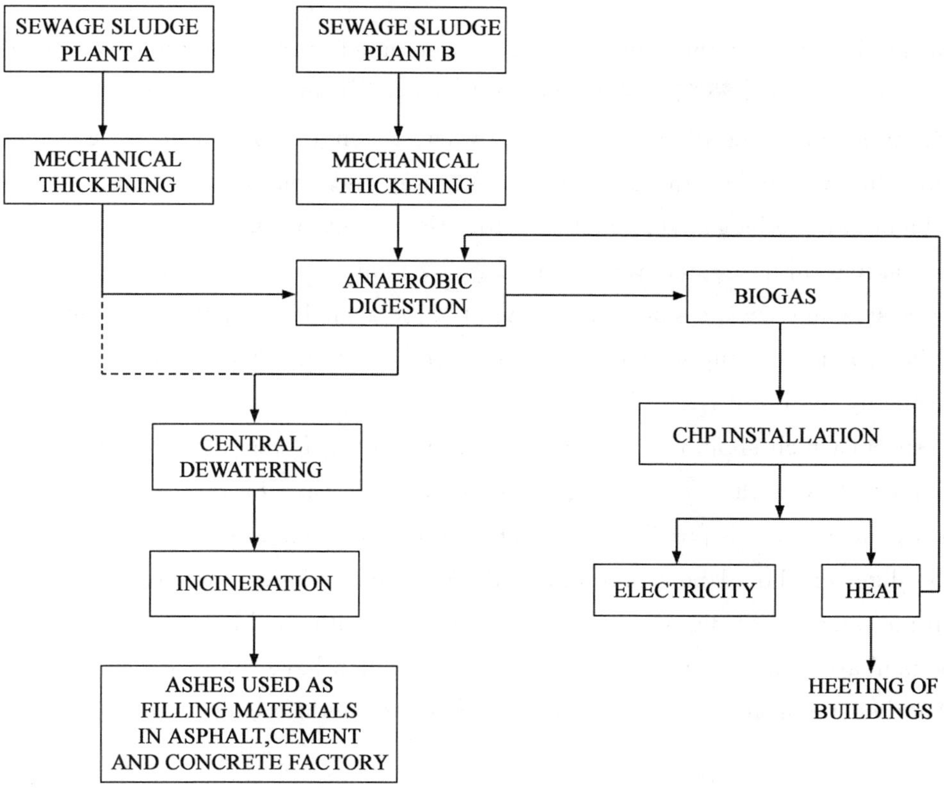

Figure 1 Traditional method of energy recovery from sewage sludge

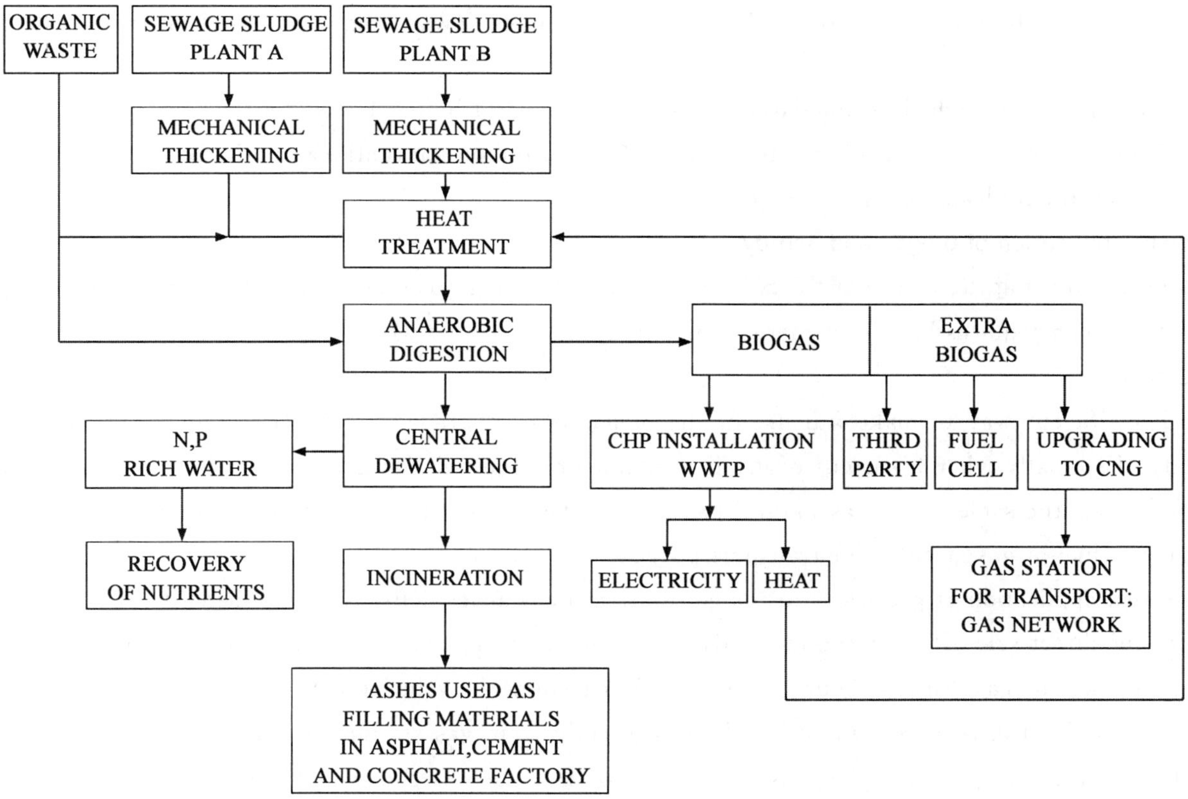

Figure 2 Alternatives of energy recovery from sewage sludge

Co-digestion

With co-digestion the digestion of the sewage sludge is combined with other types of waste in the same digester (biogas reactor system) at the STP. Other biomass resources can be livestock manures, food waste from restaurants, or other organic industrial wastes. On the one hand, the organic wastes is stabilized and controlled well and on the other hand the energy and emission problem is solved. Due to the addition of extra biomass, the biogas yield-and therefore the energy output-increases.

In China, enough agricultural organic waste (manure), food waste (full of fat) from restaurants, industrial organic wastes and sewage sludge are thought to be available as potential biomass sources from which biogas can be produced to supply the renewable energy. The market potential of this technology is considered as reasonably high.

There are a few aspects which should be paid special attention to. Make sure that infrastructure for co-digestion is available. Only if the existing digesters in the STP have extra space in terms of hydraulic residence time or organic load, it is possible to collect additional organic wastes from others. The materials that come into the digester should be monitored well like characteristics, qualities and quantities of the materials. Preventing overload of digester, buffering and pre-treatment like screening and heating of the sources should be paid attention to. Especially fatty sources which have the tendency to clock piping need pre-treatment. Another important aspect is a proper mixing of the difference sources in the digester.

Heat treatment

Heat treatment is another alternative to treat more sludge and finally producing more biogas and energy. It is to recycle the heat back to the heat treatment before anaerobic digestion in order to increase the process temperature of sludge treatment (up to 120-130℃) with a higher pressure at 5-6 bar. However, the higher temperature rise by the heat improves the efficiency of sludge dewatering. Applying heat treatment results in:

- Lower viscosity which enables to operate the digestion at higher DS content;
- More breakdown of organic matters; 15%-40% more than conventional method;
- Which finally leads to smaller digesters.

Efficient application of biogas and energy

In addition to using the biogas at the STP itself, the extra biogas can be sold to the third party, like paper industry which requires biogas as energy source.

CNG gas

Another efficient way to make full use of the biogas is to upgrade it to CNG (compressed natural gas) quality. Especially when a surplus of gas is available or if the residual heat can not be applied, the green/clean gas has the same quality as natural gas and can be injected in an existing gas network or used as a substitute fuel at gas stations for transport purposes.

The direct application of green gas to the gas station for vehicles delivers several benefits. (1) One benefit is that the direct sale of gas is commercially attractive. For the application of gas in the household from a gas network, the gas price is about 0.55 €/m^3. For feeding the vehicles with gas from a gas network, the gas price is about 0.70 €/m^3 while for the gas directly from gas station the gas price is estimated to be 0.4-0.6 €/m^3. The gas price depends on several factors, such as the availability of existing gas network and gas station, the features of the end users, and the scale of the biogas treatment plant. (2) The second benefit is for the end users. The use of clean gas produces almost no particulate matters and prevents dust problems that diesel vehicles have. (3) It is also beneficial for environmental protection. Cars driving on green gas/natural gas emitted 20% less CO_2. It decreases the dependency on natural gas and fossil fuel. The replacement of fossil fuel by green gas might be a sustainable and economic option because of the increasing price of fossil fuel and depletion of fuel resources.

Figure 3 Green gases for transportation purposes (Examples in Sweden)

Fuel cell

The new technology for the future is called fuel cell or formation of hydrogen. It directly converts the biogas to electricity, heat and water through an electrochemical reaction using hydrogen and oxygen with zero pollutant emission instead of a gas engine. The hydrogen is produced from biogas while air provides the oxygen. The biogas is firstly reformed to hydrogen within the cell and then moves to the fuel cell anode through an electrolyte. Air is fed to the cathode.

Several benefits determine the application of fuel cells. (1) It is much cleaner then typical power generation and would not pollute the local community. Fuel cells have significantly low emission rates of NOx and SOx, and are well-suited to locations that are impacted by stringent air quality regulations. (2) It efficiently generates a constant supply of electricity and heat 24 hours a day by comparison with a gas engine. With alternative B2, the treatment installation for 100 000 pollution units can generate 0.1 MW of electricity which can supply roughly 225 household demands. (3) It can be selected when extra biogas has the potential to be economically used as a feedstock. (4) It allows decentralised power generation.

Fuel cells contribute to the sustainable application of energy potential. However, the long-term R&D programmes and driving forces from the government and market should be established before its practical application.

3 Phosphorus recovery from sewage sludge

In China, the market price of the recovered phosphate is to be raised due to the shortage of natural phosphorus. Therefore it is now financially attractive to recovery phosphorus from sewage sludge. In the Netherlands, there are two ways to recovery phosphorus at STP.

One way is to recover phosphorus from ashes coming from sewage sludge. Only the ashes from sludge originating from STP with biological P removal or chemical P removal with alum salts are suitable for the phosphorus recovery industry. The ashes originating from sewage sludge containing iron salts can not be accepted (Stowa, 2007). Finally the recovered phosphorus is used for the fertilizer industry.

After dewatering of the digested sludge, a filtrate flow enriched with ammonia-nitrogen and phosphorus exists. To recover phosphorus from the filtrate, precipitation of phosphorus via the formation of struvite (crystalline magnesium ammonium phosphate hexahydrate) is an efficient way. The Crystalactor® Technology (Giesen, 1999) is an option to achieve the crystallisation.

The Crystalactor® is a fluid-bed type of cyrstallizer and has been applied at full-scale in the Netherlands. The technology was installed at the Geestmerambacht (Stack, 2004). The Crystalactor® process has many advantages, such as: (1)compact installation, (2)production of high-purity phosphate pellets, containing an extremely low water content (5% to 10% moisture), (3)re-use possibilities of the pellets.

Figure 4 Left: Schematic Crystalactor®,
Right: Full-scale Crystalactor® installation at Geestmerambacht

4 Conclusions

Organic wastes, climate, and energy are crucial to the world and all issues are interrelated. Generally speaking, the current world is changing to focus on energy recovery from the wastes and has to seek for an efficient approach of organic wastes disposal with sustainable and economic concerns.

The optimal solution would differ from case to case due to the different features of each STP, the different characteristics of the biomass sources, the existing gas network, the digestion infrastructure, the market demand and the features of the end user, etc. Although some attention has to be given to the selection of an optimal solution, those problems can be solved by technical means.

It can be concluded that anaerobic digestion plays an important role in the reduction of organic matters and energy generation. Co-digestion of sewage sludge combining with other biomass in the same digester system is an efficient solution for increasing yield of biogas and further reduction of CO_2 emissions. Heat treatment is crucial for further breakdown the organics and increasing the sludge treatment capacity. Different opportunities are available to recovery P from sewage sludge.

References

Kalogo Y., Monteith H., (2008): *State of Science Report: Energy and Resource Recovery from Sludge*. Global Water Research Coalition, pp. 13-31.

Smith D. A, Clark K. W., (1995): *Wastewater Treatment and Sludge Management-Energy Reference Guide*. Report prepared by NYSERDA, 2034-ERTER-MW-93.

Shizas J. I., Bagley D., (2004): *Experimental determination of energy content of unkown organics in municipal wastewater streams*. Journal of Energy Engineering, ASCE, 130(2), pp. 45-53.

Giesen A., (1999): *Crystallization Process Enables Environmental Friendly Phosphate Removal at Low Costs*. Environmental Technology, 20. pp. 759-775.

Guo H, Koornneef E., Lue B., (2009): *Dutch Sludge Treatment Approach: Beneficial for China?* Proceedings of IWA Sludge 2009 Conference.

Stark K., (2004): *Phosphorus recovery-Experience from European countries*. Proceedings of Polish-Swedish seminars, Stockholm, June 6-8, 2004 Integration and optimization of urban sanitation systems.

Stowa, (2007): *P-Recovery of Iron Poor Sludge From Sewage Treatment Plants (in Dutch)*. STOWA (2007), report no. 31. ISBN 978.90.5773.380.2.

通过提升沼气工程运营人员的能力改善沼气工程的运行性能

S. Prasitpianchai[1], C. Sawatdeenarunat[2], U. Kantha[2], N. Tovivek[2]

(1. 德国国际合作机构,曼谷 10330;2. 清迈大学,清迈 50200)

【摘要】本篇文章中有关评估沼气工程绩效的内容来自于 Nakornping 能源与发展研究所(ERDI-Nakornping)与德国国际合作机构(GIZ)合作开展的用农村废弃物生产的甲烷转化为能源项目(M2P 沼气项目)影响评估的一部分。该项目的目标是通过提升工人的能力来提高沼气产量。项目开展期间针对 3 个目标群体,即养猪厂、木薯加工厂和棕榈油加工厂组织了 12 次专题讨论会。在针对每个目标群体开设专题讨论会结束之时项目利用 3～5 个月的时间进行培训活动的效果评估。根据本项目进程,该篇文章主要对养猪厂的结果进行论述。此项评估是以调查问卷和参数测量的形式进行,然而在项目过程中发现调查问卷的效果并不理想,很难从中分析出有用的数据。参数测量的方法虽然面临诸多困难,如因为目前猪的死亡率高而禁止进入养殖厂和前往偏远站点的高额费用,但是该方法效果显著。项目团队通过在北部省份开设的 3 个专题讨论会中,从 66 家养猪厂中挑选了 7 家作为调查对象。1～3 号养猪厂的结果表明高产气量的可能性微乎其微,故 1～3 号被排除在分析范围外。4～7 号养猪厂的结果表明在培训前和培训后 COD 的平均去除率为 86%,但单位 COD 沼气产量培训后比培训前从 0.17 m^3/kg 上升到 0.30 m^3/kg,即沼气产量增加了 76%。

【关键词】沼气工程绩效监测;能力建设;运营人员培训;沼气产量增加

IMPROVE BIOGAS PLANT PERFORMANCE THROUGH CAPACITY BUILDING FOR OPERATORS

S. Prasitpianchai[1], C. Sawatdeenarunat[2], U. Kantha[2], N. Tovivek[2]

1. German International Cooperation (GIZ); 2. Energy Research and Development Institute - Nakornping (ERDI-Nakornping), Chiang Mai University.

Contact: Uthen Kantha, Energy Research and Development Institute-Nakornping (ERDI-Nakornping), Chiang Mai University, 239 Huaykeaw Road, Ampur Muang, Chiang Mai, 50200 Thailand, Tel.: +66-53-942007-9, E-Mail: pr@erdi. or. th, uthenkuntha@hotmail. com

EXECUTIVE SUMMARY

The biogas plant performance assessment in this paper is a part of the impact assessment in the Methane-to-Power from Agricultural Waste Biogas (M2P Biogas) Project, which is a collaboration between the Energy Research and Development Institute-Nakornping (ERDI-Nakornping) and the German International Cooperation (GIZ). The project aims to increase biogas yield through capacity building in human resources. Twelve workshops were organized nationwide for 3 target groups in pig farms, tapioca factories and palm oil mills. Assessment in effectiveness of the training activity is carried out 3-5 months following completion of the workshop in each target group. For the given timeframe, this paper focuses only on the results from pig farm. The assessment was conducted using questionnaire and measuring parameters. The questionnaire method was proved ineffective and the results could not be interpreted for meaningful data. The measurement of parameters method, although had faced several difficulties such as restrictions of farm access due to current widespread of swine disease and high cost of travel to remote

sites, is more effective. The team has selected 7 farms out of 66 farms from 3 workshops in the northern provinces. The results from farm 1-3 did show very high gas yield that could not be possible. Therefore, results farm 1-3 were excluded from this analysis. The results from farm 4-7 showed that the COD removal efficiencies were 86% on average for both before and after the training. The biogas yield was increased from 0. 17 to 0. 30 m^3/kg per removed COD, respectively a 76% increase of the gas yield was achieved.
Key words: Biogas plant performance monitoring; capacity building; operators training; biogas yield increase

1 Introduction

Biogas technology has been promoted and developed in Thailand for more than 20 years. There are currently more than 2 000 biogas plants all over the country. Most of the biogas plants are wet systems using substrates from livestock manure and industrial wastewater. Perhaps, the success of biogas technology in Thailand owing largely to utilization of simple construction and low cost materials that makes investment cost affordable for farmers. Most common technologies employed are Covered Lagoon and UASB that are a dug out of channels and covered with plastic sheet. The simplicity of the construction has obscured the real problem in the operation of the system. Poor operation has led to many problems such as low gas yield, plant failure and even serious accidents that could cause loss of lives and assets. Therefore, ERDI-Nakornping and German International Cooperation (GIZ) has jointly implemented the Methane-to-Power Biogas from Agricultural Waste (M2P Biogas Project) aiming to increase technical knowledge and operational skills of the biogas plant operators. The project has developed training courses for operators at difference levels and sectors in order to meet the needs of the target audiences. The courses are divided into 3 modules targeting at pig farmers, tapioca factories, and palm oil factories. This paper will review only on results and impact assessment of the operators from pig farms.

1.1 Background

The team had organized 10 workshops nationwide in the provinces where there are concentrations of pig farms. *Figure 1* shows distribution of biogas plants in Thailand. Pig farms biogas plants are mostly concentrated in the north and western provinces. In the eastern and northeastern region, there are biogas plants in tapioca factories while the plants in the south are usually in palm oil mills.

The training module for pig farms was developed and revised 3 times to finally meet the needs of the farmers. The 2 day course comprised of theory, workshop and on-site training. Participants in each course range from 10-50 persons. There were 164 farms participated in the training in total. Size of the digester is in the range of 30-5 000 m^3. The most common size is 100 m^3 (for around 600 pigs). It is anticipated that after the training, farmers should be able to operate their plants at higher efficiency and therefore increase at least 10% of the gas yield. The higher amount of gas yield would resulted in more production of electrical and mechanical power for on farm usages and therefore avoid around 32 000 t CO_2 equivalent per year. To verify this hypothesis, the project had made a survey with questionnaire and parameters measurement.

1.2 Research objectives

This research paper is a part of the impact assessment on the project activity in training for biogas operators. The objective of this work is to verify the impact of training activity by means of an increased

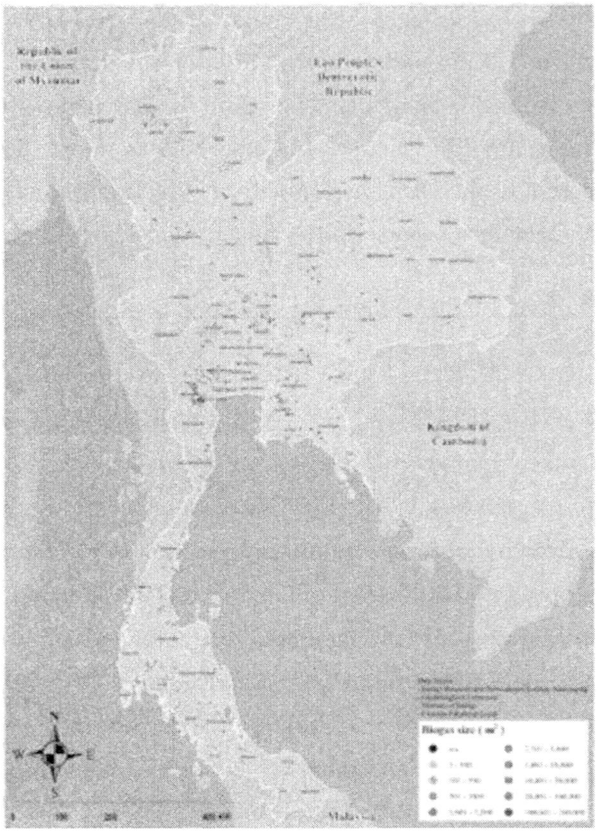

Figure 1 Biogas plants distribution in Thailand 2011

gas yield and enhanced COD removal efficiency.

2 Methodology

Data acquisition was made by 2 methods.

(1) Questionnaire: during the training, farmers were taught how to fill out the form according to their biogas plant types, meter reading and to collect data of their gas consumption, pump operating hours (for quantity of wastewater). The raw data were calculated into gas yield. Out of the 164 farms, only 12 had returned the questionnaire.

(2) Measurement of biogas plant characteristics: the team started the training workshop from the Northern provinces and move down the locations of workshops further to southern provinces. Therefore, only pig farms in the Northern provinces are in the suitable timeframe for parameters measurement. The project had sampling 7 pig farms from 66 farms that attend workshops in the north. Measurements in other region are being conducted; however, are not yet available during the time of this paper. Five parameters were collected as shown below:

- Inlet flow rate of wastewater to digester (m^3/d);
- COD of inlet and outlet wastewater (mg/L);
- Amount of gas production (m^3/d);
- Temperature of headspace biogas (℃);
- Pressure (cmH_2O; only for fixed dome)。

2.1 Biogas plants

There are 3 types of biogas plants commonly used in small pig farms:

Figure 2 CD-Junior reactor

Figure 3 Inlet wastewater sampling

(1)CMU-CD (Chiang Mai University-Channel Digester) was developed by ERDI-Nakornping in 1995. At present there are more than 700 plants all over the country, mostly in swine farms. The digester is design with simple construction and low cost materials that are affordable for farmers. The COD removal efficiency of this reactor is in the range of 80%-85% and gas production is 0.3-0.5 m^3/kg COD removal.

(2)CD-Junior is similar in construction with CMU-CD but in a smaller scale. The digester was modified to have capacity for small livestock farms in order to lower the investment cost for farmers while maintain COD removal rate and efficiency at the same level with CMU-CD.

(3)Fixed dome is a conventional anaerobic digester and has been installed to many livestock farms in Thailand for more than 20 years. This type of digester is usually recommended for small farms that have less than 60 LU.

2.2 Sociological background of operators

The operators of biogas plant in small pig farms are mostly the farmers themselves and foreign labours. Farmers are educated at the level of primary to secondary school. Foreign labours are from neighbouring countries and illiterate. Therefore, the training course was design to be given in simple language with graphical explanations. The course contents focus mainly on good practices in biogas operation, simple daily tasks, check list and troubleshooting. Key emphasis was given to safety operation.

2.3 Factors influencing biogas yield

There are several factors that could have impact on gas yield. The data acquisition during difference time of the year could be deviated. The team recognise that many of these factors are uncontrollable and therefore there could be deviation in the results.

Table1 Influential factors and impact

Factors	Influential impact
Size of livestock	Characteristics of manure are varied depending on the pig size such as manure from fattened pigs near slaughter weight has a higher concentration of solid while manure from sows is more liquidate.
Feed ingredients and formulation	Pig feed ingredients are formulated to provide high nutrition. Some of the formula contain slow degradable biomass that result in low gas yield. Some of independent small scale pig farms that are not in the contracted farming with agricultural conglomerates sometime feed their pigs with food waste and market waste. Manure from these farms usually contains high amount of insoluble organic.

Factors	Influential impact
Season	Wet and dry seasons are also critical condition during the time of data acquisition. Rain water mix with manure can dilute the substrate and reduce the amount of gas production.
Location	Rainfall, level of ground water, and temperature are varied in difference parts of the country. These climatic conditions can have impact on gas yield such as biogas plants in the northern region have lower gas yield in cold months while plants in the southern part are effect from long period of rainy months.

3 Results from questionnaire

Questionnaire is the lowest cost method for data collection; however, there are risks of inaccuracy and low collection rate. The team had given the questionnaire to all 164 farms that attend the training course. The farmers were requested to record daily wastewater and gas consumption. Measurement of wastewater is done by counting the number of hours in pump operation and radius of pipe. Gas could be used for mechanical and electrical generation. Farmers were taught to read the nameplate of their engines and record the number of hours the engines operate. The record forms should be sent by post 3 months after farmers attend the training. There are only 12 farms that return the form and only 3 forms contain useful data. Questionnaire method is proved having ineffective result in this sociological group.

4 Results from measurement

Upon completion of training in the northern region, the team has selected 7 farms for actual measurement based on size and type of biogas plants and availability of existing data prior to the training.

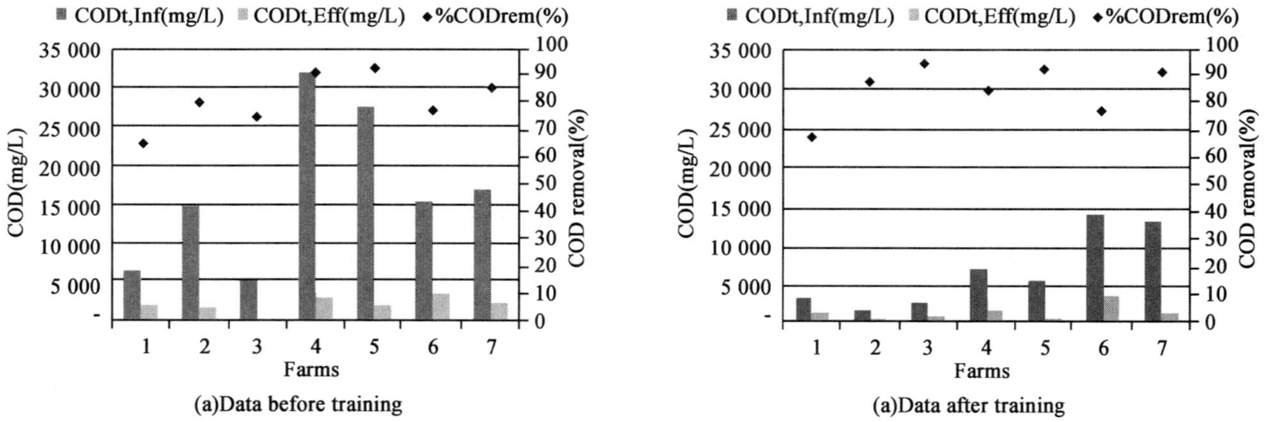

Figure 4 Comparison of influent and effluent COD of wastewater before and after training

Data from farm 4 and 5 showed discrepancy information on influent COD that was high in the first measure (31 900 mg/L and 27 600 mg/L) but very low in the second measure (6 700 mg/L and 5 000 mg/L). This was from influential factor in the size of pigs. During the first measurement, pigs in farm 4 and 5 were at slaughter weight and therefore there were higher amount of wastes. The second measurement was conducted 9 months afterward when there were new batch of pigs and they were only in the fattening weight. Farm 6 and 7 had influent COD in the range of 13 000-16 000 mg/L. Effluent COD in all plants were below 3 500 mg/L and is usually treated next stage in open ponds to further reduce COD and generally be used as fertilizer on site. Farm 1-3 showed low COD particularly in the second measure. This could be from factor in frequent rainfall during the time of measurement.

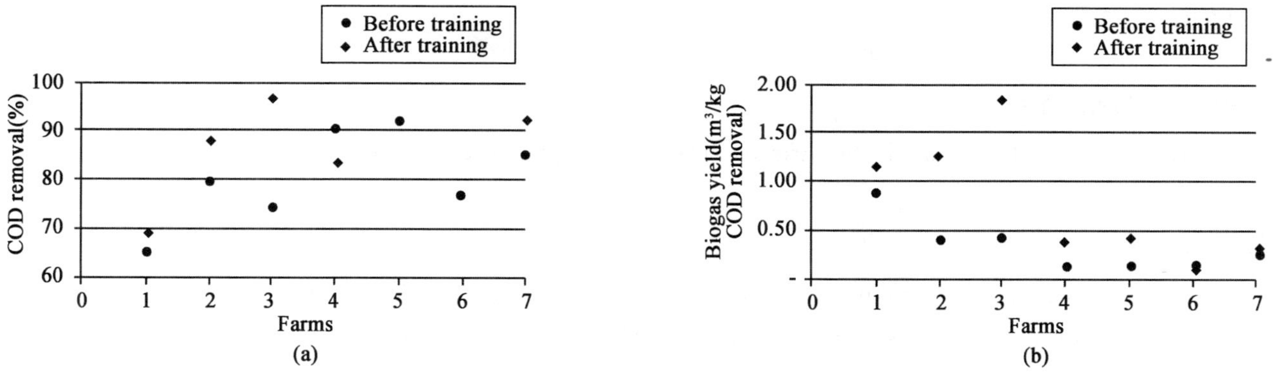

Figure 5 Comparison of (a) % COD removal and (b) Biogas yield

Percentage of COD removal and biogas production per kilogram COD removal (m^3 gas/COD removal) are used as the key indicators of the performance of biogas plants in this paper. Measured results from Farm 1-3 showed irregularly high gas yield particularly in the second measurement or after the training. The gas yield data were shown more than 1 m^3/kg COD removed where biogas production from pig manure is generally around 0.5 m^3/kg COD removed. There are several factors that could contribute to these errors. For example, Farm 3 had medium gas yield in the first measurement but very high gas yield in the second measurement. Primarily assumption is that the plant had organic overloading that was shown increase from 2.14 to 3.19. During the time of the first measurement, pigs were at the slaughter weight (100 kg). Overloading of manure influents could slow down the reaction in fermentative acidogenesis so gas yield was only 0.34 m^3/kg COD removed. In the second measurement, the farm had started a new batch of pigs, piglets produce less amount of waste but higher in rapid digestible biomass due to feed formula. Once the plant was no longer overload and bacteria receive high nutrients, the fermentation process was more effective and produced more gas. The team had also aware that there could be error in the process of data collection or lab testing.

Therefore, results from Farm 1-3 are excluded in this analysis. Results from Farm 4-7 shows the COD removal rates were around 86% on the average for both before and after training while there are significantly increase in the gas yield. Averaging results from these 4 farms, gas yield was increase from 0.17 to 0.30 m^3/kg COD removal. Farm 6 has lower gas yield, collected data show the influent flow rate was increased from 27 to 45 m^3/d which could be from excessive water usage in barns cleaning and higher rainfall. High water flow rate could flush out bacteria from the reactor, dilute the system and result in lower gas production.

5 Conclusions

The increase of gas yield and the COD removal efficiency are measurable impacts and the most apparent way to verify the effectiveness of the training course on the farm operator's behavior. However, there are several factors that could influence the performance of biogas plant particularly in pig farming where there are cycles of pigs at varying weights and ages on farm and therefore produce inconstant flow of wastewater. Although, some results show irregularity that need attention to analysis the source of problem, the overall results show an increase of the gas yield but COD removal efficiency remains unchanged as it has stayed already at a high value. It can be concluded that the performance of biogas plant can be improved by training of the operators. The gas yield has increased 76% which was higher than the original assumption of 10%.

Acknowledgements

The authors would like to thank the farm owners for allowing access to farm premise for data collection.

References

Aggarangsri P. ,Teerasountornkul T. , (2011): *Practical Design and Efficiency of Large-Scale Biogas Digesters for Swine Farms in Thailand.* Journal of Sustainable Energy & Environment Special Issue, pp 51-55.

Wongsapai W. , Thienburanathum P. and Rerkkriengkrai P. ,(2008): *Biogas Situation and Development in Thai Swine Farms.* Proceeding of International Conference on Renewable Energy and Power Quality (ICREPQ,08), 12-14 March 2008, Santander, Spain.

中国大中型沼气工程的沼气潜力及其对缓和气候变化所做的贡献

C. Brauner[1], B. Raninger[2], R. Villa[1], 董仁杰[3]

(1. Cranfield 大学,Cranfield,MK43 0AL;2. 德国国际合作机构,北京 100125;3. 中国农业大学,北京 100083)

【摘要】中国的煤炭消耗量、氮肥消耗量、固废产量以及二氧化碳和甲烷排放量均居世界首位,同时中国还处于城市化进程最快的时期。目前,中国正在努力应对这些问题以确保能源供给与社会、经济、环境的可持续发展。

厌氧发酵技术对上述问题有着重要的影响,并对可再生能源有着积极的贡献。

本文评估了不同种类的生物质资源,并据此对其理论沼气潜力进行了评价:如城市垃圾(生活垃圾、市政污泥)和农业废弃物(农作物剩余物、畜禽粪便)。本研究只计算了大中型沼气工程的沼气潜力,对于填埋式、分散式污水处理系统和小型户用沼气池不予考虑。

目前预测的中国理论沼气潜力为 2.9×10^{11} m^3/a,其数量占能源总需求的 6.9%。到 2030 年,加上以边际土地上的能源作物为原料,沼气潜力可以进一步增加至 4.394×10^{11} m^3/a。即沼气工程发电机的装机容量潜力可以达到 71.4 GW,从而为可再生能源做出深远的贡献。

厌氧发酵的副产品沼渣沼液是一种绿色肥料。如果对所有潜在的生物质原料进行厌氧处理,产生的沼渣沼液将能够取代近 30%的无机肥,并且超过中国目前生态肥料的需求量。

选择厌氧发酵工艺作为处理指定原料的主要方法可以使减排潜力达到 4.75×10^9 t CO_2 当量。到 2030 年,其甚至可以达到 7×10^9 t CO_2 当量。

从目前和未来可利用的生物质资源预测中可以看到虽然还有许多资源没有被充分开发利用,但也还是可以从中看出发展可持续沼气能源的政策目标(到 2010 年占能源总量的 6.6%,到 2020 年占 10%)是可以实现的,并且可以做得更好。

【关键词】沼气潜力;大中型沼气工程;温室气体减排;能源作物

BIOGAS POTENTIAL OF MIDDLE- AND LARGE- SCALE BIOGAS PLANTS AND ITS ESTIMATED CONTRIBUTION TO CLIMATE CHANGE MITIGATION IN CHINA

C. Brauner[1], B. Raninger[2], R. Villa[1], DONG Renjie [3]

1. Cranfield University, England and University of Natural Resources and Life Sciences, Vienna; 2. Deutsche Gesellschaft für Internationale Zusammenarbeit (GIZ) Gmbh; 3. China Agricultural University

Contact: Christiane Brauner, MSc, Cranfield University, England and University of Natural Resources and Life Sciences, Vienna, Austria, E-Mail: Brauner_christiane@yahoo.de;

Prof. Dr.-habil Bernhard Raninger, GIZ, FECC of Ministry of Agriculture No. 55 Nongzhan Beilu, Chaoyang, Beijing 100125, PRC, Tel.: +86-10-6508-7080-200, E-Mail: Bernhard.Raninger@giz.de, www.biogas-china.org;

Prof. Dr. Raffaella Villa, Centre for Environmental Science and Technology University, Cranfield University, College Road, Cranfield, Bedfordshire, MK43 0AL, England, E-Mail: r.villa@cranfield.ac.uk;

Prof. Dr. DONG Renjie, Energy Engineering and Low Carbon Technology, China Agricultural University, Beijing 100083, China, E-Mail: rjdong@cau.edu.cn

EXECUTIVE SUMMARY

China comes first on the world list as to coal and nitrogen-fertilizer consumption, solid waste production and CO_2 and CH_4 emissions. At the same time the country has to cope with one of the most rapid periods of urbanization in history. China is currently making many efforts to reduce and even to cope with these problems in order to ensure its energy supply and a sustainable social, economic and environmental development.

Anaerobic digestion technology could greatly impact on all the above-mentioned problems and positively contribute to the renewable energy sector.

An assessment of biomass resources was performed to evaluate the theoretical biogas potential from: urban wastes (household waste, municipal sewage sludge) and agricultural wastes (crop residues, animal manure). This study calculates only the biogas potential of feedstock, which is used in middle- and large-scale biogas plants. Landfill gas, decentralized wastewater treatment systems and small household digesters are not included.

A theoretical current annual biogas potential of 290 billion m^3 was estimated, which could account for 6.9% of the total energy demand. Furthermore, the annual potential could increase to 439.4 billion m^3 by 2030 by including the feedstock energy crops from marginal land. Biogas plants with an initial capacity potential of 71.4 GW could, therefore, be installed and contribute more to the renewable energy targets as so far planned.

Digestate, a by-product of the process, is a sustainable green fertilizer. By treating the whole assessed feedstock, the digestate generated would replace nearly 30% of inorganic fertilizer and exceed China's current demand for bio-fertilizer.

By using the anaerobic digestion process as the main treatment option for the selected feedstock, the current total reduction potential of CO_2 equivalents is 4.75 billion tons. By 2030 even 7 billion tons of CO_2 equivalents could be saved.

The current and future forecasts of the biomass availability highlights that the resources are far from being fully exploited, but they also show that the policy targets to develop sustainable biogas energy (6.6% by 2010 and 10% by 2020) could be easily reached and could be much more ambitious.

Key words: Biogas potential; middle and large scale biogas plants; green house gas mitigation; energy crops

1 Introduction

At present, over 1.3 billion people live in the People's Republic of China (PRC), which is approximately 20% of the world population, with a growth rate nearly of 0.5% (Index Mundi, 2011).

China's economy is also growing rapidly; it is the second largest producer of electricity after the United States. Its energy demand is also the highest in the world (The World Bank, 2010). As indicated by the National Energy Administration (NEA), its primary energy consumption overtopped 3.07 billion tons of standard coal equivalent in 2009, i.e. an increase of 30% since 2005 (EMCA, 2011).

The PRC, therefore, ranks first in the list of the world's highest emitting countries (EIA, 2009). However, with its low-carbon development strategy, the government of China has set an ambitious target for non-fossil energy consumption, which should increase to 15% by 2020. This would contribute to a reduction in the carbon dioxide emissions per gross domestic product (GDP) of 40% to 45% by 2020 in comparison with the year 2005 (EMCA, 2011).

In order to achieve these goals, the primary energy consumption should not exceed 4.2 billion tons within the period of China's 12th Five-Year Plan (FYP) (2011-2015) (EMCA, 2011).

In the last decades, renewable energy such as biomass, solar and wind energy as well as hydroelectricity has become a very important factor for energy supply and for making the climate change less severe (Tang et al., 2010).

Bioenergy uses bio-organic materials, i.e. biomass for energy generation. Therefore, it includes materials of vegetable and animal origin as well as their waste products. However, before using them as renewable energy source, they have to be treated. Transportation, mechanical treatment and deposit must be done in such a way as to be able to coordinate the biomass produced with the energy demand (FNR, 2005).

The following bio-energy technologies find application in China: biogas, syn-gases from straw and stalk gasification, electricity from biomass power generation, and various solid and liquid bio-fuels, such as bio-ethanol (Wu et al., 2009).

This study deals with the technology of anaerobic digestion focusing on the main feedstock of China which is suitable for biogas generation, such as agricultural waste, i.e. crop residues and animal manure, urban waste, i.e. the native organic fraction of household waste and municipal sewage sludge, and as well as energy crops from marginal land.

1.1 Background

1.1.1 Biogas development in china and current performance

In former days, China's biogas industry mainly focused on small rural household biogas digesters using livestock and poultry manure as a feedstock. Since 1950 over 35 million household scale anaerobic digesters with an annual biogas output of about 12 billion m^3 have been installed and this number should be increased to 80 million units by 2020. It is doubted that all of them are in good operation.

On the basis of national renewable energy legislation (China Renewable Energy Law 2006), the aims to be achieved are changing from the environmental benefits of digestion to gaining energy out of this feedstock using adopted technology. In recent years, the country has put a focus on building medium and large-scale biogas plants (MLSBP), located around large cities, which use diversified feedstock for biogas production (GIZ, 2011a).

Until 2009 about 34 000 small-scale biogas plants and 22 900 medium-and large- scale biogas plants (Fermenter $> 50\ m^3$), with 3 717 large-scale installations (Fermenter $> 300\ m^3$) were built. A biogas-driven power generation with a capacity of 42 MW electricity was established (GIZ, 2011a).

In spite of the rapid developments and investments made by the government during the period from 2003 to 2010, the Chinese biogas sector faces some major problems. Their standard cannot compete with international performance as insufficient specific biogas is produced, i.e. not enough is produced and thus, there is no or weak process controlling and monitoring. Furthermore, the PRC does not have relevant know-how regarding the design and operation of the pre-treatment equipment (MPRA, 2010). So far, just three plants have been connected to the electrical grid and only seven units of them are registered 'Clean Development Mechanism' (CDM)-Projects whereby only two can generate some CDM carbon credits based on certified emission reduction (CER). China has enhanced the building of plants by financially supporting them, yet the running of the plants as well as the use of gas and digested, have not been taken into account. (GIZ, 2011a)

1.1.2 Policy targets concerning renewable energies with special focus on biogas energy

Since a further increase in population, urbanization, industrialization and residential energy consumption is predicted the country has to cope with enormous challenges that are addressed with large-scale policy programmes and renewable energy development (The World Bank, 2010 and NDRC, 2007a).

In the "Medium and Long Term Development Plan for Renewable Energy in China" the National Development and Reform Commission (NDRC) states that by 2010, China is willing to achieve 10% of the total primary energy consumption from renewable sources, which was managed with success. By 2020 this share will even be increased to 15% (NDRC, 2007b).

China will dispose of more than 360 Gigawatts (GW) of renewable energy by 2020, provided all the aims concerning renewable power are achieved (Martinot et al. , 2007).

Concerning these renewable energy policy targets, the biogas energy sector was increased from 19 billion m^3 (including the weak household digesters) by 2010 and should rise to 44 billion m^3 by 2020.

Due to its expansive territory and diverse landscape, China disposes of a rich array of renewable energy and thus its biomass and organic waste resources are ranked highest in the world (Martinot et al. , 2007). Therefore, there is a good chance of replacing fossil fuel, such as coal and natural gas, mineral oil and its derivates, and nuclear energy (especially after the Fukushima disaster), by using this feedstock as an energy source. By 2050 1 billion tons of coal, which is one-third of China's current consumption, could be replaced by biomass-based energy (China Daily, 2011). In order to achieve these ambitious targets a capacity target of 13 million kilowatt hours per year of biomass power generation has been set for the next five years (Reuters, 2011).

Concerning biogas energy, 10 000 large-scale biogas plants on livestock farms and 6 000 industrial organic sewage treatment plants will be built by 2020, with an estimated production of 14 billions m^3 biogas per year; a capacity of up to 3 GW electricity shall be generated (NDRC, 2007a). The detailed targets of the biogas development, which are still to be set in the 12th Five Year Plan, even exceed the previous targets just mentioned.

1.1.3 The 12th Five-Year Plan (2011-2015)

The Five-Year Plans (FYP) of the People's Republic of China is a series of economic and social development initiatives, drafted by the Government of China. In the 11th FYP (2006-2010) China, for the first time, set some goals concerning the sustainable development of an environment and energy policy. The 12th FYP (2011-2015) puts even more emphasis on sustainability, such as 'Green Development', and 'Low Carbon Development'. Market mechanisms such as the planned Emissions Trading Scheme, which have never been used before should lead China to more successful results. At present, groups of professionals are working on detailed implementation concepts, which will be released in autumn 2011 (GIZ, 2011b).

Objectives of the biogas sector development until 2020 and later

China's current Five-Year Plan foresees a growth of the biogas sector.

The following paragraph lists the aims concerning development in the biogas sector:

(1) Diversification of bio-energy based on local circumstances with great emphasis on large-scale plants, which utilize biogas for electricity and heat supply.

(2) In 2020 the total amount of all kinds of bio-energy utilization will be over 40 million tons of coal equivalents, and 50 million units of household digesters should be developed. The installed capacity of biogas power generation will be 2 GW, and 12 billion kWh will be produced.

(3) The target of the 5-FYP is to build 10 000 small and 1 500 middle and large scale biogas plants (MLBGPs, >300 m^3 digester volume per year in 31 provinces and 2 000 counties till 2015. From current 2% of the breeding farms 20%-30% shall be equipped with biogas plants. Feedstock will be

manure, straw and other bio-waste (MOA, 2011).

(4) Biogas and especially bio-methane to be from increasing importance for the rural renewable energy supply and nutrient cycles. The use to households through compressed bio-methane (CBM) in local gas grids will be supported and, biogas upgrading shall be applied for projects between 50 and > 2 000 m^3 biome than per day (MOA, 2011).

(5) Problems of low productivity and low lifetime of Biogas plant shall be solved. Biogas plants shall mostly follow international state of the art technology with high performance based on a sustainable business concept (MOA, 2011).

(6) 55 MOHURD: fermentation plants with a total capacity of 10 000 t/d for restaurant and kitchen waste from households are to be built in 33 cities. This should be a good starting point for separate collection of the biogenic fraction of household waste in the future.

(7) Improving gas grid constructions and supply of public service systems. Thus, the large-scale biogas project and large-scale biomass gasification project could provide gas to additional 3 million households in 2015. By 2015, 200 "Green Energy Demo Counties" and 10 000 "New Energy Demo villages" (an average of about 300 households) will be built. The government will subsidize infrastructure constructions, including gas grid, gas storage and purification (NDRC, 2011).

To sum up one may say that the 12th FYP does not consider the development of biogas at first sight but considering the impact of the anticipated measures, there are great opportunities for the industrial scale biogas sector and its development, which is supposed to boost during this period (NDRC, 2011).

1.2 Research objectives

This paper is divided into five parts, which are listed below:

Part 1 assesses, based on present data, the amount, distribution, availability and the status- quo of the biogas energy potential of the PRC's abundant biomass resources.

Part 2 assesses, based on present arising, the energy potential in 2030.

Part 3 gives insight into how much of China's total energy demand could be met by the estimated biogas energy potential and could thus replace primary energy sources. In addition, present and future policies are seen in the light of biogas potential in order to demonstrate the current standing and resource utilization of biomass.

Part 4 deals with the German biogas sector and compares it with that of China.

Part 5 deals with the contribution of biogas energy to climate change mitigation. Calculations of saved amounts of Greenhouse gas (GHG) emissions are made using the total available biomass feedstock for biogas generation.

2 Methodology

2.1 Introduction

The aim of this paper is to give an extensive and detailed presentation of all selected kinds of feedstock for the generation of biogas and to estimate some kind of maximum biogas production potential, which is the result of research and discussions with various experts.

As already mentioned in the previous chapter, this study assesses the current and the future feedstock such as, agricultural waste including crop residues, animal manure and energy crops from marginal land for biogas energy, urban waste including urban household waste and municipal sewage sludge.

It also estimates to what extent biogas could meet China's energy demand and how much GHG can be saved.

The data collection was done using published international papers and Chinese governmental reports. The data was scanted and different for each feedstock identified. Therefore an individual method of feedstock estimation as well as biogas potential calculation had to be chosen for each kind of feedstock. In addition to the collecting and selecting of data, basic statistical analyses were made. Microsoft Excel has been used to perform the calculations and draw graphics.

2.2 Current feedstock levels, distribution and biogas potential

2.2.1 Urban household waste

Information Source:

The Sino-German RRU-BMW Project (Resource recovery and utilization of bioorganic municipal waste) conducted from 2005 to 2009 (Raninger et al., 2006; Raninger et al., 2009) collected data on the estimated potential of organic household waste in China as well as the quantity, quality and composition of Municipal Solid Waste (MSW). This project was carried out in four communities in Shenyang and was taken as a basis for this paper in making assumptions of the current amount of the biodegradable fraction within household waste.

Table 1 Summary of data used (Urban Household Waste)

Data	Source
Total Urban MSW: 360 million tons	Raninger, 2011
Organic Fraction within MSW: 80%	Raninger et al., 2006
Biogas Potential: 100 m^3/t FM	
Total Solid (TS) content: 35%	
Volatile Solid (VS) content: 80%	Steffen et al., 1998
Waste in Cities	
Urban Population China: 621 860 000	NBSC, 2010a
Generation of MSW: 0.9 kg/c/d	Raninger et al., 2006
MSW Generation: 200 million tons	
Waste in Counties	
Population in Counties: 626 223 092	
Generation of MSW: 0.7 kg/c/d	Raninger, 2011
MSW Generation: 160 million tons	

Assumptions:

The current collected amount of Biodegradable Municipal Waste (BMW) is mainly landfilled at present. It could, however, become a highly valuable resource for biogas production.

On the basis of the European Union Landfill Directive of 1999, landfilling of bio-organic wastes is going to be prohibited in Europe (65% reduction until 2016) (The Council of the European Union, 1999). Since 2005 only waste material consisting of up to 5% bio-organic matter (3% total organic carbon) has been disposed of in Germany and Austria (BMJ, 1991). It is, therefore, expected that earlier or later China will not allow the landfilling of organic waste either. This study assumes that all the generated organic waste can be used in order to gain biogas energy out of it rather than being lost in landfills. It has to be mentioned that due to various reasons Landfill gas collection does not be effective at landfills in China, which excluded this way to generate reasonable amounts of Landfill-bio-gas, it works somehow in industrialised countries. (Raninger, 2010)

2.2. Municipal Sewage Sludge

Data used to estimate the current production and the total biogas potential of sewage sludge are summarized in the following *Table*.

Table 2 Summary of data used (Municipal Sewage Sludge)

Data	Source
Population in cities and counties: 1 248 083 092	
22% of the total amount are treated: 80 000 t/d, Total: 133 million t/a	Zhao et al., 2006
Biogas potential: 0.027m^3/d. c	Lindeburg, 2011
TS content: 20%	Raninger, 2011

2.2.3 Crop Residues

Information Sources:

In order to obtain the current available amount of crop residues, China's main eight kinds of crops were chosen and the amount of arising has been taken from the 'China Statistical Yearbook 2010' (NBSC, 2010b) (see *Table 4*). Numbers of arising and energy potential of agricultural by-products such as rice hull, corncob and bagasse has been taken from the 2nd Volume of the 'Study of Potential and Constraint of the Biomass Sector in China' conducted by the Europe-China Clean Energy Centre (ECCEC, 2011). Information about the current distribution have been also taken from ECCEC as well as from the study by Zhong et al. (2003).

Data Recalculation:

All the results achieved from the current situation as well as from energy potentials are based on recalculation of original data. On the basis of indices for straw: crop conversions and the energy content within the crop residues taken from different studies, calculations were made estimating the biogas potential.

Table 3 Summary of data used (Crop Residues)

Biogas Potential: 300 m^3/t FM	Raninger, 2011

Current situation, biogas potential:

Table 4 Productivity of crop straw in China in 2009

Crops	Yield of Crops (×10^4 t)*	Index of straw: crop**	Yield of residues (×10^4 t)	Proportion (%)	Biogas potential (×10^9 m^3)
Rice	19 510.3	0.623	12 155	16.3	36.5
Wheat	11 511.5	1.366	15 725	21.1	47
Maize	16 397.4	2	32 795	44.1	98
Beans	1 930.3	1.5	2 895.5	3.9	8.7
Tubers	2 995.5	0.5	1 497.8	2	4.5
Oil bearing crops	3 154.3	2	6 308.6	8.5	19
Cotton	637.7	3	1 913.1	2.6	5.7
Sugarcane	11 558.7	0.1	1 155.9	1.6	3.5
Total	67 695.7	—	74 445.9	100	222.9

Note: * NBSC, 2010b, ** Wang, 1994

Current utilization and availability of energy usage

Figure 1 shows the current utilization of crop residues. However, according to Shi (2011), approximately 60% of the total crops of straw could be used for energy generation.

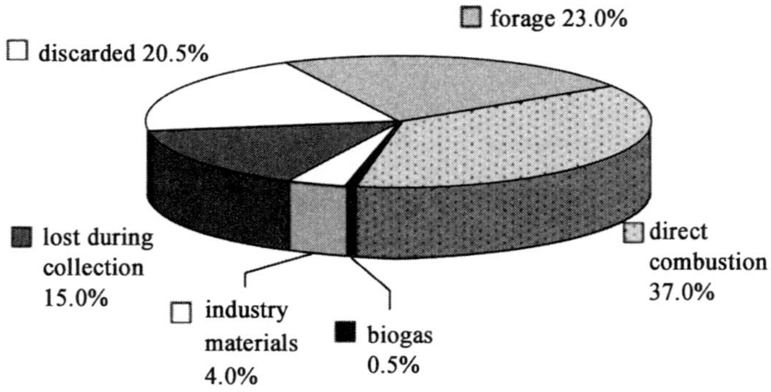

Figure 1 Current utilization situations of crop residues in rural China (Liu et al., 2007).

2.2.4 Animal Manure

Information Sources:

In order to estimate the amount of animal manure generated in one year, this study focuses on China's three main kinds of animals kept and bred on farms, such as pigs, cows and poulty.

Data on the current number of animals were used on the basis of the 'China Rural Statistical Yearbook 2010'.

Information about the distribution of large and medium-sized farms was taken from the ECCEC (2011).

Data Recalculation:

In order to estimate the biogas potential, the conversion table from the PPTA report from the Asia Development Bank (ADB, 2009) was used. It provides data about the generated amount of manure (kg/d/c), the chemical oxygen demand (COD) (kg/d/c) as well as its converting factor to biogas used in China. The conversion is as follows: 1 kg COD = 0.23 m^3 biogas. Data on the collectable amount of manure used from Shi's study (2011) are considered and included in the calculations.

Table 5 Average production of animal manure and biogas in 2009

Animals	Population on hand ($\times 10^4$)*	COD (kg/d/c)**	Biogas potential ($\times 10^9$ m^3/a)
Pig, total	46 996	0.6	23.7
Cow, total	10 726	5	45
Poultry, total	533 000	0.02***	8.9
Total	590 722	—	77.6

Note: * NBSC, 2010c, ** ADB, 2009, *** Layer chicken

Current utilization and availability for energy usage

According to Cui et al. approximately 35%-40% remains untreated, 50% is returned to the field directly, 5% are composted, and only 3% is going to methane fermentation (Cui, 2005; MOA, 2011).

Beside the 3%, which is already in use, a utilization factor of 43% is taken for the calculations, assuming the huge untreated amount can be fully made use of.

2.3 Future feedstock levels and biogas potential

The year 2030 was chosen to show the further increase in amount and biogas energy potential of each feedstock in the future.

2.3.1 Urban household waste

Assumptions:

In order to predict the future potential of China's urban household waste for biogas production, the following assumptions are made.

As a result of the rapid urbanization, 1 billion people will live in urban areas by 2030 (McKinsey&Company, 2008). Therefore, only 50% of the counties' population are still remaining in the counties, which results in 313 111 546 people.

There are contradictory predictions in various studies as to the amount of organic waste produced by people in the future.

According to a study made in the USA, the organic fraction in the MSW will diminish, as people will use more ready-made package foods and microwave ovens (Daskalopoulos et al., 1998). The World Bank estimates that even if a larger amount of packaging waste, paper products and plastics and a smaller amount of coal ash for heating compared to gas will be used in households, the amount of the organics will still be more than 50% of the waste stream by 2030 (The World Bank, 2005). The changes in lifestyle and growth in wealth predicted for the future will entail an increase in money spent for food (Ru et al., 2010). Materials suitable for recycling are recovered by cheap labour and have a market value in China also this trend most likely will remain (Raninger, 2009). Therefore this study assumes that the organic fraction of MSW will remain nearly the same.

2.3.2 Municipal sewage sludge

Assumptions:

In order to estimate the amount generated only the rapid urbanization, i. e. the population will rise to 1 billion people by 2030, is considered in these calculations (McKinsey&Company, 2008).

2.3.3 Crop Residues

Assumptions:

The following graph shows a clear continuous rise in the production of total crops in the last 10 years. Therefore, a further increase in the future can also be assumed.

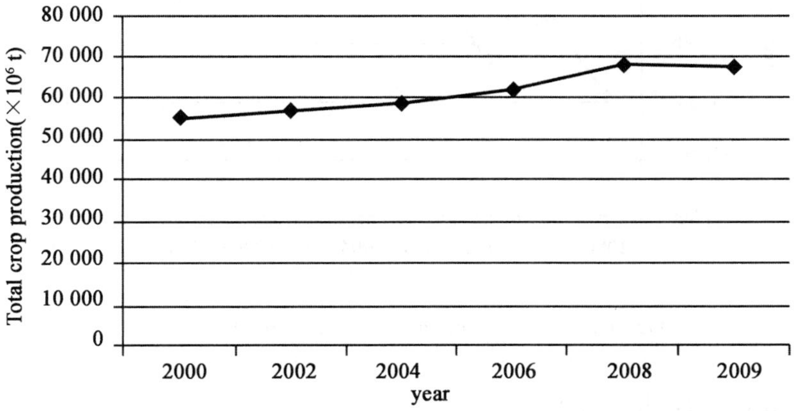

Figure 2 Total production of eight different kinds of crops from 2000 to 2009 in China

Due to the regional and climatic situation, and the traditional intensive use the increase of crops will be limited. But according to many studies, one of the most influential factors for future development of farmland is national land use planning (Hu, 2010a).

There will be also a rise in the population in the future and changes will take place concerning the land use, which would entail a continuous decrease in the entire farmland area as in the farmland area for each person. The global demand for food will increase.

The Government of China sees to it that there will be no shortage of food by making sure that arable areas are never less than 1.8 billion Chinese mu, i. e. 120 million hectares. By 2020 arable land should still be about 1.805 or 1.818 billion Chinese mu. (Hu, 2010a)

It can also be predicted that mechanized farming will increase whereas manual work in the fields is going to decline. This would result in a decrease in 'collection loss' and direct returns to the fields (Liu et al., 2007). Therefore, more feedstock will be available for energy production and the organic fertilizer, which is a highly valuable product gained out of the biogas process, could further be used in the fields in order to ensure sustainable agriculture.

Liu et al. (2007) calculated the crop residues increase rate since the 1970s whereby the average rate has been about 1.2% in the last twenty years. This seems to be a constant index and is also used for prediction until 2030.

2.3.4 Animal manure

Assumptions:

On the basis of the 'China Rural Statistical Yearbook' from 2000 until 2010, calculations on the situation of three different kinds of livestock such as pigs, cows and poultry during the last 10 years are made in order to predict the increase of animal manure. The numbers have been added together in order to calculate the increase in percentage over these years.

The following figure shows the continuous increase in arising of livestock for the last 10 years. It is assumed that the average rate of 12% between the years 2000 and 2009 will stay the same.

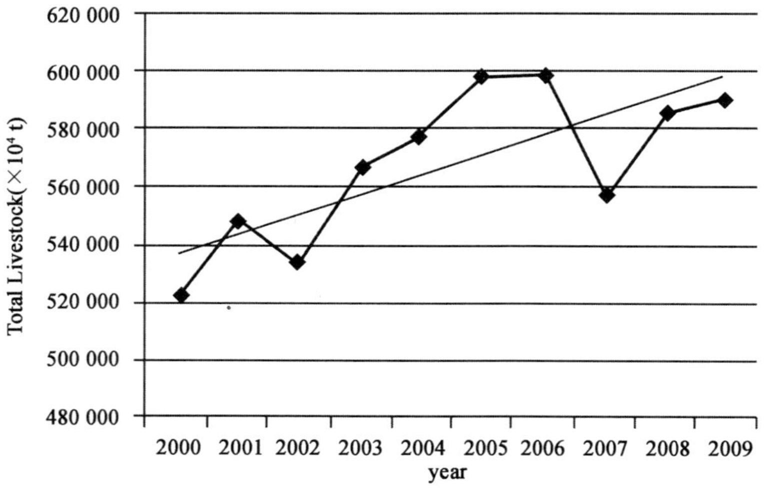

Figure 3 Total arising of livestock from 2000 to 2009

2.3.5 Energy crops from marginal land

Information source:

This part of the study is mainly based on one of the latest studies of 'China's Biomass Resources' published in 2011 and conducted by Shi Yuanchun.

Assumptions and data recalculation:

Evaluation of biogas potential is performed in the following steps:

Step 1: Marginal land appropriate for growing energy plants was analysed. The crucial results were the size, composition and distribution of marginal land in China.

Step 2: Exploration of energy plants that are suitable for biogas production was made and what amount and species of plants can be grown per hectare were taken account of in the calculations in order to get data about how many tons of plants are available for the entire generation of energy.

Step 3: The Biogas potential of the amount of different kinds of plants has been estimated.

Step 4: A utilization factor of 40% by the year 2030 is estimated in this study.

Definition of 'marginal land'

In order to estimate the available size of marginal land, one must first define what is meant by marginal land. However, no reliable source concerning the definition of marginal land has been found as data vary very much, and researches choose different kinds of species for their evaluation.

'Marginal land' has been used in this study as 'the land which has poor natural conditions, has not been used for agriculture currently but has production potential and development value to produce a certain amount of biomass. Such land is not fit for farmland temporarily, but can grow some kinds of plants with high adaptability' (Hu, 2010a).

The marginal land, which is used for growing energy plants, consists of woodland (shrub land and sparse forest land), grassland and land without vegetation including shoal/bottomland, saline and alkaline land as well as plain land and unused abundant agriculture land (Zhuang et al., 2010).

Size and composition of marginal land suitable for energy plants

The estimated figure concerning the size and composition of marginal land is based on factors such as land use data, terrain data (like elevation and slope), meteorological data (including precipitation data and air temperature data) and soil characteristics.

The following table illustrates the percentages of marginal land, 'wasted marginal land', which is still uncultivated land and cultivated land, called 'current marginal land'.

Table 6 Marginal land resource suitable for producing biomass feedstock in china

Type		Area ($\times 10^4$ hm^2)	Percentage
Wasted marginal land	Total	8 491	51.9%
	Suitable for farmland	2 787	17%
	Suitable for forest	5 704	34.9%
Current marginal land	Total	7 883	48.1%
	Marginal farmland	2 000	12.2%
	Marginal forest	5 883	35.9%

Note: Modified according to Shi, 2011.

Biogas potential

In order to compare the overall outputs, the yield per hectare is a better measurement, due to different climate conditions. *Table 7* shows the distribution, the unit output of major biomass plants as well as their biogas energy potential.

Table 7 Potential energy output of biomass plants of China's marginal land

Type of marginal land	Area ($\times 10^4$ hectares)*	Suitable energy plants and area percentage*	Energy Output (tce /(t · hm^2))*	Energy Potential (10^4 tce)*	Unit output (t/hectares FM)*	Biogas energy potential ($\times 10^9$ m^3 VS)
Suitable for farmland	2 787	Sweet sorghum, 35%; Sweet potato, 15% (north);	2-3	22 140	60-80 (stalks) 15-20	205 25

Type of marginal land	Area ($\times 10^4$ hectares)*	Suitable energy plants and area percentage*	Energy Output (tce /(t · hm^2))*	Energy Potential (10^4 tce)*	Unit output (t/hectares FM)*	Biogas energy potential ($\times 10^9$ m^3 VS)
Current marginal land	2 000	Cassava, 35% (south)	4-6		20-30	53
		Sorghum, 15% (south)			60-80 (stalks)	63
Barren mountain	5 004	Energy 60%	3.3	9 908	6.5	59
		Oil plant forest 20%	1.8	1 801	4.0 (seedlings)	12
		Energy crop 20%	3-5	4 003	6.5	20
Desert	700	Xerophyte shrub	2.6	1 820	4.0**	8.4
Firewood forest	175	Firewood forest	4.2	735	4.0***	2.1
Oil plant forest	343	Oil plant forest	1.8	617	4.0 (seedlings)	4.1
Shrub wood	5 365	Shrub wood	2.6	13 950	4.0	64
Total	16 374	—		54 974	—	515.6

Source: * Shi, 2011

Notes: ** oil plant forest, *** shrub wood

The following data were used to calculate the biogas potential.

Table 8 Summary of data used (Energy crops from marginal land)

Data	Source
Biogas potential: 300 m^3/t VS	Braun et al., 2010
Reference Plant: Miscanthus	
TS content: 30% (m/m FM)	Jewell et al., 2003
VS content: 67% (m/m DM)	

2.4 Contribution of biogas to the total energy demand; potential versus production; including policy targets

This part covers the potential contribution to the total energy demand in 2010 as well as the next 20 years of the current and of the future arising in feedstock, calculated in the first two parts of the study. Moreover, a comparison is made between the biogas energy targets as specified in China's Renewable Energy policy targets for the next 10 years and the current energy potential of the feedstock which is already used in China, such as municipal sewage sludge and agricultural biomass waste.

Table 9 Summary of data used (Biogas energy potential)

Data	Source
1 m^3 of Methane: 9.97 kWh	FNR, 2011a
1 m^3 of Biogas: 1.8 kWh electricity	

2.5 Germany's biogas sector

Information source:

In 2005 a biogas feedstock assessment was carried out by the 'Fachagentur für Nachwachsende Rohstoffe' (FNR, 2005), which has been used to compare the biogas market sector between the two countries: the People's Republic of China and the Federal Republic of Germany.

2.6 GHG Savings-contribution of biogas to climate change mitigation

In the last part of this thesis calculations are made, estimating the saved amounts of CO_2 equivalents by using the total available biomass feedstock for biogas generation.

Assumptions and Calculations:

There are three ways to avoid GHG emissions by generating biogas out of an AD process, which are summarised below. Additionally CO_2 emissions can also be saved if a certain amount of crop residues is no longer burnt on open fields.

Table 10 shows a summary of the results from Part 1 of the study. These data as well as the figures from *Table 11* have been used to estimate the saved amounts of emissions in CO_2 equivalents.

Table 10 Amount, energy potential and methane content of the different kinds of feedstock and amount of crop residues

Type of Feedstock	Available Amount ($\times 10^6$ t/a)	TS content (%)	Biogas potential ($\times 10^9$ m^3/a)	CH_4 content, assuming 60% of biogas ($\times 10^9$ m^3/a)
		CURRENT LEVELS		
Urban Household Waste	291	35	29.1	
Municipal Wastewater Sludge	133	20	12.3	
Crop Residues (Straw)	581	40	215	
Animal Manure	658	4	33.4	
Total	1 663	—	290	174
	37% of total			
Crop Residues	215			
		FUTURE LEVELS		
Urban Household Waste	327	35	33	
Municipal Wastewater Sludge	141	20	13	
Crop Residues (Straw)	588	40	218	
Animal Manure	737	4	37.4	
Energy crops (40%)	680	30	138	
Total	2 473	—	439.4	264
	37% of total			
Crop Residues	218			

Table 11 Global warming potential$_{100}$ of gases

Gas	Value (in kg CO_2 equivalents)	Source
CO_2	1	IPCC, s. a.
CH_4	23	
N_2O	296	
NH_4NO_3 as N	7.2	Williams et al., 2006

2.6.1 Gaining electricity out of a renewable energy source; replacing coal

This section estimates the amount of GHG emissions, which can be avoided by replacing coal by energy gained from the anaerobic digestion process producing the same amount of power.

As a first step, the amount of electricity produced out of the total amount of biogas energy potential of the total feedstock is calculated below.

This calculation assumes that the heat of the CHP process is not used. Therefore, the calculation, which is stated below, slightly underestimates the amounts of saved CO_2 emissions.

$$\text{Methane } (m^3) \times \text{Calorific Value of Methane } (MJ/N\ m^3) \times \text{Efficiency of CHP}$$

Furthermore, calculations are made as to the sum of emissions that are released by generating the same amount of power by using coal as energy source. As a last step these two numbers are compared.

Table 12 Summary of data used (GHG Savings-Calculation No. 1)

Data	Source
CV of CH_4: 36 MJ/Nm^3	FNR, 2011a
CHP Efficiency (only electricity): 35%	FNR, 2005
Emission intensity of conventional coal: 1.186 CO_2/MWh	Wang et al., 2009

2.6.2 Choosing anaerobic digestion instead of landfilling as a waste treatment option

In this section, only two kinds of feedstock, namely urban and agricultural biomass waste are looked at. Energy crops are planted with the purpose of generating energy from them. Therefore, landfill is no option for this resource.

First, emissions generated by landfilling as well as during the anaerobic digestion process are calculated and in a second step are compared. The additional amount of emissions, which are generated by landfilling, can be avoided by choosing the more environmental friendly option of anaerobic digestion.

This calculation assumes that not all the landfill sites are controlled. Rather than collecting methane and using it as a renewable energy source, the gas is venting into the atmosphere. Furthermore, it has to be mentioned that methane emissions of landfill sites will take longer to emit into the atmosphere than during an AD process, even Chinese easy decayable bioorganic waste fraction converts very fast into leachate and some biogas (Raninger, 2010).

For these calculations the following equations have been used:

$$\text{Emissions of Landfill} = 2.46\ \text{t}\ CO_2\ \text{equivalent/t of waste (The World Bank, 2005)}$$

$$\text{Emissions of Anaerobic Digestion} = \text{Methane}\ (m^3)\ (CO_2/CH_4)^{*} \times (\text{kg}\ CO_2/Nm^3\ CO_2)^{**} \times \text{GWP of}\ CO_2$$

* This formula stands for the amount of CO_2 received, when CH_4 is burnt. It depends on the units of methane. Because of the stoichiometric ratio for combustion, which is stated below, the ratio between Nm^3 (normal cubic metre) of methane and carbon dioxide is one.

** As gas yields are calculated in m^3 and tons of CO_2 are to be calculated, this figure must be converted. This is, thus, indicated as: '(kg CO_2/Nm^3 CO_2)'. Thereafter, it has to be further multiplied by the density of each gas.

Stoichiometric ratio for combustion:

$CH_4 + 2O_2 = CO_2 + 2H_2O$

Table 13 Summary of data used (GHG Savings-Calculation No. 2)

Data	Source
Density of CO_2: 1.977 kg/m^3	O'Leary, 2000

2.6.3 Replacing inorganic fertilizer

In 2009 China's consumption of nitrogen (N) fertilizer resulted in a total of 36.9 million tons, being 35% of the total world's use. (FAO STAT, 2011) and the per capita consumption is about 3 time of the world's population average.

In order to ensure the same amount of plant yield, only 25%-30% of organic fertilizer can be added to the inorganic fertilizer (11 million tons). (Roelcke, 2011), but the plant availability of the total nitrogen in AD slurry (mainly Ammonia) is 60% (Wendland. 2011). The total N content in the total feedstock even overtops this demand (see *Table 14*), which exceed China's demand.

Assumptions:

The calculation concerning the emissions saved by replacing the inorganic fertilizer-by using the end-product of the AD technology i. e. the digested-by an organic fertilizer is rather simple.

The process of generating inorganic fertilizer, mainly Ammonium Nitrate, is highly energy consuming. The nitrogen (N) of the e. g. food waste, if landfilled, would then be completely lost. This is not true of the crop residues as the alternative can be left on the ground so that they still supply N for the next cultivaticn. But as farmers mainly burn it on the open fields, it can be assumed to be mainly lost.

The emissions, which are generated while the fertilizer is produced, are the so-called embedded emissions i. e. 'indirect emissions' of the inorganic fertilizer. These are measured by the amount of active compounds they contain (kg of N). From each kg of N contained in the fertilizer, it takes 5.039 5 kg CO_2 equivalents, as a lot of gas has to be burnt.

In this study it is assumed that the bio-fertilizer from AD is as well absorbed by the crops as the inorganic fertilizer and that therefore they are not insignificant N losses on the organic fertilizer.

Calculations:

In order to achieve the same amount of plant yield when using inorganic fertilizers, the specific quantity of organic fertilizer, which can be added to the inorganic fertilizer, is calculated.

The generated amount of emissions, by producing the same amount of inorganic fertilizer can therefore be saved. For these calculations the following equation is used:

$$N/\text{Feedstock (t)} \times \text{GWP of } NH_4NO_3$$

The following table lists the data, which are used for the calculations.

Table 14 Total Solid (TS) and typical values of Nitrogen (N) content of the feedstock

Feedstock	TS content (%)	N content (%)	N content ($\times 10^6$ t)		Source
			Current Levels	Future Levels	
Food Waste	35	4.14	4.2	4.7	Dong et al., 2011
Crop Residues	40	2.79	6.5	6.6	
Animal Manure	4	1.36*	0.36	0.4	
Energy crops	30	1.2**	—	2.4	
Urban Sludge	20	11 kg N/t	0.3	0.31	Defra, 2010
Total	—	—	11.36	14.41	—

Notes: * Pig Manure, ** Sugar beet

2.6.4 Using crop residues as feedstock rather than burning them on fields

Assumptions:

According to Liu et al. (2007), 37% of crop residues are burnt directly on open fields. By instead treating this amount in a clean, sustainable way in an AD plant, huge amounts of GHG can be avoided.

Calculations:

This chapter covers the calculation of the amounts of GHG emissions, which are generated by burning straw on open fields and which at the same time is the amount saved by using straw as a feedstock for an AD plant. The amount of GHG gases generated by burning straw is calculated using the methodology published by the Intergovernmental Panel on Climate Change (IPCC) in '2006 IPCC Guidelines for

National Greenhouse Gas Inventories'. The formula used by IPCC was slightly changed for the purpose of this paper. It should also be mentioned that indirect green house gases like carbon monoxide (CO) and nitrogen oxide (NO_x) were neglected. Therefore the result is slightly conservative as CO as well as NO_x however have negative impacts on the climate and lead to atmospheric and surface effects (ITS, 2005).

Estimation of GHG emissions from fire:

$$L_{Fire} = B \times C_f \times G_{ef} \times 10^{-3}$$

Where:

L_{Fire} = Amount of greenhouse gas emissions from fire, tons of each GHG e. g., CH_4, nitrous oxide (N_2O), etc.;

B= Biomass available for combustion (t);

C_f= Combustion factor, dimensionless, for maize residues: 0.80;

G_{ef}= Emission factor, g/kg dry matter burnt, for;

CO_2:1515

CH_4:2.7

N_2O:0.07

Therefore, the results for each GHG are as followed:

Current Levels:

$CO_2 \rightarrow 215 \times 10^6 \times 0.80 \times 1\,515 \times 10^3 = 261 \times 10^6$ t

$CH_4 \rightarrow 215 \times 10^6 \times 0.80 \times 2.7 \times 10^3 = 464\,000$ t

$N_2O \rightarrow 215 \times 10^6 \times 0.80 \times 0.07 \times 10^3 = 12\,040$ t

Future Levels:

$CO_2 \rightarrow 218 \times 10^6 \times 0.80 \times 1\,515 \times 10^3 = 264 \times 10^6$ t

$CH_4 \rightarrow 218 \times 10^6 \times 0.80 \times 2.7 \times 10^3 = 471\,000$ t

$N \rightarrow 2O \rightarrow 218 \times 10^6 \times 0.80 \times 0.07 \times 10^3 = 12\,208$ t

By converting the GHG into tons of CO_2 equivalents and adding them up, the amount of tons of CO_2 equivalents can be calculated.

3 Results

3.1 Current feedstock levels and biogas potential

3.1.1 Urban Waste

The part entitled 'urban waste' of this study focuses on the resource of organic household waste which is over 70% of the biodegradable fraction within municipal solid waste (MSW) generated from households in China (Liu, 2011a, Raninger 2009). In addition it also includes the feedstock of municipal wastewater sludge generated from domestic as well as from industrial sources.

Urban household waste

It is predicted that by 2030 the country will generate twice the amount of MSW that the United States produces (The World Bank, 2005). The huge amount of organic fraction within MSW can be used for biogas production and thus solve problems not only of energy supply but also with a health and environmental impact (ECCEC, 2011). The waste generated in China is due mainly to the growth of urban population and increased wealth. It can be said that because of economic activity, the population in the city produces twice or three times more waste than the population in the countryside, no matter what their income is (The World Bank, 2005).

Present solid waste management

In 2010 there were about 360 million t of MSW generated in 654 cities of China. Approximately 200 millions of them are generated in major cities. The remaining 160 million t of MSW are produced in counties (Raninger, 2011).

Nowadays, 72% of the total generated MSW in cities is collected, from which nearly 80% is landfilled. 18% is incinerated and 2% is composted (Li, 2011). 80% of the landfilled waste is put in poorly developed dumping sites. Leachate and landfill gas emissions end up as local and global ecological pollution (Raninger, 2011).

Approximately 11% of methane emissions caused by humane beings can be traced back to landfills (Global Methane Initiative, 2010). In China incineration and composting are not very successful either as there is little technical and biotechnological knowledge, low standard of equipment, unclean input material or a low calorific value of 3 500 kJ/kg for fresh matter of MSW as the bioorganic fraction consists of a lot of water. Coal is to be, therefore, often used for co-firing (Raninger, 2011).

Achievements

In its 12th FYP, China is addressing these problems with a proposed continuous reduction of pollution from waste disposal. 'Resources conservation', 'Renewable energy' and 'Environmental protection' have become an issue in state policies.

In the larger cities of China, the government has also started to educate people in waste reduction and source segregation. For example, 600 demo-communities in Beijing are provided with color-coded waste bags and waste bins where 'recyclables', 'kitchen waste' and 'other waste' can be collected separately (BMAC and OCSCCC, 2010) and 100 communities in Shanghai have started to separate dry and wet waste. By the end of this year there will even be 1 009 ecologically aware communities and by 2015 they will have their own property management (Liu, 2011b).

Current levels and biogas potential

China's urban population which are living in major cities produces approximately 163 million tons and the population in counties 128 million tons of biodegradable waste per year. In total, 29.1 billion m^3 of biogas could be gained in biogas plants from this amount.

Municipal Sewage Sludge

Sewage contains a mixture of contaminants such as organic matter (COD, BOD) including nutrients, nitrates and phosphates, also pathogens, toxic chemicals, heavy metals, and inert matter such as sand, debris. Removing these contaminants is the main reason to treat wastewater. In addition, biogas can also be gained from the sludge separated from the wastewater. The wastewater is generated from domestic as well as from industrial sources (Gray, 2010).

Present wastewater management

At present, China disposes of 2 630 treatment plants. Nevertheless, 60% of the generated amount of wastewater sludge is being landfilled; the rest is burnt or composted. It is also used as construction material, for example in brick production.

Current levels and biogas potential

In total, 146 million t of sewage sludge is produced per year, which results in an amount of 12.3 billion m^3 biogas.

3.1.2 Agricultural waste

This part of the study focuses on crop residues as well as on animal manure, which are the main resources occurring in the agricultural sector.

Crop Residues

Crop residues, such as straws stalks, and husks, are defined as 'vegetative organic materials', which remain on the field after harvesting. This term includes field residues i. e. crop straw as well as process residues i. e. by-products.

In addition to coal, oil and natural gas, straw, which is available on a large scale in China, can be seen as the fourth biggest energy resource as it contains 40% of carbon in its fibre (ECCEC, 2011). It can be seen as a highly valuable resource not only because energy can be generated from it but also because it maintains soil fertility, prevents soil erosion and produces sustainable crops. Moreover, besides being used as feed for livestock it is also of great use for cooking and heating in rural households, in the Northern part of China where no firewood is available (Xie, 2011).

Distribution

50% of the straw resources are distributed in 10% of the country. The highest amount of nearly 180 million t, which is approximately a quarter of the total straw amount, is produced in the north of China, whereas Tibet has the lowest amount with 0.1% (ECCEC, 2011).

The north-eastern as well as the Southern provinces of China are the regions where most of the crop straw is produced, such as in Heilongjiang, Hebei, Henan, Shandong, Jiangsu, and Sichuan. However, there are different types of straw, i. e. maize in Heilongjiang and Liaoning, maize and wheat in Hebei, rice and wheat in Jiangsu, rice and sugar cane in Guangdong, and maize and cotton in Sinkiang (Zhong et al., 2003).

Current levels

The predicted amount of straw was about 744 million t and the total energy output was 480 billion m^3. The three major food crops of maize, wheat, and rice are the main source of crop residues, with an amount up to 328 million t, 157 million t, and 122 million trespectively, accounting for even more than 80%.

Further potential raw materials for biogas energy generation are the agricultural by-products such as rice hull, corncob and bagasse. In 2006 the output of rice hull, corncob and bagasse was 36.5, 72.7 and 24.9 million t respectively, from which 23, 55 and 2.6 billion m^3 biogas could be generated (ECCEC, 2011).

Biogas potential

Nearly 447 million tons, which are 60% of the useable amount, of the whole generated amount can be put into biogas energy generation. Including the by-products, it comes to a total amount of 215 billion m^3.

Animal Manure

Animal manure is a very suitable feedstock for the technology of anaerobic digestion and it is widely used throughout in China. If this resource were used for gaining energy, there would be much less urban and rural environmental problems, such as pollution of air, water and soil; in particular, pollution of drinking water and ecological agriculture could be avoided (ADB, 2009).

In China there are somehow two methods of feeding domestic livestock. Families and small-sized farms practise the traditional method, namely 'natural feeding' and breed mainly sheep, horses and ducks. The manure from this type of farming is difficult to collect as it is distributed on grass and ponds or through household biogas digesters. Farms with some pigs and cows can also have such 8-12 m^3 passive digesters or the dung is used as fertilizer directly.

On the other hand, there is the centralized feeding in medium (breeding belts around the cities) and large-sized farms for cattle, pigs and poultry (ADB, 2009). The manure of these animals is easier to collect as they are held in a confined area (Li et al., 2001). Therefore, this section estimates the amount and energy potential of animals arising in centralized feeding farms.

Distribution

Due to the different climatic situation in the various regions of China, the amount of livestock also varies from province to province. Large and medium-sized farms in the provinces of Henan, Guangdong, Shandong, Hebei, Liaoning, Hunan, Fujian, Jiangsu, and Zhejiang produced the highest amount of manure in 2006, namely up to 63.6% of the whole country (ECCEC, 2011).

Current Levels

In 2006 in China 519 million pigs, which is 49% of world production, with a 3.5% annual increase were raised. The total generated and collectable amount of manure accounts up to 3 billion t in 2009, 3.5 billion t forecasted for 2015 with a COD three times higher than the industry emissions (ADB, 2009).

Biogas Potential

The biogas yield of the available amount for biogas generation adds up to approximately 33.4 billion m^3.

3.1.3 Total Current Biogas Energy Potential

The total biogas potential, including urban and agricultural biomass waste sum up to 290 billion m^3.

3.2 Feedstock arising and biogas potential until the year 2030

Due to the predicted increase in urbanization, the highest increase in the occurrence of waste will be in cities. It will reach 263 million t of organic fraction in major cities and 64 million t of biodegradable fraction in counties. Resulting in 327 million t in total, i.e. 33 billion m^3.

The resource of wastewater sewage sludge will also undergo an enormous increase of about 6%, resulting in a biogas potential of 13 billion m^3.

As the increase in crops is limited by factors like land use changes as well as by the regional and climatic situation, this fact will not change a lot in the future; only 1.2% is assumed. Therefore, a biogas yield of 218 billion m^3 is predicted.

On account of the rapid economic development and the increased living standards, the demand for meat as well as for dairy products is rising. This has led to a steady grow in the numbers of commercial livestock farms from about 2.4 million in 2002 to about 4.3 million in 2006, with an improved quality and increased productivity (ADB, 2009).

Assuming the average rate of 12% during the period from 2000 and 2009 will stay the same, a biogas yield of approximately 37.4 billion m^3 would be the result by the year 2030.

Therefore, due to an increase in urbanization and change in people's lifestyle, the highest rise will concern the resources of household waste as well as animal manure.

3.2.1 Energy crops from marginal land

A fifth kind of feedstock, the energy crops from marginal land, is also included in the forecast calculations of biogas potential for the year 2030. Using energy crops for energy production is a relatively new field, especially for the biogas generation. This study assumes a utilization factor of 40% of the total energy crops amount for 2030.

Bioenergy produced from energy plants will play an increasingly important role in future energy supply and as China has fairly limited cultivated land resources, bioenergy development will mainly rely on the exploitation of marginal land. This section focuses on the assessment of marginal land resources and biogas potential of energy crops in China (Zhuang et al., 2010).

While China has now some ten years experience in biogas from livestock manure waste water, in the last few years they focused on the fermentation of straw, but the country has also just started to do some research on the highly promising field of energy crops from marginal land (Hu, 2010a). In addition to the energy production economic development in remote and depressed rural area can be improved too. Research studies also prove that besides greenhouse gas emission abatement, significant eco-environmental benefits can be achieved. By planting bio-energy plants on marginal lands, the eco-function of soil/water conservation and wind erosion protection can be improved (Li et al., 2009).

Land utilization

China's total surface area is approximately 9.600 million km^2 (TCG, 2011). In contrast to other countries, China has a large population but limited arable land (Hu, 2010a). *Figure 4* shows the proportion of land and how it is used.

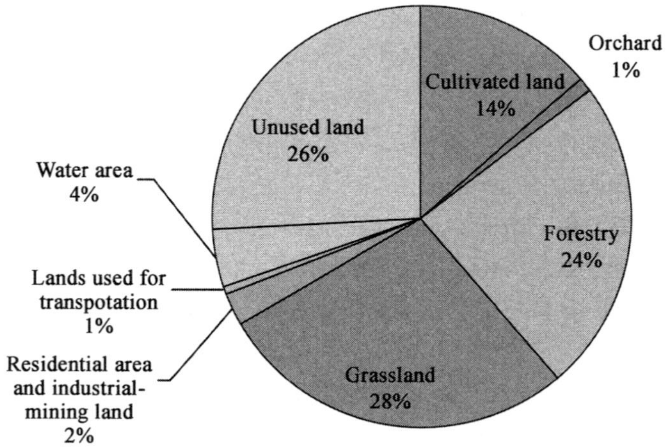

Note: The data base on the investigated land area of 9.507 million km^2 of China (Hu, 2010a).

Figure 4 Classification of land utilisation in China

As bioenergy industries should not compete for land with the production of food, feed and fiber, the plantation of energy crops will be restricted to marginal land (Tang et al., 2010).

Size and composition of marginal land suitable for energy plants

Shi (2011) estimates that around 163.74 million hectares of China's total marginal land might be suitable for growing energy crops. 84.91 million ha, which account for 51.9% of the total area, are still not used. 70.8% of the total marginal land is well suited for forestry, 29.2% for agriculture (Shi, 2011).

Distribution

Marginal land resources can be found in two main areas of China:

(1) Southwest China as well as Yunnan, Guizhou, Sichuan Province and Chongging City. The area of marginal land in this region is about 38.7 million hectares, accounting for 39.1% of the territory of China. In this area, the cultivation of energy plants is supported by sufficient water and sunshine.

(2) Inner Mongolia and Northeast China with a good quality of soil and low slopes (Zhuang et al., 2010).

Energy plants

Energy plants are defined as any plant containing biomass, which serves as energy source. China disposes of a great biodiversity (Guo et al., 2011). The main energy plants suitable for growth in China are Energy corn, sweet sorghum, cassava, sweet potato, sugarcane, sugar beet, jerusalem artichoke, *Pennisetum clandestinum*, and *Panicum virgatum* (Hu, 2010a). The characteristics of the terrain (like elevation and slope) and soil as well as meteorological factors (including precipitation data and air temperature data) should be considered when choosing the type of plant to be cultivated.

Biogas potential:

The total energy potential of energy crops planted on the total available area of 163.74 million hectares come to nearly 345 billion m^3 biogas. As this study assumes a utilization factor of 40%, 138 billion m^3 will be used by the year 2030.

3.2.2 Total future biogas energy potential

Figure 5 summarizes the current biogas potential and shows as well the expected changes up to the year 2030. Any kind of feedstock will increase in the future. The ranking will remain the same, i. e. crop residues are on the top of the list followed by the new but highly valued feedstock of energy crops and animal manure. Urban waste represents the smallest amount and, therefore, also in energy potential. Nevertheless it is more likely to be used than for example the energy crops.

In summary it may be said, the future biogas potential will increase to 439.4 billion m^3 by 2030 with a total increase of 52%, respectively.

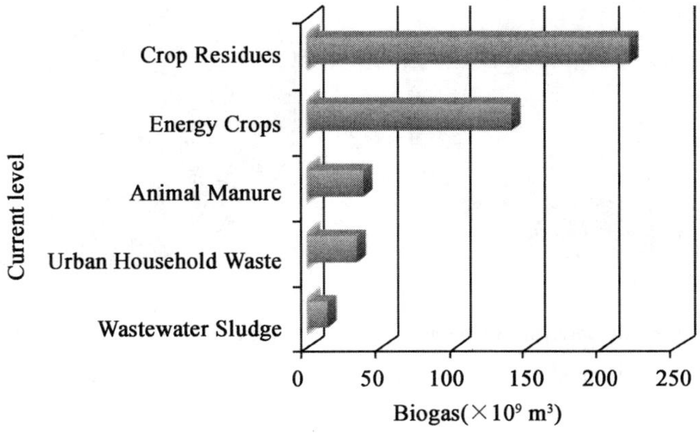

Figure 5 Current situation of municipal, agricultural biomass waste and energy crops and the increment until 2030

3.3 Contribution of biogas energy to china's energy demand

China being the most populous country is the largest energy consumer in the world. It is ranked first in global coal production as well as in coal consumption amounting to 3.5 billion tons in 2009, which makes nearly half the world's consumption with a 180% increase since 2000. Since 2009 the country has no longer been exporting but rather importing coal (EIA, 2011).

According to the 'State Grid Energy Research Institute', in 2010, China's energy demand sum up to 91 000 petajoule (PJ), and will further increase to 189 000 PJ by 2030 (Hu, 2010b).

3.3.1 Contribution of biogas

In total, the current biogas potential totals 290 billion m^3. Assuming an average methane content of 60 %, this would result to an amount of methane of 174 billion m^3/a. Assuming all the available biomass feedstock is used as an energy source a total biogas energy potential of about 6 246 PJ/a could contribute to 6.9% to the overall energy demand. The following figure represents the respective contribution of each feedstock.

Energy potential

This part deals with the energy potential generated out of the total estimated biogas potential of 290 billion m^3 biogas, which can be used for electricity and/or heating purposes.

Electricity production

With a conversion factor of a Combined Heat and Power (CHP) plant of 35%, an available energy potential of around 607 TWh/a can be calculated. With 8 500 full load hours per year, this would correspond to an installed power capacity of 71.4 GW. In addition, through co-combustion, the wastewater sewage sludge can be used in existing coal-fired power plants.

Heat Generation

With a average conversion efficiency for the exclusive provision of heat of 96%, the potential availability of heat results in 5 996 PJ/a. Otherwise by gaining heat through a CHP plant with a conversion efficiency of 58%, heat of 3 623 PJ/a could be generated.

Due to the predicted strong increase in total energy consumption in China over 5%, which is approximately 9 463 PJ of the enormous energy demand in 2030 (189 000 PJ) could be met by biogas energy utilization.

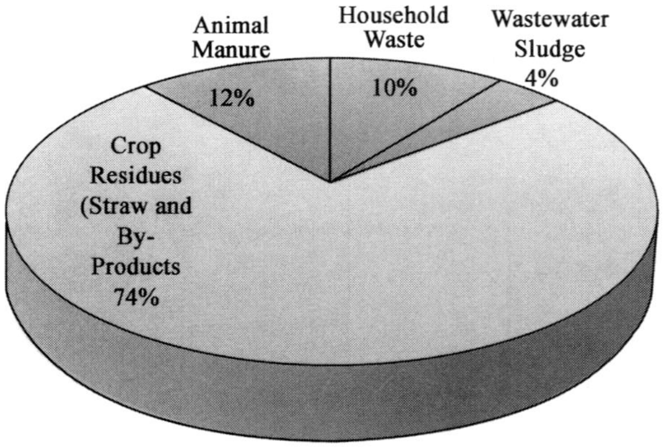

Figure 6 Contribution of the current potential of biogas energy to total energy demand

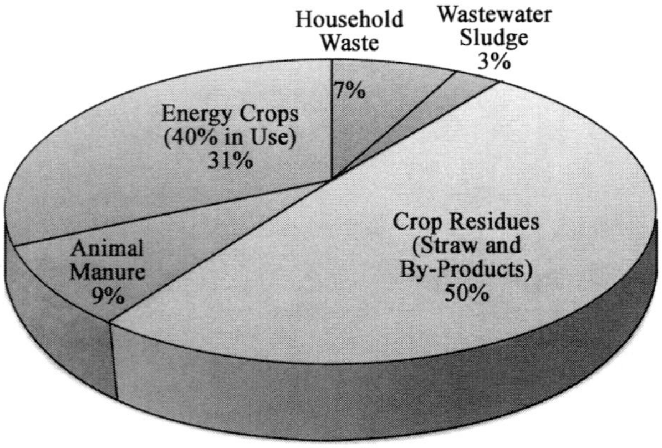

Figure 7 Contribution of the future potential of biogas energy to total energy demand

3.4 Potential versus production; including policy targets

Given the amount of biomass waste available, the use of biogas is still relatively low although China has made efforts to support the production of biogas by setting appropriate goals until 2020. The current performance of biogas-plants and grid-connection restrictions will not allow to reach even these goals.

Policy Targets

Part 1 of this study estimates a current potential of up to 290 billion m^3 of municipal sewage sludge and agricultural waste, the two kinds of feedstock, which are already mainly in use in China.

However, according to the renewable energy targets achieved by the NDRC in 2010, only 6.6% of the available feedstock was used for energy supply on the market. The use of feedstock will increase to 10% until 2020.

Production

The PRC already invested large sums in the construction of biogas plants in the last years, especially in those treating agricultural waste. The numbers of small, medium and large sized plants treating agricultural waste feedstock rose rapidly from 26 586 in 2007 to 39 510 in 2008 and to 56 534 in 2009, which results to a total biogas yield of 764.92 million m^3 out of agricultural biomass waste (GIZ, 2011a). However, this is less than 1% of the total agricultural resources of about 248.4 billion m^3.

As animal manure has a biogas potential of 33.4 billion m^3 resulting in 60 GWh electricity, it would not have been a problem to contribute about 14% to the demand for electricity in households of China (440 GWh) (NBSC, 2011) by transforming the amount of animal manure from industrial livestock farms into biogas.

Another example of a lost opportunity in terms of biomass resources in China is the sewage sludge.

Less than 1%, which is 100 million m^3, of the total amount of municipal wastewater sludge (12.3 billion m^3) was used in biogas energy systems in 2010.

3.5 China compared with Germany, the European leader in biogas production

According to the FNR, in 2005 Germany had a total biogas potential of 22 to 25 billion m^3, including also landfill sites. The following figure shows how the different kinds of feedstock contribute to the total biogas potential.

By using the whole substrate mix, a total energy potential of 360 to 520 PJ/a could be gained, which would contribute between 3.9% to 5.6% to Germany's energy consumption of 9 288 PJ in 1999. (FNR, 2005)

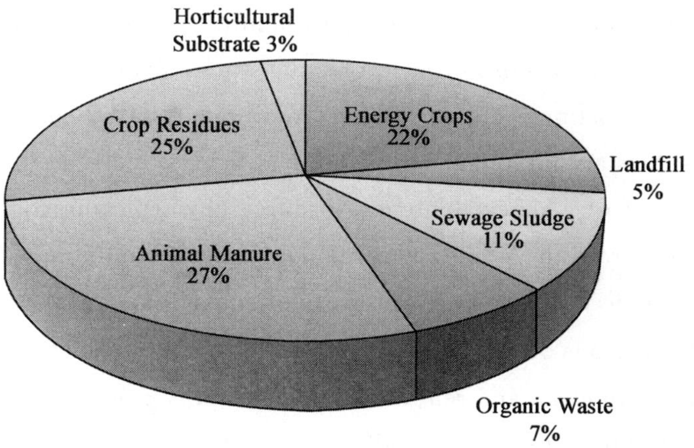

Figure 8 Germany's total biogas potential (FNR, 2005)

Germany's biogas potential shows a similar contribution of each feedstock to the total energy demand. The main provider of feedstock is the agricultural sector in both countries.

Even though Germany's biogas potential is small compared to that of China as it consists of smaller areas (e.g. only an area of 2 million hectares which is suitable for growing energy crops was estimated) and having fewer inhabitants, over 5 900 anaerobic digestion plants have already been installed in Germany with an established power generation capacity of 2 300 MW electricity, which is a 20 fold increase compared to 2001. In 2010 12.8 billion kWh, which corresponds to 2.1% of the country's energy demand, and 12.6% of the energy supply of renewable energies (FNR, 2011c)?

Since the beginning of this year, 52 bio-methane plants have been connected to the natural gas grid and many more projects are under construction and in planning (FNR, 2011d).

3.6 GHG emissions and savings

According to the 'United Nations Framework Convention on Climate Change' (UNFCCC)-secretariat (1994), over 74% of total GHG emissions is generated by China's energy sector, which made up to 7.5 billion t of CO_2 equivalents in 2009 (IEA, 2009), followed by 15% emitted from the agricultural, 7% from the industrial and 4% from the waste sector (UNFCCC-secretariat, 1994), resulting in nearly 10 billion t of GHG emissions in 2009.

Therefore, replacing fossil fuel sources by energy generated from clean, renewable resources turns out to be a big opportunity to contribute to climate change mitigation. In addition, avoiding CO_2 emissions also reduces airborne pollutants, such as toxic heavy metals, ozone-forming chemicals and sulphur dioxide, which is responsible for acid rain (Li et al., 2009).

Contribution of Biogas to Climate Change Mitigation

This part of the study aims at estimating the amount of saved tons of carbon equivalent of CO_2 emissions due to the production of biogas energy and organic fertilizer gained from the feedstock of urban and agricultural biomass waste as well as of energy crops from marginal land whose potential was carefully estimated in Part 1 of this study. In addition, it shows which role biogas technology could play in order to achieve the ambitious targets of the national policies concerning reductions in GHG emissions.

Calculations and Results

There are three ways to avoid GHG emissions by generating biogas from an AD process.

First, energy is produced out of renewable resources.

Secondly, by assuming the waste is otherwise going to landfill or open lagoons and thirdly, the second highly valuable end-product of the process of anaerobic digestion, namely the produced digestate replaces inorganic fertilizer, whose production requires energy. Additionally, CO_2 emissions can also be saved if a certain amount of crop residues are no longer burnt on open fields.

- Generating electricity out of a renewable energy source; by replacing coal. In total, 2.2×10^{12} MJ electricity can be gained from the waste, which amounts to 611 million MWh.

According to Wang et al. (2009), the emission intensity of conventional coal results in 1.186 CO_2/MWh. Therefore, the amount of emissions of CO_2 equivalents, which can be avoided by replacing China's number one energy source i. e. coal with clean and renewable biogas energy amounts to 725 million t.

- Choosing anaerobic digestion instead of landfilling or storage in open lagoons as a waste treatment option.

Emissions generated by the two waste treatment options are compared and the saved amount of GHG emissions was calculated; this amount to 3.76 billion t of CO_2 equivalents.

- Replacing inorganic fertilizer: By replacing inorganic fertilizer by using the end-product of the AD process, the digested, 81 200 t of CO_2 equivalents can be avoided.
- Using crop residues as feedstock rather than burning them on fields.

Instead of burning a huge amount of straw and agricultural by-products on the field, this high valuable feedstock is going to be treated in an AD plant. Therefore, emissions of approximately 261 million t of CO_2 equivalents, which are generated during the burning process, can be saved.

Results:

Therefore, the current total reduction potential of CO_2 equivalents is 4.75 billion t and it will rise to nearly 7 billion t until the year 2030.

4 Conclusions

The following table summarises all the findings regarding China's and Germany's biogas potential on the basis of the energy demand as well as policy targets of these countries.

Table 15 Chinese biogas energy potential and policy targets in relation to the total Energy demand and German performance figures, 2010-2050

Year	Final energy consumption (PJ/a)	China					Germany
		Renewable Energy targets	Biogas Energy targets	Biogas Potential			Biogas
		% Energy consumption	Energy consumption	bn Nm^3	% Energy consumption	Policy target % potential	% Energy consumption
2010	91 000	10 >9 bn Nm^3	19 bn Nm^3	290	6.9	3.1	Achieved 1.5

Year	Final energy consumption (PJ/a)	China					Germany
		Renewable Energy targets	Biogas Energy targets	Biogas Potential			Biogas
		% Energy consumption	Energy consumption	bn Nm^3	% Energy consumption	Policy target % potential	% Energy consumption
2020	131 000	15-16 >24 bn Nm^3	44 bn Nm^3	365	10	6.6 <10**	Target 5.5
2030	189 000			440	5***		
2050		26-30					Potential up to 10*

Note: NDRC, Medium and Long Term Development Plan for Renewable Energy in China (Beijing, September 2007);
* Potential in 2020 = 2050 von 0.5 PJ (FNR, 2011)
** Target Biomass = 18.5% within Renewable Energy
Target Biogas all (NDRC 2007)
State Grid 2010
*** CHINA: 2030 > 9.000 PJ, > 20× Germany

With regard to the specific aims of the present study the following conclusion and recommendations can be drawn:

- Due to its large and diverse areas and the huge population, the country has the availability of abundant biomass resources. This study estimates a current annual biogas potential of 290 billion m^3, which could cover 6.9% of the total energy demand. A biogas-driven power generation with a capacity of 71.4 GW electricity could be established. Furthermore, the resources will constantly grow and an amount of 439.4 billion m^3 by 2030.
- In order to use the energy resources in the most efficient way, the use for heating energy should be promoted in particular. The heating energy can contribute to the economic returns of an anaerobic digestion plant.
- The agricultural sector has the greatest potential, such as crop residues, followed by energy crops from marginal land and animal manure. Animal manure currently used as the main feedstock for biogas production should be co-digested with diversified feedstock especially with straw.
- As to the highly valuable feedstock of energy crops from marginal land, the country is, besides of a vivid bio-ethanol production to be blended with petrol (10% bio-ethanol in many provinces) still at the starting point of doing research in this very promising field. So far, there is neither an agreed definition of the term 'marginal land' nor have assessment standards been implemented concerning suitable land resources, energy crop varieties, feedstock availability and related energy potential. Therefore, research in this highly valuable resource should be supported.
- Concerning the future of the biogas potential, the ranking will still remain the same, i. e. crop residues are on the top of the list followed by energy crops and animal manure. Urban waste represents a smaller amount and therefore also in energy potential. But it should be mentioned, due to the rapid urbanization, urban waste will increase the most, namely over 12% by 2030 and 2050. It is not only a great challenge for China to gain energy from this waste but also to cope with these huge amounts of waste in the cities in terms of environmental pollution. Therefore municipal solid waste management needs to be enhanced.
- Biomass resources offer a realistic chance to ensure energy security and sustainable social and economic development. However, the resources are still not used in a sustainable and efficient way and are instead wasted. High quantities (35%-40%) of animal manure are untreated and huge amounts of crop residues are still burnt on open fields, which lead to pollution (Liu et al., 2007). A

more environmental-friendly management in this sector would also be required.

- Although the country is rich in biogas feedstock, the current and future forecasts concerning the availability of biomass not only mean that the resources are far from being fully exploited but they also show that the policy targets to develop sustainable biogas energy are not ambitious enough; 6.6% of it was used by 2010 and 10% will be used for 2020.
- Notwithstanding this, an enormous increase in both biogas production and utilization has been recognized during the last years. From 2007 to 2009, the numbers of plants has increased by 53%. However, only three of them have been connected to the electrical supply network so far and only seven units are so-called CDM-Projects whereby only two can generate some CDM carbon credits based on certified emission reduction (CER).
- In spite of the discernible trend of a further increase in biogas energy production in the future, the necessary know-how as well as technology concerning design, construction and performance of the plant as well as pre-treatment of the feedstock and maintaining of the whole process are often still missing. The limited capacity of technical service for project operation creates a bottleneck for the rapid development of biogas projects at present. It is, therefore, important not only to enhance the utilization of biogas feedstock and to extend energy supply but also to encourage the competent authorities to invest into maintenance and control of the process of existing plants. A further step would be promoting the industrialization of biogas energy development.
- Besides methane, which is used to generate energy from the biogas, the second product, the digestate, is also of great potential. By treating the whole feedstock, the generated amount would exceed China's demand for bio-fertilizer. Therefore, not only money but also a huge amount of GHG emissions can be avoided by replacing inorganic fertilizer.
- The study estimates an amount of 4.75 billion tons CO_2 equivalents, which could be saved by using anaerobic digestion as the main treatment option and it could rise to nearly 7 billion tons until the year 2030. Energy could be produced at lower costs-provided low-priced carbon is sold on the world market-which would entail an increase in the bio-energy development(Li et al., 2009)

China disposes of a remarkable potential for biogas energy, both at present and in the future. Germany, being a much smaller country and disposing of less biomass feedstock than China, has nevertheless made any efforts to promote its biogas energy sector in recent years. China should, therefore, take Germany as an example and derive more benefit from its biomass resources as well as from the generation of clean and sustainable energy.

The biogas sector represents a promising option for using renewable energies, which could make further contribution to a sustainable energy supply as well as to mitigating greenhouse gases in the future.

The responsible authorities and decision-makers should be motivated to further promote the production and the use of biogas as well as maintain and control existing plants.

This study is meant to make a contribution to this development.

References

ADB Asian Development Bank, SINOC - Hassal, Raninger B. et al., (2009): *PRC: Preparing the Integrated Renewable Biomass Energy Development Project-Final Report*. Project Number: 40682 Beijing, China.

BMELV, (2011): in 1st governmental consultation meeting & 6th Forum for commercial and technological cooperation,

28 June 2011, Berlin, Germany.

BMAC, Beijing Municipal Administration Commission, (2010): *Lu Wa in action-Guideline for waste reduction and waste sorting in pictures*, Beijing Press, Beijing, China(in Chinese).

BMJ, Bundesministerium der Justiz, (1991): http://bundesrecht.juris.de/bundesrecht/abfablv/gesamt.pdf (in German).

Braun R., Weiland P. Wellinger A., (2010): *Biogas from energy crop digestion*. http://www.ieabiogas.net/_download/energycrop_def_Low_Res.pdf.

China Daily, (2011): *Sustainability through renewable energy*. http://www.chinadaily.com.cn/opinion/2011-05/10/content_12476300_2.htm.

CREIA, Chinese Renewable Energy Industries Association, (2011): 12th *Five-Year Plan for Renewable Energy Development-Draft*. Beijing, China.

Cui Z., (2005): *The Diversity and Applied Approaches of Biomass Resources in China*. Biomass-Asia Workshop 2005, Powerpoint presentation, http://www.biomass-asia-workshop.jp/biomassws/01workshop/en/program.html.

Daskalopoulos E., Badr O., Probert S. D., (1998): *Municipal solid waste: a prediction methodology for the generation rate and composition in the European Union countries and the United States of America*. Resources, Conservation and Recycling 24, pp. 155-166.

Defra, Department for Environment, Food and Rural Affairs, (2010): *Fertiliser Manual RB* 209, 8th Edition, UK, pp. 75.

Dong B., Chen L., Zhang Y., Zhao L., Wan X., Gao X., Zhang X., Luo J., (2011): *Effect of pig manure addition on biogas production of combined two- phase anaerobic digestion*. Vol. 2, Institute of Energy and Environmental Protection, Chinese Academy of Agricultural Engineering, Beijing, China.

EC, European Commission, (2010): *Biodegradable waste*. http://ec.europa.eu/environment/waste/compost/index.htm.

ECCEC, Europe-China Clean Energy Centre, (2011): *Study of potential and constraint of the biomass sector in China*, (unpublished report).

EIA, Energy Information Administration, (2009): *International Energy Outlook* 2009. http://www.eia.doe.gov/oiaf/ieo/pdf/0484(2009).pdf.

EIA, Energy Information Administration, (2011): http://www.eia.gov/countries/cab.cfm? fips=CH.

EMCA, ESCO Commitee of China Energy Conservation Association, (2011): *China's primary energy consumption will be kept to between* 4 *to* 4.2 *billion tons of standard coal by* 2015. http://www.emca.cn/BG/en/newsinfo.aspx? cid=10&id=184.

EPA, United States Environmental Protection Act, (2011): *Methane*. http://www.epa.gov/methane/.

FAO STAT, Food and Agriculture Organizaiton of the United Nations, (2011): *statistical database*. http://facstat.fao.org/site/575/DesktopDefault.aspx? PageID=575#ancor.

FNR, Fachagentur für nachwachsende Rohstoffe, (2005) *Handreichung-Biogasgewinnung und-nutzung*. Gülzow, http://www.big-east.eu/downloads/FNR_HR_Biogas.pdf (in German).

FNR, Fachagentur für nachwachsende Rohstoffe, (2011a): *Faustzahlen Biogas*. http://www.biogasportal.info/daten-und-fakten/faustzahlen/ (in German).

FNR, Fachagentur für nachwachsende Rohstoffe, (2011b): *Einheimische Bioenergie: Was kann sie 2050 leisten?* http://www.nachwachsenderohstoffe.de/uploads/media/Wuerfel_2011_gross.jpg (in German).

FNR, Fachagentur für nachwachsende Rohstoffe, (2011c): *Biogasanlagen in Deutschland*. http://www.biogasportal.info/daten-und-fakten/biogasanlagen/ (in German).

FNR, Fachagentur für nachwachsende Rohstoffe, (2011d): *Biogaseinspeisung in das Erdgasnetz*. http://www.biogasportal.info/daten-und-fakten/biogasanlagen/biomethan-anlagen/ (in German).

GIZ, Gesellschaft für Internationale Zusammenarbeit, (2011a): *China Biogas Sector-Historical Overview*. http://www.biogas-china.org/index.php? option = com_flexicontent&view = category&cid = 18&Itemid=35&lang=en.

GIZ, Gesellschaft für Internationale Zusammenarbeit, (2011b): *Klima- und Energieziele des 12. Fünfjahresplans-Chinas grüne Herausforderung an Europa*? Klimaschutzprogramm, Beijing, China (in German).

Gray J., (2010): *Sustainable Sewage Design*. http://www.sustainablebuild.co.uk/sustainabledesignsewage.html.

Guo D. G., Zhang X. Y., Shao H. B., Bai Z. K., Chu L. Y., Shangguan T. L., Kun Y., Zhang L. H., Xu G., Sun J. N., (2011): *Energy plants in the coastal zone of China: Category, distribution and development*. Renewable and Sustainable Energy Reviews, 15, pp. 2014-2020.

Global Methane Initiative, (2010): *Global Methane Emissions and Mitigation Opportunities*. http://www.globalmethane.org/documents/analysis_fs_en.pdf.

Hu L., (2010a): *Potential and Regional Distribution of Marginal Land for Energy Crops Planting in China*. Chengdu. Powerpoint presentation, December 2010.

Hu Z., (2010b): *Low Carbon Electricity: Integrated Resource Strategic Planning + Smart Grid*. Seminar zum deutschen Energiekonzept und dem 12. 5-Jahresplans der VR China, Powerpoint presentation, 3 August 2010, Beijing, China.

IEA, International Energy Agency, (2009): *Key World Energy Statistics*. http://www.iea.org/textbase/nppdf/free/2009/key_stats_2009.pdf.

Index Mundi, (2011): *China Population Growth Rate*. http://www.indexmundi.com/china/population_growth_rate.html.

IPCC, Intergovernmental Panel on Climate Change, (2006): *IPCC Guidelines for National Greenhouse Gas Inventories Vol. 4-Agriculture, Forestry and Other Land Use, Chapter 2: Generic Methodologies applicable to multiple land-use categories, Equation 2.27-Estimation of Greenhouse gas emissions from fire*. http://www.ipcc-nggip.iges.or.jp/public/2006gl/vol4.html.

IPCC, Intergovernmental Panel on Climate Change s. a., Technical Summary, http://www.ipcc.ch/ipccreports/tar/vol4/english/pdf/wg1ts.pdf.

ITS, Institute of transportation studies, (2005): *Project 4: The Indirect Global Warming Potential of NOx Emission*. http://hydrogen.its.ucdavis.edu/research/track4/tr4pr4.html.

Jewell W. J., Cummings R. J., Richards B. K., (2003): *Methane fermentation of energy crops: Maximum conversion kinetics and in situ biogas purification*. *Biomass and Bioenergy*, 5(3-4), pp. 261-278.

Khaleey Times online, (2011): *China's CO_2 emissions rose over 10 pct in 2010*. http://www.khaleejtimes.com/DisplayArticle08.asp? xfile=data/environment/2011/June/environment_June13.xml&section=environment.

Li J., Zhuang X., DeLaquil P., Larson E. D., (2001): *Biomass energy in China and its potential*. Energy for Sustainable Energy, 5(4), pp. 66-80.

Li X., Huang Y., Gong J., Zhang X., (2009): *A study of the development of bio-energy resources and the status of eco-society in China*. Renewable and Sustainable Energy Reviews, pp. 1-6.

Li X., Yuan H., (2011) *China Biogas Industry in New Period - challenges and future development*. Center for Resources and Environmental Research, Beijing University of Chemical Technology, Beijing, China. (in Chinese).

Li Y., (2011): *Grundsituation der unschaedlichen Behandlung der Siedlungsabfaelle in China und die Entwicklungsschwerpunkte im 12. Fuenfjahrplan*. GTZ und NDRC waste workshop-12th Five-Year Plan, Powerpoint presentation, 25 May 2011, Beijing, China (in German).

Lindeburg M., (2011): *Civil Engineering Reference Manual of the PE Exam*. 12th edition, Professional Publication Inc., USA.

Liu H., Jiang G. M., Zhuang H. Y., Wang K. J., (2007): *Distribution, utilization structure and potential of biomass resources in rural China: With special references of crop residues*. Renewable and Sustainable Energy Reviews, 2008, 12, pp. 1402-1418.

Liu J., (2011a): Chief Engineer, China Urban Construction Design & Research Institute, Personal conversation, 6th July 2011, Beijing, China.

Liu J., (2011b): *Current situation and new progress of household waste in China*. Sino-German Environmental Policy Programme Workshop on the 12th Five-Year Plan on Municipal Solid Waste Management, 25 May 2011, Beijing, China.

Luo S., (2011): *Ecoagriculture in China: A Systems Approach*. Journal of Resources and Ecology, 2(1), pp. 8-14. (in Chinese)

Martinot E., Junfeng L., (2007): *Powering China's Development-The Role of Renewable Energy*. Worldwatch Report, Report number: 13: 978-1-878071-83-5, Washington.

McKinsey & Company (2008): *Preparing for China's urban billion*. http://www.mckinsey.com/mgi/publications/china_urban_summary_of_findings.asp.

MOA, China Ministry of Agriculture, Science and Education, Biomass Department, (2011): *verbal information by Dir. Li Shaohua, to German Biogas Stakeholders in Germany*.

MPRA, Munich Personal RePEc Archive, (2010): *Rural biomass energy 2020: People's Republic of China*. http://mpra.ub.uni-muenchen.de/24987/1/MPRA_paper_24987.pdf.

NBSC, National Bureau of Statistics of China, (2010a): *China Statistical Yearbook* 2010, *Total population by urban and rural population and birth rate, death rate, national growth rate*. http://www.stats.gov.cn/tjsj/ndsj/2010/indexeh.htm.

NBSC, National Bureau of Statistics of China, (2010b): *China Statistical Yearbook* 2010, *Agriculture*, 13-15 *Output of Major Farm Products*. China Statistics Press, Beijing, China.

NBSC, National Bureau of Statistics of China, (2010c): *China's rural yearbook* 2010, 7-42 *The increase or decrease of living animals at the end of the year*. China Statistical Press, Beijing, China, pp. 193. (in Chinese)

NBSC, National Bureau of Statistics of China, (2011): *China Statistical Yearbook*, *Energy*, 7-6 *Electricity Balance Sheet*. China Statistics Press, Beijing, China. (in Chinese)

NDRC, National Development and Reform Commission, (2007a): *Medium and Long Term Development Plan for Renewable Energy in China*. http://www.martinot.info/China_RE_Plan_to_2020_Sep-2007.pdf.

NDRC, National Development and Reform Commission, (2007b): *China's National Climate Change Programme* 2007. Beijing, China.

NDRC, National Development and Reform Commission, (2011): *China 12th FYP on (agricultural) Biogas*

Sector Development. in BioEnergy Expo Conference on Biogas Industrialization in China Conference, 19 May 2011, Qingdao, China.

NPC, The National People's Congress of the People's Republic of China, (2011): *China prepares to end GDP obsession*. http://www.npc.gov.cn/englishnpc/news/Focus/2011-% 0903/07/content_1635453.htm.

OCSCCC, Office of Capital Spiritual Civilization Construction Commission, (2010): *Lu Wa in action-Guideline for waste reduction and waste sorting in pictures*, Beijing Press, Beijing, China. (in Chinese)

O'Leary, D., (2000): *Properties of Carbon Dioxide*. http://www.ucc.ie/academic/chem/dolchem/html/comp/cO2.html.

Raninger B., LI R., Bidlingmaier W., et al., (2006): *Bioorganic Municipal Waste Resource Recovery and Utilisation Project in Shenyang-RRU-BMW Summary Report'*. Proceedings of the 1st ICEEE Int. Conference on Bioorganic Municipal Waste and Landfill Management, Shenyang, P. R. China, pp. 1-120.

Raninger B., Yang Guangzhong, R. Li, Feng Lei, (2009): *The RRU-BMW Project - Four Years Pilot Tests on BMW Source Separation in Shenyang, China*, Biomass and Organic Waste as sustainable Resources, Proceedings of the Int. Conference ORBIT 2009, Biogas Engineering and Application, Vol. 1, China Agriculture University Press, ISBN 978-7-5655-0190-6, Beijing, China, pp. 35-44.

Raninger B., Li R., Chen X., Xu H., (2010): *Biomass Wastes-to-Energy in China, Biogas from Landfills or Anaerobic Digestion?* Depotech 2010, Montanuniversität Leoben, Abfallwirtschaft, Abfalltechnik, Deonietechnik und Altlasten, Hersg. Lorber, Adam, Aldrian, et. al,. ISBN 987-3-200-02018-4, pp. 105-111.

Raninger, B., (2011): *Biogas from landfill in China? ... implication on solid waste management*. Biogas technology: Status and trends in Asia, Powerpoint presentation, 24 May 2011, Bangkok, Thailand.

Reuters, (2011): *China to focus on developing biomass, solar power-paper*. http://www.reuters.com/article/2011/08/01/china-energy-renewable-idUSL3E7J103A20110801.

Roelcke M., (2011): Professor at Technische Universität Braunschweig, E-Mail Contact, 8th August 2011.

Ru X., Lu X., Li, P., (2010): *The China Society Yearbook*. Vol. 4, Brill, Leiden, The Netherlands.

Shi Y., (2011): *China's resources of biomass feedstock*. China Engineering Sciences, Beijing, China, pp. 16-23 (in Chinese).

Steffen R., Szolar O., Braun R., (1998): *Feedstocks for Anaerobic Digestion*. http://www.adnett.org/dl_feedstocks.pdf.

Tang Y., Xie J. S., Geng S., (2010): *Marginal Land- based Biomass Energy Production in China*. Journal of Integrative Plant Biology, 2010, 52 (1), pp. 112-121.

TCG, Travel China Guide, (2011): *Country Description of China*. http://www.travelchinaguide.com/essential/country.htm.

The Council of the European Union, (1999): *Council Directive* 1999/31/*EC of the European Parliament and of the Council of* 26 *April* 1999 *on the landfill of waste*. http://eur-lex.europa.eu/LexUriServ/LexUriServ.do? uri=OJ:L:1999:182:0001:0019:EN:PDF.

The World Bank, (2005): *Waste Management in China: Issues and Recommendations*. Urban Development Papers, Working Paper No. 9., s. l.

The World Bank, (2010): *Asia Sustainable and Alternative Energy Program ASTAE*. Annual Status Report 18, Washington, D. C.

UNFCCC - secretariat, United Nations Framework Convention on Climate Change - secretariat, (1994): *Emissions summary for China*. http://unfccc.int/files/ghg_data/ghg_data_unfccc/ghg_profiles/application/pdf/chn_ghg_profile.pdf.

Wang Q. H. ,(1994): *Investigation of rural straw utilization in China*. Rural Eco-environment, 10(4), pp. 67-71. (in Chinese)

Wang H. , Nakata T. , (2009): *Analysis of the market penetration of clean coal technologies and its impacts on China's electricity sector*. Energy Policy,37,338,pp. 351.

Wendland M. , Bayerische Landesanstalt für Landwirtschafft (LfL), (2011): *Verbal information about ongoing AD slurry application research*. Weihenstephan Germany.

Williams A. G. , Audsley E. ,Sandars D. L. ,(2006): *Determining the environmental burdens and resource use in the production of agricultural and horticultural commodities, Main Report*. Defra Research Project IS0205. Bedford: Cranfield University and Defra.

Wu C. Z. , Yin X. L. , Yuan Z. H. , Zhou Z. Q. ,Zhuang X. S. , (2009): *The development of bioenergy technology in China*. Energy 35, pp. 4445-4450.

Xie G. , (2011): Professor at China Agricultural University. Personal conversation, 24th June 2011, Beijing, China.

Zhao L. , Wang G. ,Mang H. P. , (2006): *Chinese Biomass Energy Potential - Concept for Sino German Cooperation, Sustainable Development of the Bioenergy Sector, Inception Study*. Part 1, Summary of Current Biomass Energy Situation, Beijing, China.

Zhong H. P. , Yue Y. Z. ,Fan J. W. ,(2003): *Characteristics of crop straw resources in China and its utilization*. Resources Science 2003, 25, pp. 62-66. (in Chinese)

Zhou, Y. , (2005): *Advance in the study of clean straw biomass as energy*. Shaanxi Chemical Industry, 34(10), pp. 595-597 (in Chinese)

Zhuang D. , Jiang D. , Liu L. ,Huang Y. ,(2010):*Assessment of bioenergy potential on marginal land in China*. Renewable and Sustainable Energy Reviews, 15, pp. 1050-1056.

POLLUTANT GENERATION COEFFICIENTS IN ANIMAL PRODUCTION

CHEN Haiyuan, WU Shubiao, GUO Jianbin, DONG Renjie, CHEN Li, PANG Changle

China Agricultural University

Contact:Prof. Dr. PANG Changle, China Agricultural University, College of Engineering, Qinghua East Road No. 17, Haidian District, 100083 Beijing, PR China, Tel.: +86-10-62737885, E-Mail: pangcl@cau.edu.cn

EXECUTIVE SUMMARY

Pollutants generation coefficients of livestock breeding industry are the important basic data in the field of environmental assessment. However, the reported coefficients have a significant difference. This study analyzes the features of different domestic livestock pollutant productions and suggests new pollutants generation coefficients. It simplifies the methodology of pollutant assessment on the farm. The suggested pollutant generation coefficients were verified using the related livestock data of a big city. And the results are as follow: Among all kinds of livestock, cattle are the heaviest in feces and it accounts for 61.2% of the total feces in the city. As for layer 13.1%, pig 12.3%, broiler 6.3%. The calculated amount of pollutants applying the simplified methodology fit well to the previous calculation.

Key words: Feces; pollutant generation coefficients; estimation

畜禽养殖业产污系数的确定

陈海媛,吴树彪,郭建斌,董仁杰,陈理,庞昌乐*

(中国农业大学能源工程与低碳技术研究室,北京 100083)

【摘要】 畜禽养殖业产污系数是环境领域重要的基础数据,当前在产污系数的选取上差异很大。本文通过分析国内不同畜禽的产污特点,将畜禽养殖业的产污系数进行合理整合,系数简化,只需统计各畜禽的存(出)栏量,便可得出产污量,并结合某大城市的养殖业进行验证和预测。案例预测结果表明:在畜禽种类中,根据确定的猪、鸡和牛的产污系数(COD、TN、TP),估算出该市产生粪污量最多的是牛,占该市主要畜禽粪便产生量的61.2%,其次是蛋鸡13.1%、猪12.3%、肉鸡6.3%,与目前文献中计算方法得出的结果基本一致。

【关键词】 畜禽粪污;产污系数;估算

1 引言

随着规模化畜禽养殖的迅速发展,畜禽粪污中的N、P已远远超出当地的土壤负荷[1],而未经合理处理的粪污排入水体,对河流和海洋造成了严重污染[1]。

畜禽养殖业污染物排放量主要取决于畜禽养殖废弃物(包括粪便、工艺污水、垫草垫料等)产生量。而废弃物产生量与畜禽种类、养殖规模密切相关,除此之外,还受养殖废弃物处置及处理工程和技术(如粪污的厌

* 通讯作者:庞昌乐,男,博士,副教授,中国农业大学能源工程与低碳技术研究室,北京市海淀区清华东路17号 100083, pangcl@cau.edu.cn,010-62737885

氧发酵技术和堆肥技术)的影响。为估算畜禽养殖业污染物排放量,国内开展了许多研究,进行不同畜禽种类产污系数的确定,但是在产污系数的选取上却存在着很大的差异(见表1)。

本文根据国内畜禽养殖业的现状,重新确定了畜禽养殖业的产污系数,并以某大城市养殖业为例进行了案例分析,对该市畜禽发展所带来的潜在环境污染进行了估算,为粪污的合理利用和污染控制提供依据。

Table 1 Factors of components of livestock produced in the city

表1 不同畜禽的产污系数

畜禽品种	粪污产污系数	美国农业工程学会
猪(kg/d)	5.3[2,3],5.1[1,4,5],7.0[6],5.0[7],	5.1
肉牛(t/a)	7.7[2,3],6.9[7],12.8[6],17.5[1]	7.6
奶牛(t/a)	19.4[2,3],17.5[4,5], 12.8[6],17.5[1]	20.1
肉鸡(kg/d)	0.10 [2,3],0.08[6],0.05[1,4],0.15[1]	0.08
蛋鸡(kg/a)	53.3[2,3],54.8[6],47.8[5],27.4[1],60[8]	42.1

2 产污系数的确定

目前,畜禽养殖产污系数涉及的统计基量过多。如猪的产污系数分为保育段和育肥段,但现阶段养殖场保育和育肥段区分不明显,导致调查基量的误差较大,日常统计工作的基量来源也不统一。为了使污染源数据使用农业统计年鉴的数据,将按阶段给出的产污系数进行整合,如按保育段、育肥段折算出出栏一头猪的产污量。

(1)猪产污系数确定

猪产污系数由保育猪、育肥猪和母猪的产污系数和养殖时间综合折算确定。

$$X_{pig}=X_1\times 70+X_2\times 65+X_3/20 \tag{1}$$

式中,X_{pig}为生猪产污系数;X_1为育肥猪日产污系数,平均养殖时间为70 d;X_2为保育猪日产污系数,平均养殖时间为65 d;X_3为母猪年产污系数,仔猪产污合并至母猪中(按一头母猪年产仔猪20只计算)。

$$\text{母猪年产污系数}=\text{母猪日产污系数}\times 365+\text{仔猪产污系数}\times 45\times 20 \tag{2}$$

(2)鸡产污系数的确定

蛋鸡产污系数由育雏鸡、产蛋鸡的日产污系数和占蛋鸡的比例综合折算确定。

$$Y_{eggc}=(Y_1\times 15\%+Y_2\times 85\%)\times 365 \tag{3}$$

式中,Y_{eggc}为蛋鸡产污系数;Y_1为育雏鸡日产污系数,占蛋鸡的比例为15%;Y_2为产蛋鸡日产污系数,占蛋鸡的比例为85%。

肉鸡产污系数由肉鸡日产污系数和养殖时间折算确定。

$$Y_{bro}=Y_{1bro}\times 52 \tag{4}$$

式中,Y_{bro}为肉鸡产污系数;Y_{1bro}为肉鸡日产污系数,平均养殖时间为52 d。

(3)牛产污系数确定

奶牛产污系数由育成牛、产奶牛的日产污系数和占奶牛比例折算确定。

$$Z_{dairy}=(Z_1\times 35\%+Z_2\times 65\%)\times 365 \tag{5}$$

式中,Z_{dairy}为奶牛产污系数;Z_1为育成牛的日产污系数,占奶牛的比例为35%;Z_2为产奶牛产系数,占奶牛的比例为65%。

肉牛产污系数由肉牛日产污系数和养殖时间折算确定。

$$Z_{beef}=Z_{1beef}\times 660 \tag{6}$$

式中，Z_{beef} 为肉牛产污系数；Z_{1beef} 为肉牛日产污系数，平均养殖时间为 660 d。

根据《畜禽养殖业产污系数与排污系数手册》中的不同畜禽污染物日产污系数[9]，按上述公式，得出各类畜禽污染物(COD、TN、TP、氨氮)的产污系数，见表 2。

Table 2 Factors of components of livestock produced in the city

表 2 畜禽养殖业产污系数表

品种	COD	TN	TP	氨氮	备注
猪(kg)	36	3.70	0.56	1.80	出栏量
奶牛(kg/a)	2 131	105.80	16.73	2.85	存栏量
肉牛(kg)	1 782	70.80	8.96	2.52	出栏量
蛋鸡(kg/a)	7.18	0.50	0.12	0.12	存栏量
肉鸡(kg)	2.92	0.06	0.02	0.02	出栏量

3 产污系数验证及案例预测

3.1 产污系数的验证

现以某大城市养殖业为例，以农业统计年鉴上的畜禽的存(出)栏量和农业经济手册中饲养周期和畜禽日产污系数，对该市 2009 年畜禽产生的潜在污染物量进行估算，结果为该市主要畜禽粪尿中为 COD 为 100.4×10^4 t，氮为 5.41×10^4 t，磷为 1×10^4 t，氨氮为 105.1 t。

用本文重新确定的产污系数对该市 2002—2009 年主要畜禽粪尿成分产生量进行核算(见表 3)，计算公式如下：

$$\text{畜禽粪尿及成分量}=\text{存(出)栏数}\times\text{畜禽粪尿及成分的产污系数}^{[10]} \tag{7}$$

Table3 The Output of components of livestock feces and urine in the city from 2002 to 2009

表 3 2002—2009 年该市主要畜禽粪污成分产生量

年份	COD($\times10^4$ t)	TN($\times10^4$ t)	TP($\times10^4$ t)	氨氮(t)
2002	107.2	6.0	1.0	171.3
2003	128.3	7.0	1.2	174.1
2004	125.7	7.3	1.3	162.0
2005	148.6	7.9	1.3	170.1
2006	101.2	5.4	0.9	149.0
2007	101.4	5.4	0.9	151.4
2008	98.4	5.2	0.9	102.2
2009	100.2	5.4	0.9	105.1

由表 3 可以看出，2009 年，前畜禽粪尿主要成分的产生量与前者计算的结果基本一致，从而验证该产污系数的正确性。可以看出，该市 2005 年之前畜禽粪尿主要成分处于上升趋势，从 2006 年开始每年产生的粪尿成分基本相同，这都是在该市调整畜禽养殖业结构，由数量型向集约化、规模化的质量型的畜禽养殖业发展，使畜禽粪尿的主要成分，即潜在的污染物被有效的控制。

3.2 产污系数的案例预测

现利用本文的产污系数估算该市畜禽养殖业对环境带来的潜在污染，查看该产污系数在实际应用中是否快而精确。此外，产污核算中需要获取的基量(生猪出栏量、肉牛出栏量、奶牛存栏量、肉鸡出栏量、蛋鸡存栏量)直接从农业统计年鉴上获取,保证数据来源的一致性。

Table 4 The Output of livestock feces in the city from 2002 to 2009

表 4 该市主要畜禽粪污历年产生量 （$\times 10^4$ t）

年份	猪	肉牛	乳牛	蛋鸡	肉鸡
2002	222.85	394.48	164.04	100.06	45.76
2003	258.26	463.55	223.31	101.68	54.015
2004	274.67	477.96	271.00	94.63	48.62
2005	296.09	534.31	295.67	99.36	52.60
2006	172.40	323.3	255.04	87.02	32.00
2007	161.18	326.41	258.06	88.41	32.51
2008	185.19	374.64	252.35	59.68	34.48
2009	202.30	296.75	265.28	61.35	35.37

由表 4 知 2002 年至 2009 年,该市畜禽粪便产生量最多的牛,占该市主要畜禽粪便产生量的 61.2%,其次是蛋鸡 13.1%、猪 12.3%、肉鸡 6.3%。虽然在 2006 年后经过该市调整政策,畜禽养殖的集约化、产业化使畜禽数量有所下降,但总体是呈现出稳定增长趋势的。

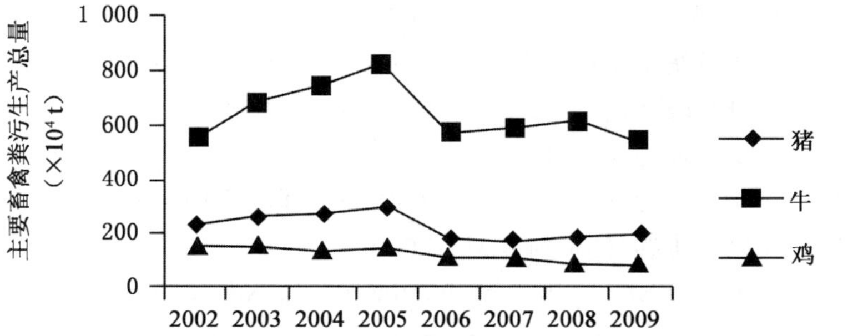

Figure 1 The Output of livestock feces and urine in Tianjin from 2002 to 2009

图 1 该市 2002—2009 年主要畜禽粪污产生总量

由图 1 可以看出,该市近几年来的主要畜禽品种的粪尿产生量是呈增长趋势的,可以明显看出牛的粪尿产生量居首位,为 3.899×10^6 t,占主要畜禽品种粪尿产生量的 37.7%,其次奶牛、猪,分别占 23.5%、21.0%,畜禽粪尿产生量最少的是肉鸡为 4.19×10^5 t,占 4.0%。受经济波动的影响,2006 年出现了较大的回落,之后除了肉牛粪尿产生量近两年一直减少,其他品种的粪尿产生量又趋于稳定增长。

在估算该市的畜禽粪污过程中,对畜禽养殖业产污系数的调整更具有操作性、方便快捷、大大减少数据处理的工作量,更适应日常环境统计与管理需求,还使之与农业统计年鉴的数据相衔接,提高准确性和一致性。

4 结论与讨论

本文通过分析国内不同畜禽的产污特点,将畜禽养殖业的产污系数进行合理整合,系数简化,只需统计各畜禽的存(出)栏量,便可得出产污量,有利于分析畜禽养殖业对环境污染的贡献。

结合某大城市的养殖业进行案例验证和案例预测,结果表明:在畜禽种类中,该市产生粪污量最多的是牛,占该市主要畜禽粪便产生量的 61.2%,其次是蛋鸡 13.1%、猪 12.3%、肉鸡 6.3%,这对加强畜禽污染防治工作起了重要的作用。

研究给出畜禽种类的产污系数主要是针对畜禽粪尿,没有考虑畜禽养殖业排出的 CH_4、N_2O 等温室气

体和 NH_3 排放，建议尽快开展对畜禽养殖业中的温室气体和氨气的产污系数的测定。

项目及基金支持

农业主要污染源产排污量削减技术评估与核算体系研究（编号：201009017）；“十一五”国家科技支撑计划“规模化沼气工程沼液、沼渣高值利用技术研究与工程示范”课题（2008BADC4B00）；北京市教育委员会科学研究与科研基地建设项目“规模化养殖场粪污双向清洁处理与利用研究”。

参考文献

[1] 刘培芳. 长江三角洲城郊畜禽粪便的污染负荷及其防治对策[J]. 长江流域资源与环境，2002，11(5)：456-460.

[2] 马林，王方浩，马文奇，等. 中国东北地区中长期畜禽粪尿资源与污染潜势估算[J]. 农业工程学报，2006，22(8)：170-174.

[3] 王方浩，马文奇，窦争霞，等. 中国畜禽粪便产生量估算及环境效应[J]. 中国环境科学，2006，26(5)：614-617.

[4] 黄沈发，陈长虹，贺军峰. 黄浦江上游汇水区畜禽业污染及其防治对策[J]. 上海环境科学，1994，13(5)：4-8.

[5] 上海市农业环境监测站. 上海郊区农业环境污染源状况[J]. 上海农业环境保护，1992，9(1)：1-2.

[6] 彭里，王定勇. 重庆市畜禽粪便年排放量的估算研究[J]. 农业工程学报，2004，20(1)：288-292.

[7] 郝兴樟. 畜牧业污染的治理方法[J]. 饲料工业，2002. 23(8)：44-45.

[8] 王新谋. 家畜粪便学[M]. 上海：上海交通大学出版社，1999.

[9] 中国农业科学院农业环境与可持续发展研究所，环境保护部南京环境科学研究所. 第一次全国污染源普查畜禽养殖业产排污系数手册（内部资料）[M]. 2009.

[10] Moore J A, Gamroth M J. Calculating the fertilizer value of manure form livestock operations[Z]. Agricultural Communications Oregon State University Administrative Service A442 EC1094, 1993：1-7.

[11] 戴锦. 分散化经营与产业化经营[J]. 农村经济，2004(12)：55-57.

[12] 贾蕊，陆迁，何学松. 我国农业污染现状原因及对策研究[J]. 中国农业科技导报，2006，8(1)：59-63.

[13] 彭新宇，张陆彪. 畜禽养殖业的环境影响及经济分析[J]. 生态经济，2007(5)：271-273.

[14] 朱兆良. 农田中氮肥的损失与对策[J]. 土壤与环境，2000，9(1)：1-6.

[15] Oenema O, Van Liere E, Plette S, et al. Environmental effects of manure policy options in the Netherlands [J]. Water Sci. Technol. 2004, 49: 101-108.

[16] 鲍艳宇，娄翼来，颜丽，等. 不同畜禽粪便好氧堆肥过程中重金属 Pb Cd Cu Zn 的变化特征及其影响因素分析[J]. 农业环境科学学报，2010，29(9)：1820-1826.

[17] 孙军德，韩泽志，初兆娴. 畜禽粪便的利用现状与发展前景[J]. 环境保护与循环经济，2008(5)：46-48.

[18] 王凯军. 畜禽养殖污染防治技术及政策[M]. 北京：化学工业出版社，2004.

[19] 北京市环境保护科学研究院. 北京市规模化畜禽养殖场污染现状与防治对策研究（内部资料）[M]. 2001.

[20] 章力建，刁其玉，王吉峰. 畜禽养中立体交叉污染及其防治对策[J]. 中国畜牧兽医，2005，32(9)：3-5.

[21] 张玉珍，洪华生，曾悦，等. 九龙江流域畜禽养殖业的生态环境问题及防治对策探讨[J]. 重庆环境科学，2003，25(7)：29-34.

沼气生产——整个底物生产过程和底物同样重要

E. Nacke, M. Spitzer

(*CLAAS KGaA, Harsewinkel 33428*)

【摘要】气候变化带来的挑战导致了欧洲能源政策的变化。因此,欧盟成员国承诺,到 2020 年,其能源消耗中的 20%将由可再生能源提供。德国政府决定把沼气发展成为可再生能源产业的主要支柱。在德国,沼气技术已经在现在的能源领域发挥着重要作用,且已发展成为重要的农业经济活动。

农业沼气生产起源于利用动物粪便。最初牛粪由于其供应方便、干物质含量高从而产沼气潜力大,成为利用最广泛的原料。直到现在,畜禽粪便依然是沼气生产的主要原料之一。

除了粪便,玉米是最具生产力的沼气植物。尽管玉米原产于墨西哥亚热带地区,但种植者已经成功使这种碳四植物适应了欧洲地中海气候,而且还可以适应德国,甚至斯堪的纳维亚和英国的气候。大约在 50 年前德国就开始利用整株玉米作为牛饲料。在过去十年间,德国已经使得新鲜玉米产量达到约 43 t/hm^2,而且大量不同品种的玉米保证其可以在各种土壤和气候条件下生长。

除了利用玉米,一些沼气工程也尝试使用高粱等其他作物作为沼气原料。高粱原产于热带和亚热带地区,需要较温暖的气候,属于甜味草种属,和玉米一样,高粱也是碳四植物,即其光合过程比谷物等碳三植物效率更高。高粱本身比玉米更耐旱,这一点在气候不断变化的区域显得更为重要。一般玉米品种的呼吸率为每吨干物质 300～400 L 水,而饲料品种高粱每吨干物质只需要 200～300 L 水。

根据地理和气候条件,每公顷高粱干物质产量为 10～25 t,因此在理想条件下,每公顷高粱能产也超过 40 000 kWh 的能量。新的培育计划希望能把干物质产量提高到 25～30 t/hm^2。

沼气工程的成本不仅取决于工程自身的设备投资和车间运行费用。其经济性很大程度上受底物的生产成本、底物的质量和发酵残余物利用效率的影响。这些在沼气生产之前或者之后的工序所产生的费用可能高达总运行费用的 50%。

确定正确收割时间最重要的指标是植物干物质含量和有机干物质含量。如果干物质含量低于 28%左右,就会产生沼液,导致大量的能量流失。目前最流行的饲料保存方法是青贮,传统的青贮塔几乎已经全部改建为地上式青贮槽,青贮槽投资低也更节省人力。采用适当方法压实青贮降低其中的空气含量。根据德国石勒苏益格—荷尔斯泰因农业部的经验数据,建议在 30%干物质含量时压实到密度为 240 kg/m^3。

对于农场和其他类似农业服务提供者来说,高效的生物质收割非常关键。全球领先的农业技术生产商 CLAAS 的服务不仅包括收割,还包括运输、后勤和数据收集分析,能提供整个底物生产过程中的机械和管理,确保沼气生产的利润。

【关键词】沼气底物生产;生物质收割技术

BIOGAS PRODUCTION-THE WHOLE SUBSTRATE PRODUCTION PROCESS IS AS IMPORTANT AS THE SUBSTRATES ITSELF

E. Nacke, M. Spitzer

CLAAS KGaA

Contact: Dr. Eberhard Nacke, Manuela Spitzer, CLAAS KGaA, Münsterstraße 33, D-33428 Harsewinkel, Germany, Tel.: +49-5247-12-1561, E-mail: nacke@claas.com

EXECUTIVE SUMMARY

Global challenges regarding climatic changes lead to a change in the European energy policy. Therefore the member states of the European Union have agreed to supply 20% of their energy consumption from renewable by the year 2020. The German government has decided to support the development of biogas as one major pillar of renewable energy production. In Germany currently biogas technology plays an important role in the energy sector and has already developed into a significant agricultural business activity.

Agricultural biogas production evolved from the exploitation of manure. Initially, cattle manure was the most widely used raw material, due to its ready availability, and its good gas-producing potential as a result of its high dry matter content. Still today, manure is one major substrate in biogas production.

Beside manure maize has developed to become the most productive crop for biogas. Using the entire maize plant for cattle feeding emerged some 50 years ago in Germany. Despite the origin of maize in the subtropical zones of Mexico plant breeders have managed to adapt this C4-plant not only to the climatic conditions of Mediterranean Europe, but as well to Germany and even Scandinavia or the Great Britain. Average fresh material yields of almost 43 tons per hectare have been reached in the last decade in Germany and a large number of varieties ensure that maize may be successfully grown in a wide range of soil and weather conditions.

Instead of using maize as substrate alternative crops like Sorghum become examined as biogas plants. Sorghum belongs to the family of sweet grasses. Just as maize it is a C4-plant, which is characterized by a more efficient photosynthetic process compared to C3-plants like cereals. Sorghum is native to tropical and subtropical regions and requires warmer climates. Due to its nature it is better adapted to water stress than maize, which may become even more important in a changing climate. Standard maize varieties have a transpiration coefficient of 300 to 400 L/t dm (litre water per ton of dry matter), whereas forage varieties of sorghum need only 200 to 300 L/t dm.

Depending on location, weather conditions and sorghum variety a yield level of 10 to 25 t dm/hm^2. So at ideal conditions an energy output of more than 40 000 kWh/hm^2 is possible. New breeding programs have the objective to reach stable biomass yields of 25 to 30 t dm/hm^2.

The costs of a biogas plant are not just determined by the investment in the installation and the operation of the plant itself. The economics of biogas production are largely affected by the production cost of the substrate, the quality of the substrate, and the efficiency of fermentation residue reuse. These process steps before and after the actual biogas production can account for as much as 50% of total operating costs.

The most important indicator for the right harvest time is the dry matter content and organic dry matter content. If dry matter content is below around 28%, fermentation effluent can form, resulting in a significant loss of energy content.

The most popular conservation method for any forage material today is silage production. Traditional storage in tower silos (harvesters) has been converted almost 100% to flat bunker silos. Bunker silos need a lower investment and offer a much better labour efficiency. Effective compaction is the appropriate method to reduce the air content of a silage clamp. Experienced data of the German federal state chambers of agriculture in Schleswig-Holstein recommend a compaction of 240 kg/m^3 at 30% dry matter content.

For farms and agricultural service providers alike, efficient biomass harvesting is vital. CLAAS as leading manufacturer of agricultural technology worldwide provides mechanisation and management solutions for the whole substrate production process to secure profitable biogas production. Not just harvesting itself but also transportation, logistics and management software including data collection and analysis of the data.

Key words: Biogas substrate production; biomass harvesting technique

1 Introduction

Reflecting the global risks of climatic change the member states of the European Union have agreed on a target to supply 20% of their energy consumption from renewable by the year 2020.

The German government has decided to support the development of biogas as one major pillar of renewable energy production. Motivation for this decision is, that biogas has proven to have the highest land use efficiency among all energy production paths from biomass.

In Germany biogas plays an important role in the energy sector and has already developed into a significant agricultural business activity. In 2010 more than 5 000 biogas plants were running with a installed capacity of almost 2 300 MW electrical energy.

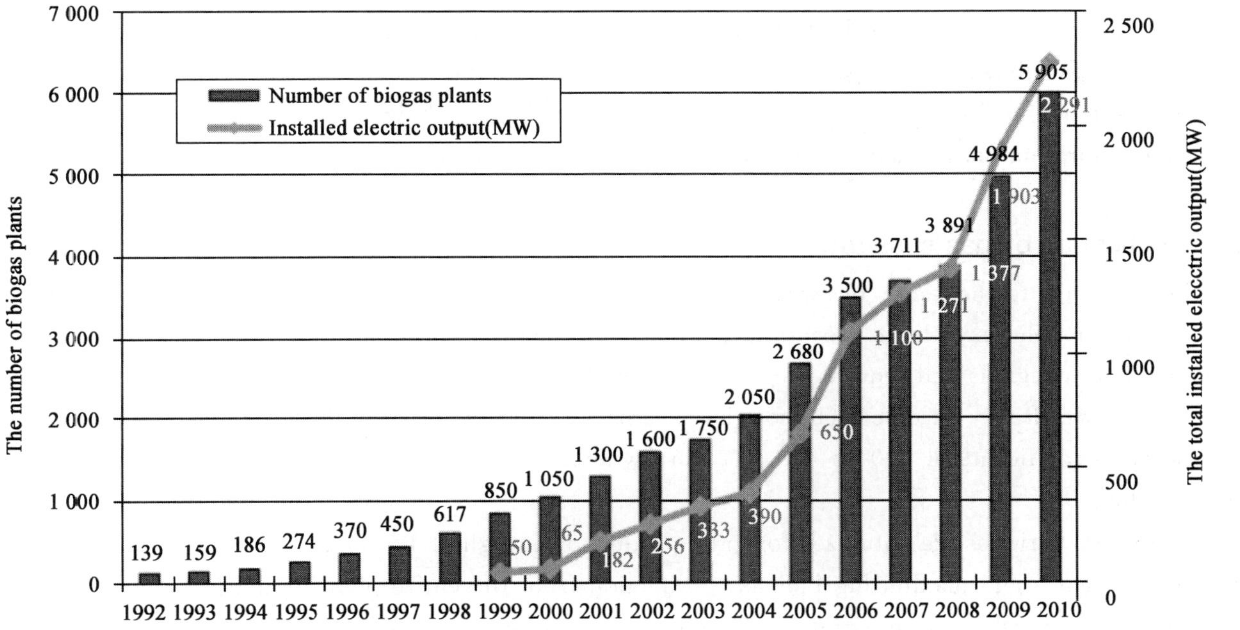

Figure 1 Development of the number of biogas plants and the total installed electric output in megawatt (MW) (as of 06/2011)

2 From animal manure to renewable raw materials

Agricultural biogas production evolved from the exploitation of manure. Initially, cattle manure was the most widely used raw material, due to its ready availability, and its good gas-producing potential as a result of its high dry matter content. Still today, manure is one major substrate in biogas production.

Besides animal waste, all arable crops or parts of these crops are very suitable to produce biogas except plants with an high lignin content. The wood polymer lignin is not broken down within the anaerobic biological process. Starch is the main energy source. A high crude fat content corresponds to a high gas yield, while sugars can be converted very quickly. As available land for biomass production is always a limiting factor the preferable choice of biomass to be used in biogas plants largely depends on the yield potential per acreage of land.

Under German conditions maize has developed to become the most productive crop for biogas. Using the entire maize plant for cattle feeding emerged some 50 years ago in Germany. Despite the origin of maize in the subtropical zones of Mexico plant breeders have managed to adapt this C4-plant not only to the climatic conditions of Mediterranean Europe, but as well to Germany and even Scandinavia or the Great Britain. Average fresh material yields of almost 43 t/hm^2 have been reached in the last decade in Germany and a large number of varieties ensure, that maize may be successfully grown in a wide range of soil and weather conditions.

Despite the fact, that maize has reached this outstanding position as fodder and biomass crop, agriculture in Germany is now seriously looking for alternatives. In some regions of Germany, maize is planted on more than 50% of the agricultural surface, which causes increasing problems. One danger are serious yield depressions by insects like the European corn borer. Additionally, the danger of monoculture maize became a public issue

Amendments to the German Renewable Energy legislation which will come into effect at the start of 2012 are reflecting these developments. There will be a maximum percentage of maize in a biogas feed ratio, which is set to 60%. Methane building potential and land use efficiency are again the main criteria to identify the best alternative crops to maize. Under German climatic conditions, attractive substrates for biogas production are traditionally grown crops like grass silage, whole crop silage, and sugar beets. As a new plant, sorghum may reach an outstanding position in the next years.

3 Sorghum as biogas substrate

Sorghum belongs to the family of sweet grasses. Just as maize it is a C4-plant, which are characterized by a more efficient photosynthetic process compared to C3-plants like cereals. Sorghum is native to tropical and subtropical regions and require warmer climates. Due to its nature it is better adapted to water stress than maize, which may become even more important in a changing climate. Standard maize varieties have a transpiration coefficient of 300 to 400 L/t dm, whereas forage varieties of sorghum need only 200 to 300 L/t dm.

Most sorghum varieties are optimized for grain production (sorghum bicolour). Adapting sorghum to more moderate climates and maximizing biomass production from the entire plant became a major challenge of breeding only recently. Grain varieties are short whereas new biomass and forage varieties reach heights between two and four meters. Sorghum bicolour varieties tiller with two or three sprouts. Sorghum sudanense varieties are building more tillers. Hybrid varieties of sorghum (bicolour × sudanense) start to combine the advantages of both species for biomass production.

Sorghum has no special requirements to the soil conditions. Ideal is a profound clay or sand-clay soil. Cold and dammed-up soils are inappropriate for sorghum. The preparation of the seedbed in spring is the same than for maize. But due to the slow growth at the beginning it is important to arrange a weeds free acreage, e. g. seedbed and soil surface. The typical vegetation period for sorghum in Europe is from Mai to September. During this time it needs a temperature sum of minimum 2 200 Celsius and the average daily temperature should be 16 C. Cold injuries occur already at 4 C.

The harvest of sorghum starts when plants reached the state of milk- or dough ripe like grains. A major issue with current varieties is to reach a dry matter content of at least 28% at harvest time to minimize losses during the ensilage process.

Experience data show, that the ideal chopping length for an optimized methane yield is three to five millimetres. Depending on location, weather conditions and sorghum variety a yield level of 10 to 25 t dm/hm^2. So at ideal conditions an energy output of more than 40 000 kWh/hm^2 is possible. New breeding programs have the objective to reach stable biomass yields of 25 to 30 t dm/hm^2. High yield potential and the request to harvest not before 28% dry matter is reached causes some issues for harvesting today. Harvesting of biomass sorghum is generally performed with a self propelled forage harvester like the CLAAS JAGUAR, equipped with a row independent header ORBIS for cutting and collection in one path. However, all sorghum varieties have a much leaner stalk than maize. Heavy rainfalls, windy periods or typical summer thunderstorms can lead to stock building. Laying material is much more difficult to collect with minimum losses. A two-stage harvest has proven to give best results. It is advisable to cut laid sorghum with a disc mover, bring it in a swath and afterwards collect and chop it with a forage harvester equipped with pickup header.

Figure 2　Maize harvest in Germany with self propelled forager JAGUAR and row independent maize header ORBIS 900 filling the loading waggon CARGOS for transportation to the clamp

4　Substrate production costs

The costs of a biogas plant are not just determined by the investment in the installation and the operation of the plant itself. The economics of biogas production are largely affected by the production cost of the substrate, the quality of the substrate, and the efficiency of fermentation residue reuse. These processes steps before and after the actual biogas production can account for as much as 50% of total operating costs (see *Figure 3*).

For a profitable biogas plant operation every part of the substrate production process is crucial.

5　Important indicators for the harvest

The most important indicator for the right harvest time is the dry matter content and organic dry matter content. If dry matter content is below around 28%, fermentation effluent can form, resulting in a significant loss of energy content.

Inorganic content, especially soil residues, must be kept to a minimum, as they do not contribute to gas production (essential minerals and trace elements are an exception to this rule). Sand and stones sink to the bottom of the fermentation tank, and can interfere with the heating systems designed to keep the substrate at the optimum reaction temperature. Removing piled sand layers from the bottom of the

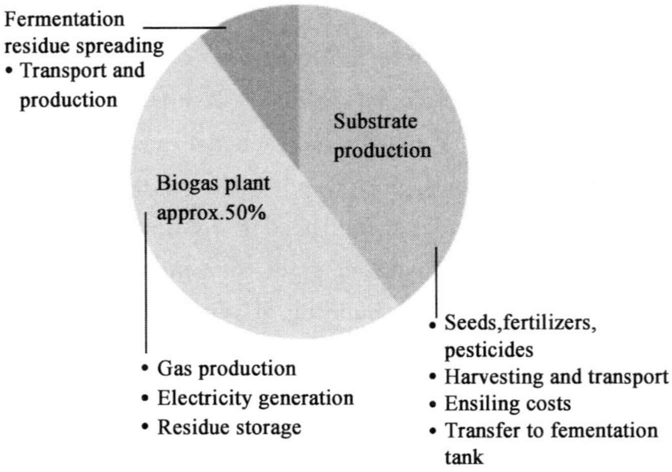

Figure 3 Breakdown of biogas production costs (%)

Figure 4 Biogas substrate production process

fermenter is very costly and considerably reduces efficiency of a biogas plant. As with animal feed, the digestibility of the biomass determines the methane yield. The bonds in lignin molecules are largely indigestible, and high gas yields can only be achieved with substrates that have been harvested at precisely the right time before plants start to incorporate higher lignin contents.

Whichever substrate is ultimately used, the way that this is preserved and prepared is absolutely crucial. Silage requirements vary between cattle feed and biogas substrate. Ruminants like cattle prefer structured material. Various studies have shown that chopping up silage more finely will, all other things being equal result in higher gas yields. Biogas bacteria need large surfaces to get best access for their dissolving activities and short chopping lengths guarantee good working surfaces for bacteria. However, energy consumption of a forage harvester is directly linked to the chopping length and there are more efficient alternatives to reach large surfaces. Reducing chopping length to five or even four millimetres is not necessary, if a forage harvester is equipped with a high performing corn cracker. Essential is not only the number of teeth, but even more a high differential speed of the rollers to increase the surface of the chopped material. Of course, for cattle feed as well as for biogas substrate, a corn cracker has to ensure, that the outer husk of every kernel of sorghum or maize is crushed or even better, that every kernel is broken into several pieces.

The choice of chop length should therefore balance these considerations against the raising fuel consumption of short cutting. A chopping length of around six millimetres seems to be the optimum for ensiling and compacting sorghum or maize. With forage rye and grass silage, a chop length of ten millimetres or longer is advisable for biological and technical reasons (faster and more effective biodegradability without the formation of floating layers).

Summing up the important factors concerning the harvest are: correct harvest time, an ideal chopping length according to the dry matter content, a hundred percent crushed kernels and an ideal compaction management on the clamp.

6 Silage quality is crucial

The most popular conservation method for any forage material today is silage production. Traditional storage in tower silos (harvest ores) has been converted almost 100% to flat bunker silos. Bunker silos need a lower investment and offer a much better labour efficiency. However, loss levels in bunker silos may reach a very high level, if the farmer is not able to reach stable anaerobic conditions in the silage clamp in a very short period. Producing high quality clamp silage is first of all time crucial. A silage clamp needs to be build up and properly sealed in a short period of time to prevent oxygen from invasion. Therefore a prerequisite for the change from tower to bunker has been the development of high capacity harvesting chains with self propelled forage harvesters, which are able to deliver large amounts of short and exactly chopped material in a very short time.

Only under anaerobic conditions a lactic acid fermentation will take place, which will lower the pH and preserve the biomass for a long time with minimum energy loss. If oxygen is not effectively be pressed out of the chopped material aerobic condition will result in the presence of detrimental microbes such as clostridia, listeria, yeast and moulds. The biochemical processes involved result in considerable energy losses in the silage clamp.

Additionally, detrimental microbes, toxic fungi, or moulding material may inhibit bacteria activity and disturb the methane building process in a biogas plant. To ensure optimum quality right up to the point where the silage enters the biogas fermenter, high-quality crops must be harvested at precisely the right moment, and properly stored in accordance with best practices in silo management.

It doesn't matter if it is maize, grass or sorghum or other substrates. The most important factor is to create an environment, which prohibits harmful organism from their activity and which avoids energy losses by aspiration. Conservation needs an oxygen free environment because Oxygen is the basis for energy consuming aspiration and a necessary environment for many, sometimes toxic fungi and bacteria. Even after good compaction of chopped material, some air-pores and channels will remain in the crop. But once the silage clamp is well sealed by a cover film, the rest of oxygen is respired within days and the existing bacteria and fungi are dying fast.

Effective compaction is the appropriate method to reduce the air content of a silage clamp. Experienced data of the German federal state chambers of agriculture in Schleswig-Holstein recommend a compaction of 240 kg/m^3 at 30% dry matter content. To reach this value it is important to consider the whole harvest chain. Only if the chain links are connected to each other a perfect clamp is possible. The throughput of the chopper and the number of transport vehicles determine the decision which compaction vehicle is appropriate. Furthermore the compaction vehicle must be able to distribute the green material in homogenous and thin layers, because only thin layers may be compacted adequately.

Throughputs of foragers vary between 50 t/h up to a maximum output of 300 t/h. Regardless if it is 50 or 300 t, this amount needs to be distributed and compacted in a minimum of time at the silo. So the compaction weight has to match the chopper performance to secure high biogas yields.

A rough rule of thumb says: take the dry matter chopper-performance in ton per hour and divide it by three. The result is the total compaction weight on the clamp. For example: to match a forager performance of 150 t Maize per hour (45 t dm content) the compaction weight on the clamp should be 15 t or more. The storage density affects the depth of air infiltration also when a silage bunker is reopened again. If it is not compacted well small channels will be left, where air can get back into the silo. Available oxygen restores the energy consuming activity of harmful microorganisms and energy is fading away by respiration. Mycotoxins can occur and inhibit the activity of methane bacteria.

7 Losses are often underestimated

Ensiling is the process by which a plant's sugars are converted into preservative acids in the absence of oxygen. Even under optimum conditions, this process inevitably results in a loss of dry mass and energy content. Below a dry mass content of 30%, silage effluent starts to form, resulting in greater losses-depending on the chop length and the silo height. And depending on the type of fermentation, how long silage is stored, how and how often silage is taken out of the clamp, and whether excess heat is produced, more energy can be lost. In total, poor ensiling can therefore result in monetary losses of over 500 €/hm^2.
Now more than ever, cost-effectiveness is of paramount importance for operators of biogas plants. And as well as equipping and maintaining the actual biogas plant, the costs of the renewable raw materials are a major factor here. Fortunately, professional assistance is available to help with every aspect of substrate sourcing, including harvesting, transporting and field preparation. Contractors and syndicates have state-of-the-art equipment with the reliability and performance to ensure fast harvesting with near-zero wastage. For farms and agricultural service providers alike, efficient biomass harvesting is vital. CLAAS Company as the European harvest leader offers a complete range of solutions that deliver this efficiency at every stage of the process, not just harvesting itself but also transportation, logistics and management software including data collection and analysis of the data.
The CLAAS technology program includes machineries for seed, soil and grain management with tractors, forage harvest technology with self propelled forage harvesters and forage machines, ensiling and residue management with 4 × 4 tractors, logistics concepts with loader wagons, loading solutions with telescopic loaders and biogas process and plant management with data management software.

References

CLAAS,(2010): *Claas Wissen Biogas*. Harsewinkel: Claas Vertriebsgesellschaft mbH.

CLAAS, (2011a): *Biogassymposium* 2011. Harsewinkel, http://www.claas.com/cc/servlet/contentblob/cl-pw/de/current/biomasse/biogas/biogas_symposium/tagungsband, property = data.pdf;jsessionid=DCD310147C0234DFE6302B894B208F09.

Fachagentur für Nachwachsende Rohstoffe e. V., (2011): *Von der Gewinnung zur Nutzung*. Leitfaden Biogas, Gülzow-Pürzen.

Fachverband Biogas e. V., (2011): *Biogas Segment Statistics* 2010. http://www.biogas.org/edcom/webfvb.nsf/id/DE_Branchenzahlen/$file/11-06-27_Biogas%20Branchenzahlen%202010_eng.pdf.

KTBL,(2009): *Faustzahlen Biogas*. Kuratorium für Technik und Bauwesen in der Landwirtschaft e. V. (Hrsg.): Darmstadt: KTBL.

KWS SAAT AG,(2007): *Kolloquium Energiepflanzen*. Einbeck: KWS Saat AG.

Moser J., (2010): *Volles Sortiment voller Energie*. Saaten-Union Energiepflanzen Aktuelle Informationen für Biogasbetriebe, Isernhagen: Saaten-Union.

Thaysen J., (2008): *Mit besserer Walztechnik höher Verdichten*! Versuch der LWK Schlewig Holstein,

CLAAS Trends 2008.

Stolzenburg K., (2009): *Realisierbare Biomasseerträ ge von Mais, Sorghum und Getreide zur Verwendung in einer Biogasanlage*. http://www.landwirtschaft-mlr.baden-wuerttemberg.de/servlet/PB/show/1297870_l1/ltz_Maistagung%202010%20-%2005%20-%20Stolzenburg%20-%20Realisierbare%20Biomasseertr%C3%A4ge%20von%20Mais,%20Sorghum%20und%20Getreide%20zur%20Verwendung%20in%20der%20Biogasanlage.pdf.

Vetter A., Heiermann M., Toews T., (2009): Anbausysteme für Energiepflanzen. Frankfurt am Main: DLG Verlag.

Wagner A., Leurs K., Büscher W., (2005): Einfluss der Häcksellänge aus Verdichtung, Silierung und Nacherwärmung. Lantechnik Agricultural Engineering 1. 2005.

Zeise K., Fritz M., (2011): Sorghum für die Verwendung in Biogasanlagen. Arbeitsgemeinschaft Landtechnik und landwirtschaftliches Bauwesen in Bayern e. V. Biogas Forum Bayern, Freising: ALB Bayern E. V.

通过沼气和垃圾衍生燃料对城市废弃物进行能量回收从而逐步取代中国填埋处理城市固体废弃物的方法

B. Raninger[1]，李润东[2]

（1. 德国国际合作机构，北京 100125；2. 沈阳航空工业学院，沈阳 110136）

【摘要】2004 年中国固体废弃物产量已跃居世界第一。预计到 2050 年大城市人口将会再增加 3×10^8 人，城市化程度将达到 70%。同时随着生活水平的不断提高，城市固体废弃物将以每年 10%的速率增长。十一五规划中，可收集废弃物处理率增加到 60%的目标未能实现，其实际处理能力比预定目标落后 2.2×10^7 t/a[1]，因此固体废弃物管理就更为重要。在十二五规划中，政府在固体废弃物管理方面的投资数额增加到 8×10^{10} 元，并采用了新的管理理念。新的固体废弃物管理办法需要根据当地实际情况，以及市政垃圾中所含的大量有机废弃物等进行进一步完善，同时沼气生产与使用的产业化也将得到快速发展[2]。但在生物有机废弃物管理方面，城市与农业部门的合作还须进一步加强。

【关键词】完善的中国固体废弃物管理；沼气；垃圾衍生燃料；气候变化；水泥产业

ENERGY RECOVERY FROM URBAN WASTES THROUGH BIOGAS AND RDF BY PHASING OUT MSW LANDFILLING IN MAINLAND CHINA*

B. Raninger[1], LI Rundong[2]

1. Deutsche Gesellschaft für Internationale Zusammenarbeit (GIZ) GmbH, Shenyang Aerospace University (SAU); 2. Shenyang Aerospace University (SAU), Institute for Clean Energy and Environmental Engineering (ICEEE), Liaoning Key Laboratory of Clean Energy (LNKLCE)

Contact: Prof. Dr. Bernhard Raninger, GIZ, Beijing & Frankfurt Eschborn, Germany, Shenyang Aerospace University, China (SAU), Tel.: +86-10-65087080, E-Mail: Bernhard.Raninger@giz.de; Prof. Dr. LI Rundong, Shenyang Aerospace University (SAU), Institute for Clean Energy and Environmental Engineering (ICEEE), Liaoning Key Laboratory of Clean Energy (LNKLCE), Liaoning, P. R. China

EXECUTIVE SUMMARY

In 2004 China has taken the worldwide lead in solid waste production. The rapid growth of megacities, till 2050 another 300 mn people will migrate and the urbanization rate will reach 70%, is besides of an increasing living standard the main reason for a continuous annual 10 m/m% MSW growth rate. Solid waste management (SWM) will play more than before an important role, as during the 11-Five Year Plan (FYP) period target's, to increase the treatment ration of the collected waste to 60%, could not be achieved and treatment capacities are lacking behind by 22 mn t/a[1]. Therefore during the upcoming 12-FYP period the investment in this sector will be doubled, reaching CNY 80 bn and new concepts adopted. New integrated SWM solutions, appropriate to the local conditions and considering better the high amount of bioorganic waste in the municipal waste, have to be approached and industrialization of Biogas

* This paper was presented at the 'International Conference on Solid Waste 2011, Moving Towards Sustainable Resource Management (ICSW2011)', Hong Kong SAR China, 05/2011.

generation and use will develop rapidly[2]. Regarding Bioorganic Waste Management the cooperation between the municipal and the agricultural sector has to be further strengthened.
Key words: Integrated solid waste management China; biogas; RDF; climate change; cement industry

1 Introduction

During the last decade the amount of collected municipal solid waste (MSW) was raised continuously in mainland China, but the waste treatment capacities could hardly follow the demand and the 'mix' of waste treatment and disposal methods could not be changed significantly. Valuable materials, suitable for material recycling, were mostly in a traditional way recovered by means of house to house and street collection, estimated amount 20m/m% of the total waste steam[3]. Beijing has in 2009, due to a better diversification of treatment and waste classification, reached the waste amount for final land filling to be reduced by 4. 8 m/m%[4]. Regarding disposal land filling appears still the most wide spread practice, accounting for more than 85% of 157 mn t/a. In 2009, 654 cities have operated 447 'Chinese standard' engineered landfills to 'manage' 57 m/m% of the total MSW [5]. The remaining estimated 28 m/m% of the collected MSW is ending up at uncontrolled landfills. If land filling is continued, from whatever reason-on purpose or lacking other solutions-China will need another 1 600 landfills in 2030 [6, 7].

But compared to western countries, a significantly different composition of MSW leads to different landfill behavior and in the typical Chinese MSW more than 80 m/m% of the organic matter is fast and easy biodegradable with a low lingo-cellulose content[6, 8]. This and the practice of landfill operation are the reasons why the expected landfill gas (LFG) collection and a certified emission reduction (CER) under UNFCCC CDM[8,9] did not realize.

2 MSW Incineration

Sine 1988 MSW incineration plants were built in China and the treatment capacity should cover about 10m/m% of the waste or 25 mn t/a in 2010[10]. But, due to various reasons, such as the low caloric value of the MSW, the insufficient flew gas emissions standard[6] and due to the lack of public acceptance, this could be not be reached

2.1 Bioorganic Waste Treatment

The MSW composting has decreased during the last ten years significantly. Ignoring biotechnological process principles, a poor hardware and mixed MSW as input material did lead to a low quality end product, not accepted by the users; though even organic fertilizers achieve good prices in eco-farming and horticulture.

The biogas generation potential of the bioorganic municipal waste (BMW) from source separation pilot tests in Shenyang in Liaoning from 2006 till 2009[11], was found high and the BMW suitable as feedstock or co-feedstock for biogas plants, and the biogas can be used for locally, but also for grid connected electric power generation, space heating, or as biogas respectively bio-methane for natural gas grid injection or vehicle use.

So far the agricultural sector has started to utilize organic wastes from animal farms and agro industry for energy production. By 2009, 3 700 large scale biogas plants with a digester volume of >300 m^3 were built and the target is to set up 10 000 middle and large scale biogas plants till 2020. Together with the biogas from industrial wastewater 14 bill m^3 shall be produced till 2020 and a grid connected power generation capacity of 3 GW should be available[12, 13]. The potential contamination of farmland by using municipal waste (MSW or BMW) as a feedstock is still seen and used as an argument not to introduce co-fermentation, even the level of organic- and inorganic pollution of BMW from households (kitchen waste from food preparation and leftovers) was found low[14].

3 Greenhouse gas emission mitigation through SWM of municipal waste

The new drivers in waste management in China are to reduce green house gas (GHG) emissions for a low carbon economy[16]. SWM has contributed to GHG emission reduction in Europe. Compared to the baseline with uncontrolled dumping the potential net savings per t of MSW are 1. 7 t CO_{2eqv} with landfilling and gas flaring, and 3. 6 t CO_{2eqv} if aerobic digestion is employed[15]. From the EU-27 GHG reduction trends1990-2006 and the projection 2010 can be seen[17] that SWM contributes the most. The EU net GHG emission from SWM are declining from a peak of 55 mn t CO_{2eqv} in the 1980s to 10 mn t CO_{2eqv} by 2020[18], as seen in *Figure 1*. Between 1990 and 2006, EU-15 greenhouse gas emissions from waste decreased by 39% and they are projected to decrease further to 45 % by 2010.

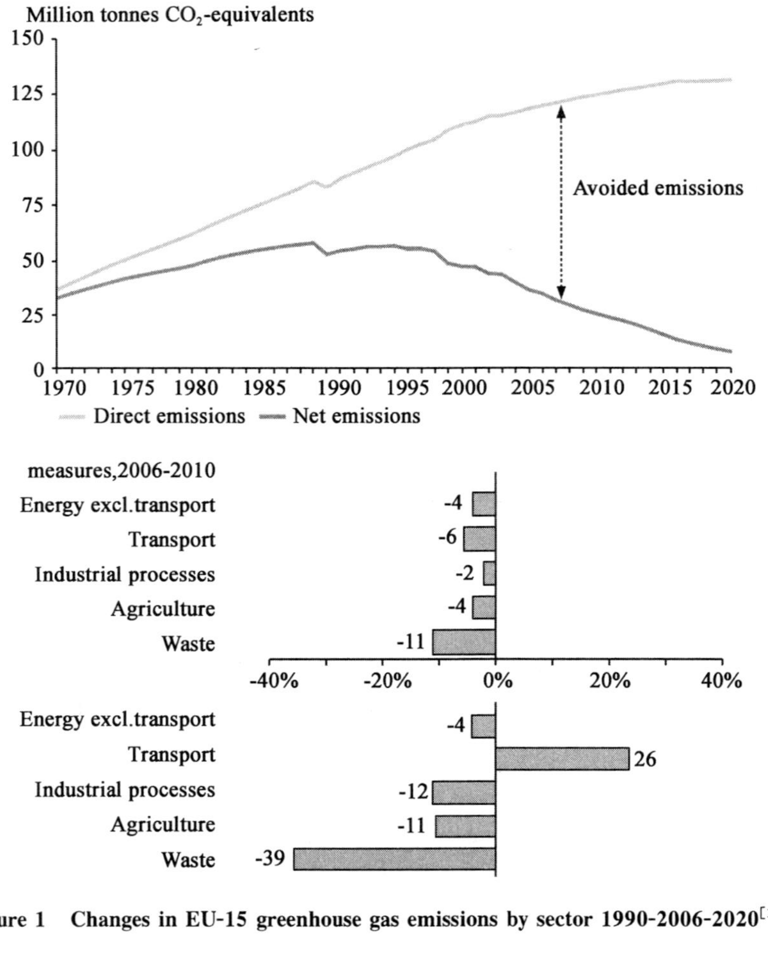

Figure 1 Changes in EU-15 greenhouse gas emissions by sector 1990-2006-2020[17].

4 Integrated solutions to approach a 'landfill free' SWM system in China

To meet the anticipated SWM targets new concepts have to be applied. This includes the consideration of the bioorganic fraction of Waste (BMW) by anaerobic fermentation and biogas use.

BMW deriving from source separation leads to a clean compost or fertilizer suitable for soil application. BMW processed in the way of mixed feedstock and sewage sludge the digester effluent after dewatering and biological drying can be landfilled or used as refuse derived fuel (RDF)[19].

To realize this concept the Chinese cement industry got the target to establish a MSW treatment capacity of 42 mn t/a in the 12-FYP till 2015. *Figure 2* shows a flow chart of such an integrated waste treatment model. The investment to built the capacity for 100 000 t/a acc. the international state-of-the-art will be about CNY 200 mn. Revenues from biogas[3], RDF, CDM and the waste fees from the cities, so far used to run the landfills, these projects are economically feasible.

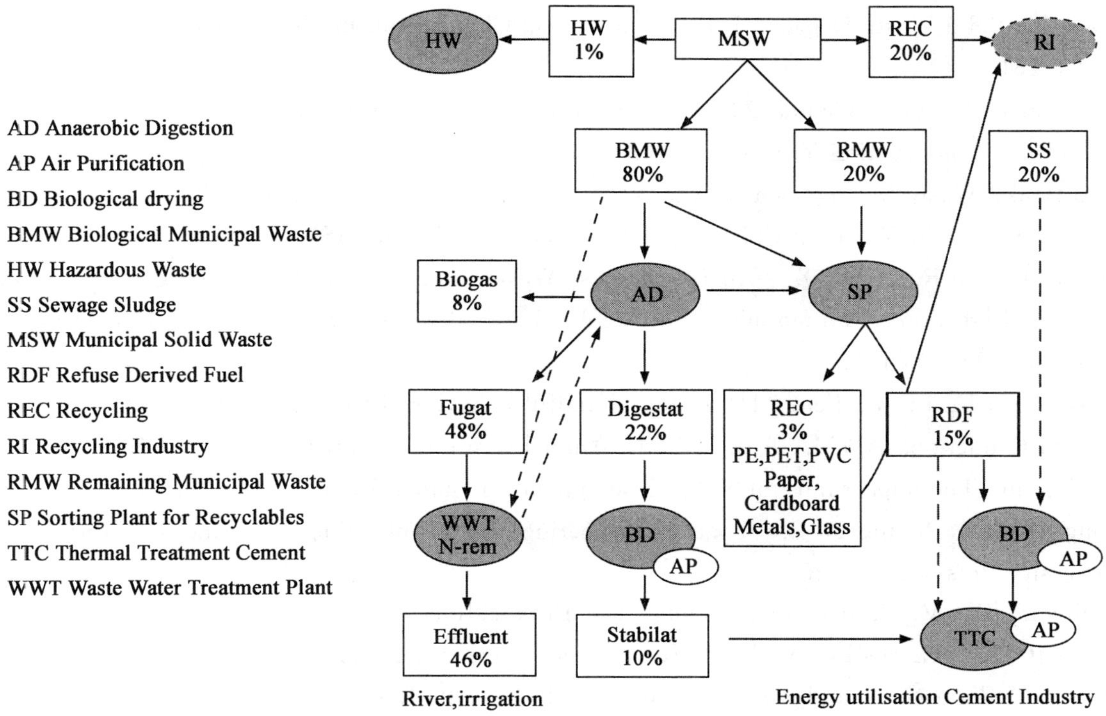

Figure 2 Integrated waste treatment model with RDF for the cement industry

5 Conclusions

From various reasons, disposing MSW by engineered (1) landfilling (incl. LFG collection and leachate treatment), by (2) composting, (3) incineration and (4) gasification, even (5) biogas production with upright fermenter systems, was not successful during the 11-FYP period[1, 4]. Therefore the existing landfill capacities are running out and to set up new landfills, in particular in the eastern coastal provinces, are due to the available space limited. 'Landfill-free 'solutions have to be approached.

The new SWM system may follow the following material streams: (1) 20-25 m/m% as secondary row material recycling prior waste collection, this will remain[3]; (2) the bio-organic matter in the residual MSW in China will stay high for the next decades[6] and 70% of the collected waste shall therefore be used to generate biogas (most likely in dry-fermentation or in combined systems); (3) the biogas will, in line with the target of the Renewable Energy Law[1, 20] supply power to the grid; (4)the fermenter residues from clean feedstock will be a valuable resource of organic matter and nutrients for soils in line with China's soil protection efforts, and (5) not clean residues from MSW will, after biological drying and stabilization either be still disposed or used as construction material additive; (6) the high calorific RDF fraction of about 20m/m%, but with 85% of the energy content[3,19] will be used as fuel for the industry, such as in cement kiln.

References

[1] China Ministry of Environmental Protection, MEP. 2010 Environmental News.

[2] Qin Shiping, NDRC. China 12th five-year plan on big biogas development. The 2nd conference on Biogas Industrialization in China, Qingdao, 2011.

[3] Raninger, B. , W. Bidlingmaier, R. Li, Lei F. Bioorganic Municipal Waste Management to deploy a sustainable Solid Waste Disposal Practice in China. Chinese Journal of Process Engineering, 2006, 6 (2):255-261.

[4] www. chinadaily. com. cn/china/2010-10/14/content_11407404. htm, 2010.

[5] China Urban Construction Yearbook 2010.

[6] World Bank. Waste Management in China. Issues and Recommendations, 2005.

[7] Xu Haiyun. All the Waste in China. Waste Management World, ISWA, 2008.

[8] Raninger B. , Li R. , Chen X. , Xu H. Biomass Wastes-to-Energy in China, Biogas from Landfills or Anaerobic Digestion? Montanuniversität Leoben, Abfallwirtschaft, Abfalltechnik, Deonietechnik, Altlasten, 2012:105-111.

[9] Raninger B. , R. Li, L. Feng, H. Zhang. Landfill behavior of Chinese MSW and its impact on CDM Emission Reduction. AWMA Ass. 's 101th Annual Conference, Portland, OR, USA, 2008.

[10] Xu Haiyun. The opportunity of WTE development in China. 2008.

[11] Dong Renjie, Raninger, B. Biogas Engineering and Application, Volume 1, China Agriculture University Press, 2010:328.

[12] Raninger, B. , M. Qian, Dong Renjie. Die Entwicklung von Grossbiogas Technologie in China. Biogas JOURNAL des Fachverbands für Biogas, Germany, 2010.

[13] NDRC. Medium & Long Term Program of the Renewable Energy Development Plan for China 2006-2020. 2006.

[14] Gehring, M. Feng LEI, Raninger B. , Li R. Quality of Separately Collected Bioorganic Municipal Waste and BMW Compost in Shenyang, China. ORBIT 2008, Wageningen, NL, 2008.

[15] World Bank. Clean Development Mechanism in China. 2nd Edition, 2004.

[16] NDRC. Chinas Policies and Actions for Addressing Climate Change. Progress Report 2010, 2010.

[17] European Environment Agency. Better management of municipal waste will reduce greenhouse gas emissions. EU, 2008.

[18] European Environment Agency. Greenhouse gas emission trends and projections in Europe 2008. Tracking progress towards Kyoto targets Assessment, 2008.

[19] Li R. , Raninger B. , Sun M. , Feng Lei, M. Gehring. Combustion Characteristics of RDF resulting from source separated MSW in Shenyang, China. ORBIT 2008, Wageningen, NL, 2008.

[20] NDRC. Renewable Energy Law, 2006.

泰国可持续利用生物质资源生产电及生物质燃料

Werner Siemers

(泰国国王科技大学(KMUTT)能源与环境联合研究所(JGSEE),曼谷 10140)

【摘要】泰国是亚洲主要的农业生产和出口国。泰国现有的能源政策除了一些常规的内容外,特别重视可再生能源的利用。其中最重要的是各种生物质能源,而风能和光伏发电等其他常规可再生能源则相对次要。因为在耕种及食品加工环节中可产生大量的生物质资源,故在不同的农业过程中都可产生电力、热能和液体燃料。

全球范围内,生物质能与民争粮已成为大家共同关心的问题。可持续的标准是生物质商品的净能量减少量和净温室气体减排量。本文对现有的和潜在的生物质产能路线进行了生命周期分析(LCA)以确定相应的能量平衡和温室气体减排量。并将这些数据与泰国传统的电力行业进行比较,确定不同利用形式的净减少量。泰国针对其常用的生物质燃料,按照欧盟法律所定义的可持续标准对这些数据进行了确认和讨论。这些案例的优点在于采用了相似的方法和数据集,使其结果在全国和全球范围内更具有可比性。文中所有的讨论也是建立在这些条件的基础上。

预计到 2022 年,泰国可再生能源产量占能源总量的 20%。实现这一目标的主要依靠生物质发电、生物质产热(热电联产)和生物质燃料。同时,政府也对生物质发电并网立法以确保其被电网接纳并保证其价格,还根据可再生能源的种类提供补贴。对不同种类的生物质燃料提供支持,保证了生物质燃料的销售价格低于化石能源的价格从而提高市场占有率。

为泰国提供不同能源利用途径所需的生物质资源的主要是中小规模农户。泰国的甘蔗种植面积约为 1×10^6 hm^2,年产量约 7×10^7 t。这些甘蔗在制糖的同时也产生了大量副产物,糖蜜废水用于生产生物乙醇,甘蔗渣也可用于可再生能源的生产。泰国有超过 8×10^6 hm^2 的农田每年生产两季或多季水稻,年产量约 3×10^7 t。稻壳已被小型电厂广泛用于发电,但秸秆仍很少用于能源生产,常在农田中被焚烧。泰国的木薯种植面积为 $1\times10^6\sim1.2\times10^6$ hm^2,年产量达 2.7×10^7 t,其中一小部分用于生产生物乙醇。油棕榈的种植面积超过 4.5×10^5 hm^2,其新鲜果实年产量约 9×10^6 t,目前有超过 1/3 的果实用于生产生物柴油。

甘蔗渣发电具有很高的温室气体减排潜力,每兆瓦时发电量能减排温室气体约 600 kg 二氧化碳当量。其他的有机物需要运输和加工,减排潜力略低。其中,稻壳的减排能力虽不如甘蔗渣但其减排潜力也很高。秸秆因需要额外的处理步骤,其减排潜力约为 480 kg CO_2 当量/MWh。糖蜜废水根据不同情况,其温室气体排放量为 23~28 g CO_2 当量/MJ 生物质燃料,远远低于欧盟制定的 35%温室气体减排标准。化石能源精炼因会增加温室气体排放而不再被接受。特别是在用木薯生产生物乙醇时,因加工厂远离原料产地而消耗了大量化石能源,只能通过提高生产效率或在精炼过程中增加可再生能源的使用比例才能降低温室气体排放量。虽然利用棕榈油生产生物柴油的过程所排放的温室气体量最低,约为 20 g CO_2 当量/MJ 生物燃料,但若按计划扩展油棕榈种植范围会引起对土地利用的变化,与之产生的负面作用也引起了大家的担忧。

综上所述,通过不同的方法,泰国还可以有更大的温室气体减排潜力。

【关键词】泰国生物质资源;生物质发电;生物质燃料;可再生能源政策

SUSTAINABLE UTILIZATION OF BIOMASS RESOURCES FOR POWER AND BIOFUELS IN THAILAND*

Werner Siemers

King Mongkut's University of Technology Thonburi (KMUTT)

Contact: Dipl.-Ing. Werner Siemers, Joint Graduate School of Energy and Environment (JGSEE), King Mongkut's University of Technology Thonburi (KMUTT), 126 Prachauthit Rd., Bangmod, Tungkru, 10140 Bangkok, Thailand, Mobile: +66-86003-2711 E-Mail: siemers_w@jgsee.kmutt.ac.th

EXECUTIVE SUMMARY

Thailand is one major agricultural producer and exporter in the Asian region. The actual energy policy in Thailand besides other general consideration places emphasis on the use of renewable energy resources. Major impact is expected from the different forms of biomass, less from the classical renewable energies like wind and solar photo voltaic (PV). Because of the large resources available at the field level and in the agro-industrial processing chain, power and heat and also liquid fuels are originating from the different agricultural based operations.

Internationally, there is great concern on the sustainability question of bioenergy use in conjunction with food and fodder issues. One criteria on sustainability is the net energy saving and net greenhouse gas reduction potential of biomass based commodities. For all present and possible biomass power routes a LCA (life-cycle-analysis) was conducted to establish the respective figures on energy balance and GHG emissions (greenhouse gas emissions). These are compared with the conventional power sector in Thailand to establish the net savings effect of the utilization patterns. Also for the prevailing biofuels in Thailand these figures are being established and discussed in conjunction wit sustainability criteria set by the EU legislation. Main advantage is seen in this respect that similar methodologies and data sets are used for a variety of cases which makes results more comparable on national and international scale. Discussion of options is then based on a better fact situation.

Thailand aims at a share of 20% for renewable energy planned for the year 2022. Major contributions to that target are based on biomass for power production, on biomass for heat production (in cogeneration) and for the implication of biofuel production from biomass materials. The government offers for the electricity production a feed-in regulation with price guarantees and grid access plus a subsidy component based on the type of renewable energy. Different forms of biofuels are supported by setting the sales price for biofuel producers and ensuring a market price below the fossil fuel alternative to encourage market penetration.

The necessary biomass for the different energy applications is produced by the Thai agricultural sector, mainly by small to medium farmers. Sugar cane is grown on some 1 million ha with a production level of some 70 million t. This production results in specific amounts of bagasse, the byproduct of sugar milling, for possible power production. Also the field rests offer a potential for heat and power. Besides that, the sugar sector also offers molasses, the byproduct of sugar refining, for the bioethanol production. Rice is

* This paper was initially presented at the Orbit 2009 CHINA. 本文已发表于 Orbit 2009 CHINA。

grown on more than 8 million ha and produces in more or two crops a total of 30 million t paddy rice per year. The rice husks are used extensively in small power plants. Rice straw still is sometimes burnt on the field and has shown no wide use as energy source. The cassava cultivation area lies between 1 and 1. 2 million hm^2, the production reaches 27 million t roots per year. A small portion of this is used by the industry for the production of bioethanol. Oil palm plantations cover some 450 000 hm^2 and produce some 9 million t of fresh fruit bunch. Nearly 1/3 of this production is currently converted and utilized as biodiesel.

The electricity production from bagasse offers high GHG savings potential in a range of 600 kg CO_{2eq} savings per MWh electricity produced. The field rests need some transport and processing effort, which reduces the net savings slightly. Also rice husk have a large savings potential, although slightly lower than bagasse. For rice straw the additional processing steps reduce the net savings more substantial to 480 kg CO_{2eq} savings per MWh. The different forms and cases for molasses based ethanol result in GHG emissions of 23 to 28 g CO_{2eq}/MJ fuel. This is far below the EU criteria for a 35% saving in GHG emissions. Use of fossil energy in refining will increase the GHG to levels which are no longer acceptable. This is true especially for cassava based ethanol. It is produced in off-site refineries with high inputs from fossil energy sources. Only high efficient production or increased employment of renewable energy in the refining process could reduce the impact. Lowest GHG emissions are found for biodiesel from palm oil products at levels below 20 g CO_{2eq}/MJ biodiesel. However, the planned expansion of oil palm plantation causes concern on possible land use changes with adverse effects.

In the conclusions it is stated that more potential exist to reduce GHG emissions in Thailand with the various options.

Key words: Biomass resources Thailand; biomass power; bio-fuels; renewable energy policy

1 Introduction

Biomass resources are the major renewable energy source in Thailand being already used to a certain extent and also being planned for further expansion for power and heat production and for biofuel production. However, up to now there remain uncertainties on the feasible potential of the different resources and the actual effects of the different alternatives. The Thai government has announced ambitious targets for a near term period up to the year 2011 and more recently adopted a national plan for the year 2022. Final alternative/renewable energy supply should reach a share of more than 20% in the final year, Energy Policy and Planning Office-EPPO (2008). Different support programmes for the individual parts of the renewable energy sector are in place, mainly a feed-in regulation for power production and a price incentive and subsidy scheme for liquid biofuels. The options for biomass power generation rely mainly on agricultural byproducts. In certain cases, e. g. for rice husk, there is already a situation reached, which uses most of the available potential. Other potentials are still not used. The biofuel production is based on first generation biofuels with feedstock also being produced for human consumption or animal fodder. In general, there is a lack of comparable criteria, whether certain developments are moving in the desired direction and whether the adopted policies are justified. One criteria being applied could be the sustainability issue, which is complicated to define as a holistic solution. To narrow the scope, the net greenhouse gas emissions and savings and the net energy balances (fossil and renewable inputs) can be taken as indictors for the degree of sustainability. This task has been undertaken for the major biomass energy supply options in the power and biofuel sector, Siemers (2009).

For the electricity production from renewable sources the sectors have been investigated and data established for the power production details, for life-cycle-analysis and for production costs. The options chosen reflect the main sources for biomass power potential (sugar sector and rice sector) in Thailand such as:

- present and future use of bagasse in cogeneration and in independent power plants;
- future use of field rests from sugar cane production in co-combustion;
- present use of rice husk in biomass power plants;
- future use of rice straw in biomass power plants.

Liquid biofuels in Thailand are produced based on first generation technologies. For bioethanol the resources used are mostly molasses, a byproduct of sugar production, and in an upcoming development the product of cassava/tapioca cultivation (starch based). Biodiesel is produced from different palm oil products, mainly crude palm oil but also residues and byproducts such as waste cooking oil and stearine. The different options under consideration are:

- Bioethanol from molasses;
- Bioethanol from raw sugar juice;
- Bioethanol from cassava roots or tapioca chips;
- Biodiesel from crude palm oil;
- Biodiesel from stearine and waste cooking oil.

The results presented here are still in an initial stage as some considerations have still to be verified and detailed. Major lack on information and analysis exist on the question on possible land use changes for the production of biofuels. But the available results can already be used to evaluate certain strategies and to define areas of further investigation. This work would be of benefit for the international discussion on bioenergy development and issues on sustainability criteria.

1.1 Background

The energy situation in Thailand is dominated by two major observations. The majority of the primary energy demand is supplied by fossil fuels and resources. Secondly, renewable energy shows a substantial share on the primary energy balance. But most of that primary energy is so-called traditional consumption of fuelwood, charcoal and agricultural residues for cooking and other small scale energy operations in rural areas. The modern portion of renewable energy such as wind, photovoltaics or biomass conversions to electricity, high quality steam or biofuels is still limited.

In the year 2007 Thailand showed energy consumption (primary sources) of a total of some 110 Mtoe (million tonnes of oil equivalent) or 4 552 PJ. Majority of the input is based on liquid petroleum products (to be imported to a large extent) with some 40% of the total. Second important energy source is natural gas at just under 30%, which is produced locally but increasingly also imported from neighbouring countries. As mentioned already, third most important energy source are the renewables but mostly traditional biomass. Coal mainly as import is used in industry, whereby lignite is used for local power generation (basically using local lignite resources). Biggest part of renewable energy is contributed by bagasse. Traditionally, all bagasse produced during the sugar milling process is used up at the sugar mills for the energy demand for production of refined sugar. Consequently, the total amount of bagasse produced in Thailand would represent in terms of total primary energy demand a figure of less than 3% on the total demand. Paddy husk (rice husks in this study) and agricultural residues are consumed as well in larger quantities. In terms of total primary energy input the rice husk use would be around 1.5% of primary energy. Modern forms of biomass energy use like biofuels (bioethanol and biodiesel together) are having a mere 1% share on the total renewable primary consumption and thus a minimal contribution to total primary energy. But in this sector increasing growth rates are observed.

Thailand's renewable energy strategy is mainly based on the promotion of biomass based resources for heat, power and biofuel. The recent government has introduced a new planning target of an overall 20% renewable contribution (it includes natural gas for vehicles) for the period of up to the year 2022, EPPO (2008). Previous targets were formulated for the year 2011 with an overall share of 8% (actual figure differs according to source and time, Amranand (2007)). As can be seen in a rough overview in *Figure 1*, the major sources for this strategy are based on biomass, either directly or in the form of biogas and liquid biofuels. Taking the biogenic content of municipal solid waste (MSW) as biomass based as well, some 95% of renewable energy is actually bioenergy.

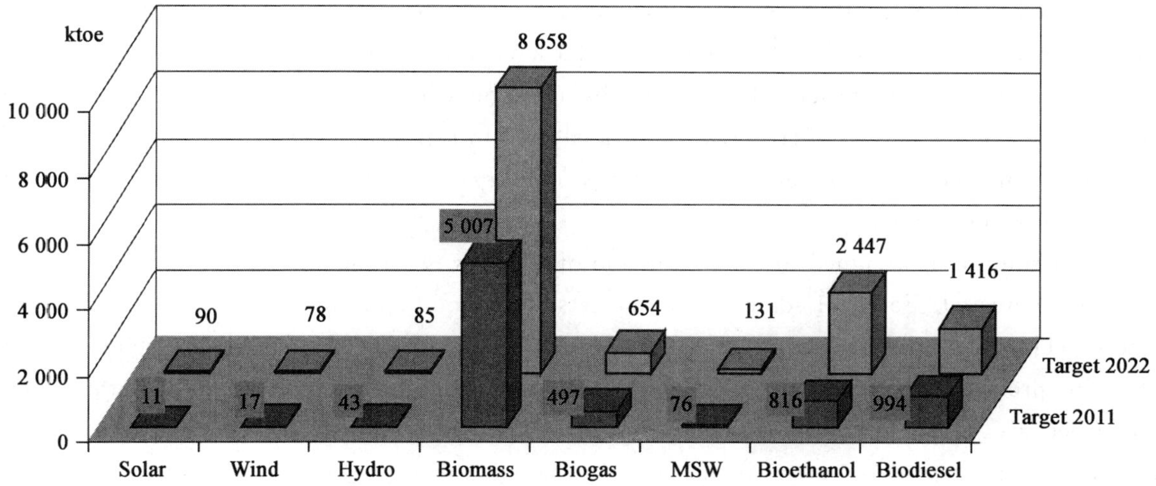

Figure 1 Political targets for renewable for 2011 and 2022

As biofuel production is concentrated on sugar and cassava inputs for bioethanol and palm oil for biodiesel, main source for heat and power are agricultural residues such as bagasse, rice husks, forestry wastes and other agricultural byproducts.

In the case of electricity generation for the final target 2 800 MW installed capacity are foreseen for coming from biomass alone. Biomass mainly is defined as solid residues from agriculture and forestry based resources. Electricity generation is defined in two ways, one way originates from industrial energy demand for any processing (in most cases agroindustry), where power plants are operated in cogeneration to produce steam as well for on-site demand. Surplus electricity then could be supplied to other industries in the neighbourhood or fed into the national grid. Second option is the operation of individual power plants also for energy needs of industry (e. g. in industrial estates), but mainly designed for electricity production for the national consumption under the feed-in regulations. In the last two years substantial increases can be seen on installed capacity. To reach the full target, however, a slightly increased exponential growth is still necessary. The other renewable energies are far less important. Contributions of hydropower (small schemes) should reach 156 MW, Wind 115 MW, Waste utilization 100 MW, biogas 60 MW and finally photovoltaics 45 MW. Besides in the biomass sector larger increases are planned especially for wind and municipal waste. Not so much in the common focus is the contribution of renewable heat to the overall targets. Based on data, however, 55% of total renewable energy is based on heat contribution. Also the majority of heat is based on biomass (the abovementioned cogeneration schemes in agroindustry and biomass boiler fuels). Smaller targets are associated with biogas production and the contribution of solarthermal energy is very small. Also in the heat sector, substantial growth rates are necessary to reach the targets. Major policy in the power sector is the feed-in of electricity under privileged conditions. Regulations are in place for grid connection and power purchase agreements between individual producers and the national grid operators. Relevant for renewable energy are two schemes, the SPP-programme and the VSPP-programme. The SPP (small power producer) allows electricity production in capacities between 10 MW and 90 MW. The feed-in tariff is based on the wholesale conventional power production and an additional subsidy, called adder, specifically adjusted to the type of renewable energy. Power purchase agreements are made with the national grid operator and final prices of feed-in are negotiated (sometimes in a bidding process) based on confirmed capacities, peak and off-peak deliveries. The VSPP (very small power producer) for capacities below 10 MW enjoy a flat feed-in tariff (usually slightly higher than in SPP) guaranteed for 7 resp. 10 years. Based on average exchange rates the adders for biomass with 0.3 THB/kWh translate to 0.65 €-ct/kWh and 0.5 THB/kWh represents 1.09 €-ct/kWh, rates which are rather low in comparison to other schemes worldwide. The wholesale tariff as the base for feed-in is announced at an average of 2.40 THB, which is transferred to European currency, around 5.2 €-ct/kWh. Interested investors seek an approval from the energy ministry and could then proceed with power purchase agreements and installations.

A specific contribution is assigned to the production of biofuels. Also here the targets differ from time to time and are readjusted. For 2011 somehow the targets were 4 million L/d on bioethanol and 3 million L/d as biodiesel. This altogether translates to a share of 30% on total renewable energy, which is produced by biofuels. The prospects for bioethanol are a bit more promising as the target seems to be in reach. For biodiesel (because also the diesel consumption in Thailand is much higher compared to gasoline consumption), higher growth rates are needed. In energy policy projections also the use of LPG and natural gas in vehicles is declared as alternative energy and added to the overall targets. Although not renewable, the use of alternative transportation fuels should lessen the dependence of Thailand on fossil fuel imports. As this contribution is substantial, sometimes different targets and percentages are discussed and presented. In place for the year 2022 now is a target of more than 20% of renewable and alternative energy. Subtracting the use of NGV in vehicles, the remaining share for biofuels translates to 9 million L/d of bioethanol and to 4.5 million L of biodiesel per day. The promotional efforts for biofuel production are summarized as a price guarantee level set by the government to which biofuels should be acquired by the blending companies. After that the government assesses the price relation to fossil fuels and adjusts prices for fossil fuel and biofuels with different level of excise tax, energy conservation fund tax and other taxes to leave biofuels competitive. In cases large differences occur between local production prices and international price levels, the government would reimburse the national operating refineries and distributors for their 'apparent' losses on the biofuel production and sales.

1.2 Research objectives

For all of the relevant conversion processes either from biomass into electricity or from biomass into biofuel, the relevant data for life-cycles should be established, sourced and verified. Major objective is to use a similar methodology for all of the different options being applied in Thailand. By this, the results can be compared and decisions being taken more reasonable on a data base, which uses similar approaches. Therefore for the emission calculation and energy balances a data bank programme is used, which need to be adopted and enlarged with the different Thailand specific data. After analysis the data can be compared and discussed. Besides the environmental issues covered by life-cycle-analysis and emission calculation also the financial situation of the different conversion routes should be assessed. For this, economic models need to be adjusted and performed. Results of economic models would be specific production cost and financial parameters like internal rate of return or net present value for the investments in the sector.

2 Methodology

One major tool used in the analysis consists of a data bank application for emission calculation. This kind of a program was developed in Germany and is in widespread use internationally. In the global emission calculation program (GEMIS) products and processes are defined and can be linked to process chains, (GEMIS). Once established, different calculations and results are available mainly on energy balance, greenhouse gas emissions, other emissions and resource use. Basic inputs are data from life-cycle-analysis along the production chain for the different commodities as in *Figure 2*. Necessary input data have been taken from available resources, reports and publications and had been verified, completed and detailed by own investigations. Input data into GEMIS are results from the LCA of the considered product. In the cases under consideration always agricultural products or byproducts are used, converted and transported before the final end use takes place. Along the production path, fossil energy is needed (for agricultural operation and transport) and also materials are needed, which imply certain energy consumption (for example N-fertiliser uses large amounts of natural gas) and an associated releases of greenhouse gases. In GEMIS products and processes need to be defined and connected together for the final assessment. Once all processes and inputs are connected, GEMIS can calculate the final result for the final end product on the base of all individual contributions (e. g. how many g of CO_2 are coming from the combustion of coal in a boiler used for the production of 1 MJ of bioethanol). The final result is a highly aggregated figure with a high number of individual sometimes minor contributions from associated processes (e. g. the energy used to produce steel, from which the boiler is manufactured which is used to supply steam to the bioethanol process). In the programme the level of accuracy and the system boundaries can be adjusted.

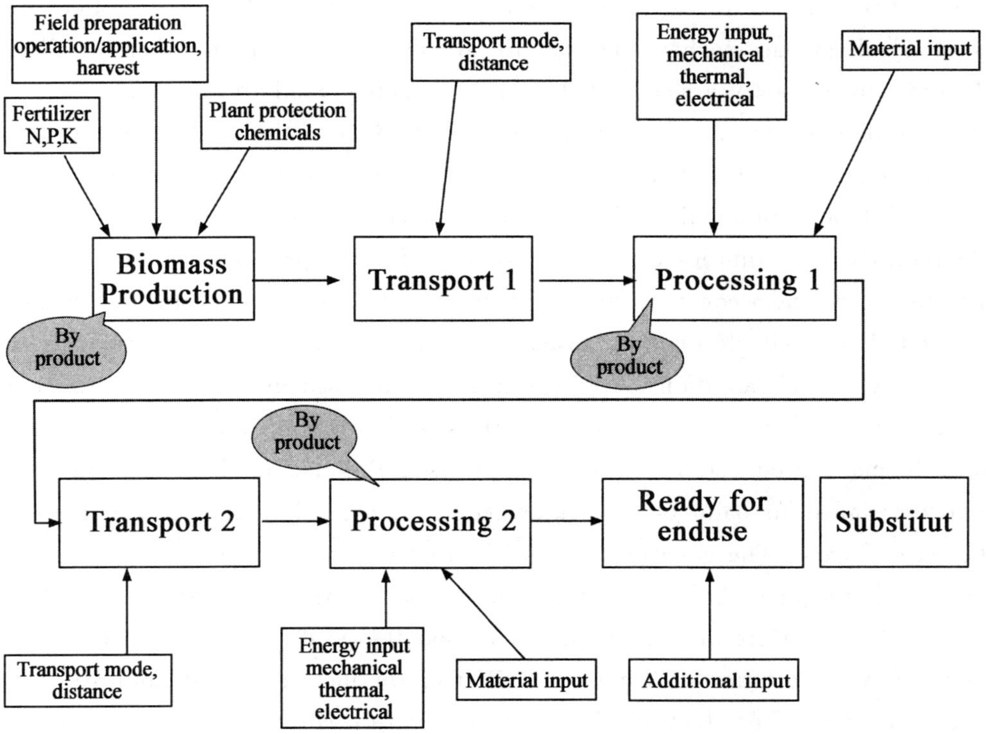

Figure 2 Principle way for life-cycle-analysis and inputs, outputs and products

3 Agricultural Production

Power and heat are mainly produced from agricultural residues or byproducts. The two major sectors are the sugar industry and the rice production. Biofuels are produced by the sugar industry as well as electricity. Other contributor to bioethanol is the cassava production. Biodiesel is based on the production of crude palm oil in oil palm plantations.

Thailand has been a traditional sugar producer in the world and is one of the major exporters of sugar to world markets. The planted area for sugar cane, only a small amount on irrigated conditions, is somehow stabilized around 1 million hm^2. Slight increases in area can be seen in history and some fluctuations occur according to market conditions and preferences by the farmers. The total production of sugar cane has been increased on average from some 40 million t in the 90s to a level of 70 million t actually. Fluctuations over the years are mainly due to changing yields in t/hm^2. The productivity level is slightly lower compared to other world market producers like Brazil. Rice is the major agricultural crop in Thailand. Rice is grown on rain-fed condition with one harvest per year nearly all over the country and in addition under irrigated conditions mainly in the Central region, where two or three harvests are possible per year. Major rice production (harvest once a year) occurs on some 8 million hm^2 and produced in the last few years around 23 million t of paddy per year. The area is relatively stable, production increases are mainly due to increased yields per hm^2 over the years. The production of second rice (two crops per year) is smaller with some 6 to 7 million t of paddies per year. Total area used for this is around 1. 5 million hm^2 (compared to the a. m. total area of 8 million hm^2). The total rice production per year then amounts to slightly below 30 million t/a. Also for rice Thailand is one of he major world producers and exporters. Cassava is also a traditional agricultural product of Thailand, mainly planted for export reasons in different forms (starch, end product, chips for fodder production) and less for domestic consumption. The area under cassava is somehow stabilized between 1. 0 to 1. 2 million hm^2 assigned to cassava. The total production of cassava in the last years in Thailand has reached some 27 to 28 million t. There had been a slight increase in productivity (t/hm^2) in the last years as well. Cassava is planted mainly in the North-East of Thailand on soils and under conditions not favourable for rice production. Out of the total production, some 25% are produced as cassava chips, which can be used locally for fodder production, but are mainly exported as ingredient to animal fodder production. Remaining production is converted to starch in local starch factories. Only recently opportunities came into existence for the production of biofuels from cassava. Some ethanol factories are under construction and planned for cassava based biofuel on the assumption that surplus cassava would be available on the market and in regions with fewer alternatives for other cash crops. Palm oil produced in plantations is a typical production scheme for the South of Thailand. As Thailand is a major exporter of sugar and cassava in the world, the contribution of palm oil products is far less compared to the major producers Indonesia and Malaysia. But also in Thailand there is a large increase in plantation area assigned to palm oil to acknowledge. Area and total production are exponentially enlarged in the recent years. The government has favoured the increase of plantation area for palm oil. This has led to a total area planted of some 450 000 hm^2 actually, still lower compared to the 1 million hm^2 each for sugar and cassava. More or less in parallel to that the production of fresh fruit bunch increased and reached in recent years a figure of 9 million t. Part of the increased production can be attributed to increased biodiesel demand. The dynamic developments in the palm oil sector are very acute, a high demand for local and exported vegetable oil, large increases in area and production levels and a political motivation for high production of biodiesel from palm. According to the latest figures some 30% of the total palm oil production would be converted to biodiesel, mainly after satisfying at least the local cooking oil demand. As could be expected, there are different opinions on the further biofuel developments. In principle it is stated, that there should be no increase in agricultural area (rather an increase in productivity), no diversion from food and fodder production and only the surplus production or byproducts should be used as feedstock for biofuel production.

4 Results and Discussion

4.1 Power production in sugar sector

Bagasse is used traditionally in sugar mills as the main fuel for the energy demand in sugar processing, both for steam and electricity. Operation can be described as low pressure steam production, use of steam for milling and sugar cooking and partially converting steam into electricity in the backpressure mode. The target in the industry was to fully utilize the bagasse throughout the season to avoid any surplus or waste problems for the mill resulting in quite low overall conversion efficiencies as there was no incentive to increase efficiency. Theoretically, there should be a substantial export potential for power available from the given amount of bagasse associated with the sugar cane delivery, if modern conversion and process technology is used. To indicate the potential, it is assumed that 1.0 TC (ton of cane) delivers 0.28 tonnes of bagasse with a NCV (net calorific value) of 7.2 MJ/kg followed by conversion of bagasse with 25% electrical efficiency (high) to deliver 140 kWh gross electricity. Demand of electricity in a high efficient mill is around 20 kWh/TC, therefore theoretical maximum surplus is 120 kWh/TC assuming the heat demand could be supplied by waste heat and/or extracted steam (and only in high demand during crushing season).

The actual situation is rather complex as a multitude of factors are influencing the final result. In order to simplify and to use descriptive and comparable figures, a model has been set up for the description of the situation in the Thai sugar industry. In analyzing data it can be shown that certain groups of sugar mills can be separated, each of it having a certain characteristic, the employed steam parameters used as the major descriptor. Accordingly, roughly 20% of the sugar milling capacity in Thailand does not export power to the grid and shows a comparably low efficiency (20 bar technology). Another 35% of production with an average of 5% electrical efficiency would export an average of some 5 kWh/TC. By increasing the system pressure from 20 to 40 bar an efficiency gain is possible to 7% overall, which also enables higher exports of 12.5 kWh/TC. This group represents some 30% of production. Higher efficiencies need high pressure systems and a shift from backpressure turbines to condensing or condensing-extracting turbines. Only recently these power plants have been established either still on-site at the sugar mill or as independent power producers attached to a sugar mill complex but also using other biomass resources (mainly as some of the CDM projects in Thailand). On average, the total export production comes up with 14.3 kWh/TC at an apparent efficiency of only 7.5% for the whole sector.

Based on this the combined modeled results for Thailand are given in *Figure 3*. The horizontal axis represents the share of the total cane production and the vertical axis the resulting export power in kWh/TC. In summary, the model reveals a total production of some 1 000 GWh with an installed capacity (virtual installed as actual capacities differ) of 275 MW and an average full load of 3 640 h (or 41% capacity utilization factor). The total theoretical potential represents the full area in *Figure 3* and would amount to 8 400 GWh of export electricity (calculated as 120 kWh/TC times a production of 70 Mt of cane). To achieve that and based on 6 500 load hours the installed capacity would be at 1 300 MW. As this figure seem not to be realistic, an intermediate approach is seen in *Figure 3* (the dark grey columns) roughly doubling the electricity production to 2 000 GWh. The final energy demand on electricity is reported as 133 200 GWh in the year 2007, DEDE (2008). The present contribution of the sugar sector therefore represents 0.75% of that demand. The near term increase to 2 000 GWh would increase the share to 1.5% of the total.

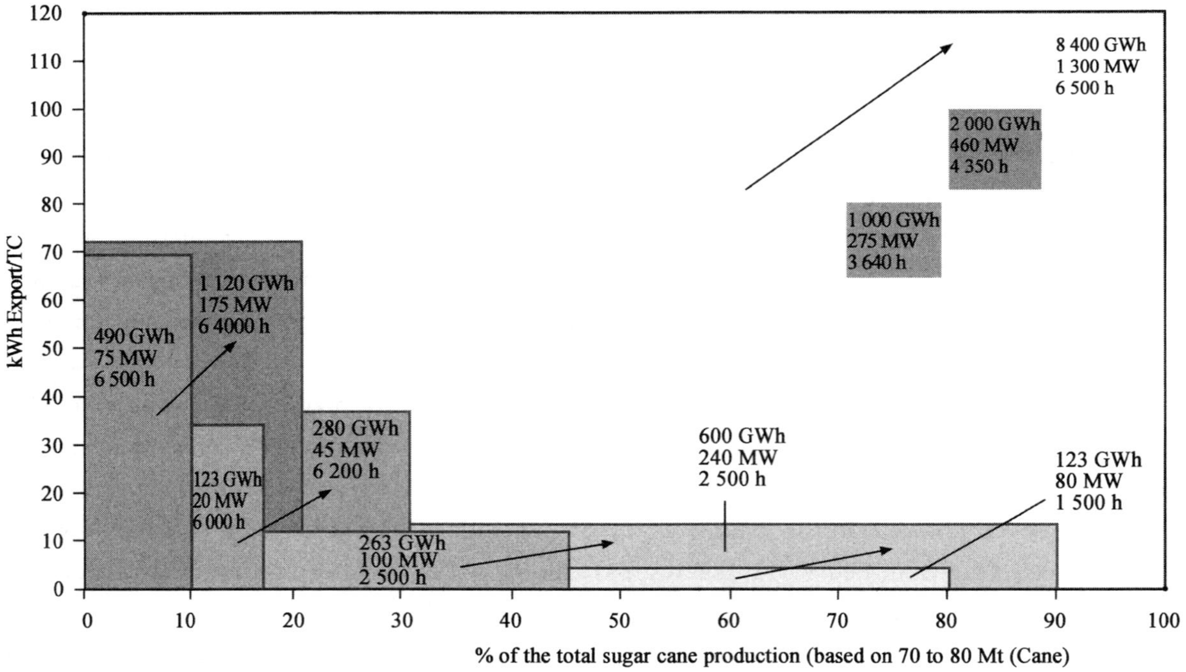

Figure 3　Present and future potential for export electricity from bagasse

After establishment of the technical background for electricity production from bagasse the data can be used as input into the emission calculation program. Three technology groups are differentiated as the 20 bar group, the 40 bar group and the advanced 70 bar group with their respective efficiencies. Bagasse as biomass fuel is regarded as emission free at the source of the power plant (renewable resource and no allocation made for agriculture and transport since a byproduct of sugar production). The only GHG emissions then being emitted are through the combustion process as methane (CH_4) and nitrous oxide (N_2O). As no reliable data are available at this stage, the default values used by IPCC have been taken, IPCC (2006) and accordingly changed to fit into the GEMIS system. In most of the cases bagasse is used as fuel in cogeneration mode. Thus, an allocation has to be made for the alternative steam production and use based on overall efficiency and technology status. Credits are then given to the electricity production for that amount of steam being co-generated in comparison to an isolated boiler steam operation fuelled by bagasse. Final results are displayed in *Figure 4*. Basically, GHG emissions are rather low at below 40 kg CO_{2eq}/MWh electricity produced. Due to the impact of efficiency, the 20 bar technology shows higher emissions compared to the 70 bar technology. The 70 bar option working in condensing mode only (not supplying steam, but operating at higher electrical efficiency) increases total GHG slightly. The present mix in Thailand of bagasse based electricity has a value of some 26 kg CO_{2eq}/MWh. These figures are to be seen against the present grid emissions, which are calculated in the range of 621 kg CO_{2eq}/MWh offering a nearly 600 kg/MWh net reduction for bagasse based electricity.

Against the well established bagasse use in the sugar sector for energy, the so-called field rest has only a minor contribution. Field rests are the remains of the sugar cane plant after harvest in the field and consist

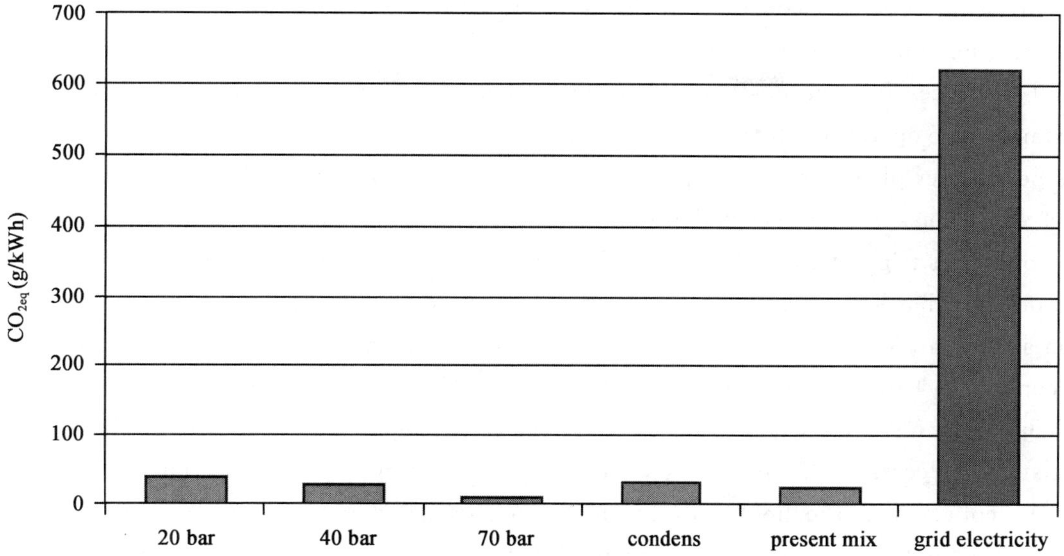

Figure 4 GHG emissions for bagasse cogeneration

of cane tops, leaves and trash. For energy utilization these resources would need a collection at field, additional transport and further handling at the power plant. Easiest way for power generation could be co-combustion in bagasse boilers at a certain maximum rate (20% seem to be not creating problems). This avoids additional power plant capacities and investments. The necessary handling efforts consume fossil energy, which results in higher GHG emissions. The final figures range from 24 to 87 kg CO_{2eq}/MWh, which is still a comfortable net gain as opposed to the present grid electricity GHG emissions. In the emission calculation a benefit for avoiding the field burning of cane was not allocated. Nearly 50% of all cane fields are burned prior to harvesting to ease harvest operation, which in this case is mainly manual labor. To change that tradition needs more interventions as can be foreseen from the energy perspective only.

4.2 Power production in rice sector

The rice sector produces two different forms of potential energy resources. These are rice husk as a byproduct of rice milling and rice straw as a resource left in the fields. The situation for these two resources is rather different and has to be analyzed separately. In recent years and triggered by high fossil fuel prices plus favourable conditions in the power sector (feed-in law and adder provision) rice husks are used in power plants for electricity production and mainly for export into the national grid. In the SPP (small power producer) scheme a total of 16 power plants uses fully or partially rice husk as fuel, EPPO (2009). The estimated installed capacity based on rice husk alone is around 140 MW. Another 9 companies are already registered under the VSPP-scheme (very small power producer) with a total installed capacity of 50 MW. These capacities can be satisfied with an amount of roughly 1. 7 Mill t of rice husks, representing 28% of all available rice husks. For the VSPP scheme another total capacity of 170 MW installed is already approved and could come into operation very soon. Waiting for approval are additionally 25 MW in SPP and 100 MW in VSPP plus planned VSPP of another 100 MW. The situation can be summarized as follows:

- Total existing capacity 190 MW;
- Likely increase to 360 MW;
- Possibly 585 MW installed total capacity.

The last figure would virtually consume all rice husks in Thailand, which is not a realistic option as there are already traditional non-energetic and energetic uses and established industrial consumers. In summary, however, all resources for rice husk would be utilised in the near future. Rice husk power plants are generally stand-alone operations with only limited demand for cogeneration or in-house consumption. As in most cases new power plants are built, the technology is more advanced and higher overall efficiencies can be assumed. Rice husk are transported (sometimes over long distances) to the power plant site. The favourable conditions triggered with high fossil fuel prices and also accompanied by inclusion of CDM in the investment decision has led to a situation, where at least local scarcities exist for rice husk and high competition arises on the use of the resources. The price for rice husk has risen from nearly nothing (an unwanted commodity) to substantial levels of some 1 000 THB/t recently (around 22 €/t). Rice straw on the other hand is not used for energy purposes besides some small scale application. Major reason is the unclear logistics and costs of the resource for any larger scale energy production. Theoretically, rice straw also has to be collected at the field site, in any form compacted to make it transport viable and then transported to the next power plant. These would involve farm operation, additional machinery for baling or briquetting and result in additional fossil energy demand. Low pressure bales are available in limited quantities and at prices which are not competitive compared to other biomass resources (1 500 THB/t or 33 €/t). At the power plant rice straw needs further processing before being combusted and converted in the boiler. The resulting efficiency should be slightly lower than for rice husks.

The results for the GHG emission calculation are shown in *Figure 5*. As expected, the fossil energy demand for transport of the rice husks and the whole logistical efforts for straw increases the total GHG emissions substantially if compared to the bagasse case, although the efficiency is higher (similar GHG emissions for CH_4 and N_2O in combustion). Rice husk power would emit a total of 67 kg CO_{2eq}/MWh, whereas the theoretical option of rice straw results in some 180 kg CO_{2eq}/MWh. This reduces the potential net GHG savings accordingly.

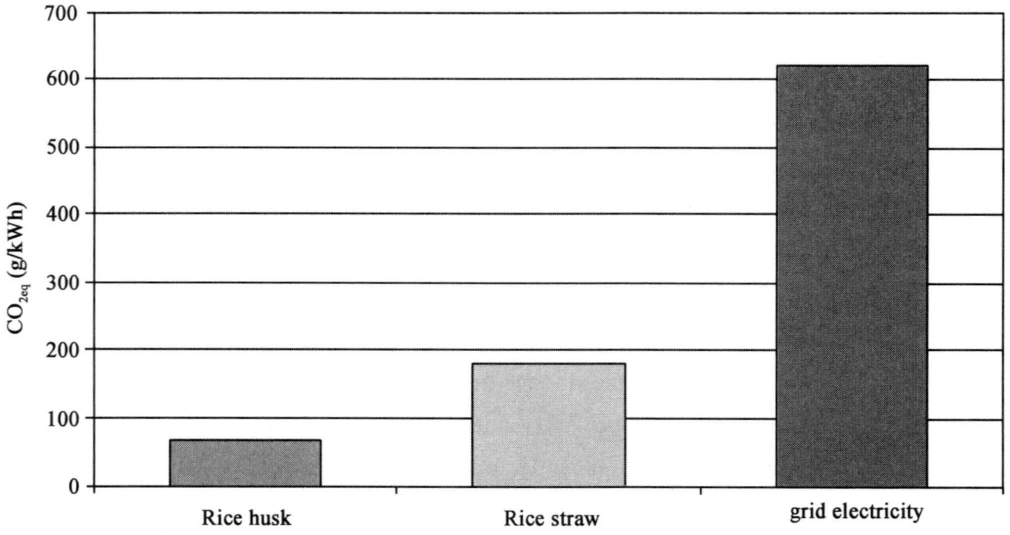

Figure 5 GHG emissions from rice husk and rice straw

4.3 Bioethanol from sugar sector

The sugar cane is transported to sugar mills after harvest and in the first step the milling/crushing, raw sugar juice with high water content is produced. This raw juice then is further processed into raw and refined sugar. A byproduct in this process is molasses, a sugar containing liquid, which can be used for alcohol production or as fodder ingredient. For one ton of cane (TC) harvested roughly 45 kg molasses are produced which could be converted to 12 L of bioethanol. Some 45% of the total molasses is presently converted to bioethanol leading to a daily contribution of around 1 million liters per day (MLPD) of pure bioethanol in the different blends on the market at the end of 2008, DEDE (2009). It is also possible to convert the raw juice (or thick juice after an evaporation step) into bioethanol (as done largely in Brazil), but this is not widely practiced on large scale in Thailand. Main reason is the agreement between farmers and sugar mills on profit sharing, which does not encourage raw juice conversion to ethanol under present agreements. Only minor supplements or non-edible sugarcane is used for the purpose of bioethanol production. A total of less than 10 refineries are in operation for the bioethanol production. In most cases the refineries are placed in the vicinity of the sugar mill, offering the big advantage to be supplied with electricity and steam for the process requirements from the sugar mill. As energy at the sugar mill is produced from bagasse, a renewable biomass resource, no grid electricity or fossil energy is necessary. The economic size of a refinery is larger than the molasses input produced at one (even larger) sugar mill. Consequently, molasses is transported from other sugar mills to these refineries. There are also off-site refineries in operation, which would buy molasses from different sugar mills for ethanol production. In this case they lack the advantage on the energy deliveries. For establishing the final results on energy and GHG emissions, the different production alternatives have been set up and calculated as shown in *Figure 6*. The base case is bioethanol from molasses at the sugar mill, which results in a final 22.5 g CO_{2eq}/MJ of bioethanol. Fossil gasoline for the comparison in this analysis stays at 93.3 g CO_{2eq}/MJ. The international discussion on sustainability criteria calls for minimum GHG reduction of complete chains of biofuels discussion. This threshold value is indicated as well in *Figure 6*. According to latest European Union regulations a savings of 35% has to be achieved either by default or calculated on the national base case, EU (2009). Thus, molasses based bioethanol is comfortably below this threshold value.

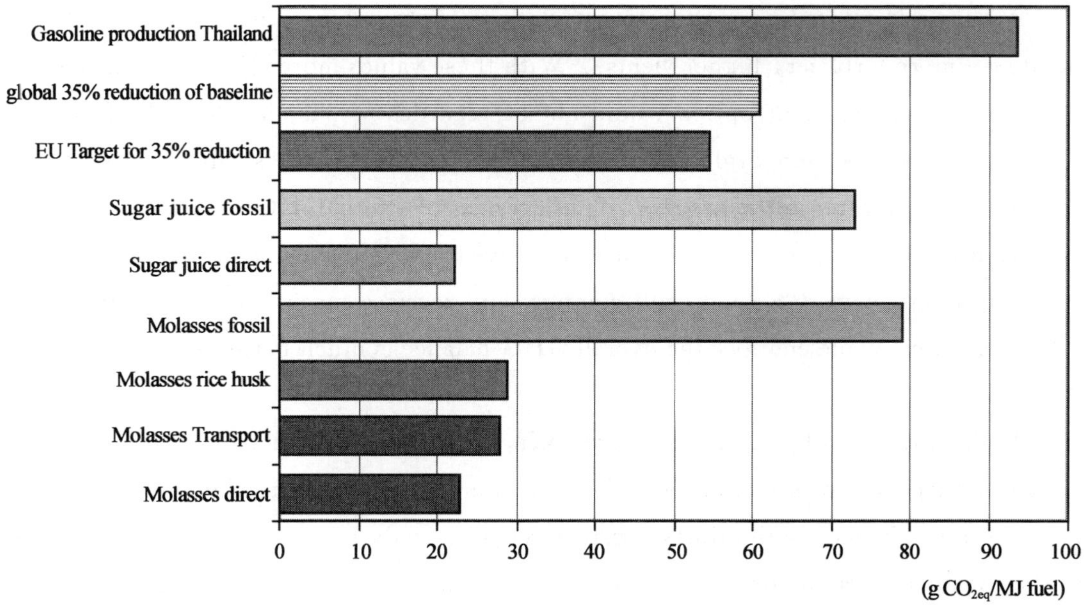

Figure 6 Results for greenhouse gas balance, sugar based bioethanol

Adding transport (for molasses delivery from different sugar mills to one central refinery) for part of the feedstock will increase the total GHG emissions to some 28 g CO_{2eq}/MJ (2 nd bar in *Figure 6*), but is still offering enough net GHG savings. The third bar in *Figure 6* depicts the case, where only fossil energy is used in the refining process (grid electricity and fuel oil) leading to some 62 g CO_{2eq}/MJ, a value too high for being regarded sustainable but still offering some GHG savings. The bar 3 represents a real case for an off-site refinery, but for the energy demand of the refinery rice husks are used and thus reducing the total GHG again to some 29 g CO_{2eq}/MWh. The bars 5 and 6 display the results for the theoretical conversion of raw juice to ethanol both for the on-site situation and the fossil energy situation. The sugar-bioethanol shows lowest results with 21 g CO_{2eq}/MJ, whereas the fossil option is too high with 57 g CO_{2eq}/MJ. Biggest contributor to emissions is agriculture in the on-site operations, transport and processing are minor steps. Only in cases, where refining consumes fossil energy, this processing step causes the highest relative emissions and fossil energy demand.

4.4 Bioethanol from cassava sector

After a conversion of the starch content of the cassava root or chip into sugars in an additional process of hydrolysis, bioethanol can be produced from cassava feedstock as well. Less than 5% of the present production is actually converted to bioethanol in Thailand, although plans for further expansion are made and a number of licensees awaits better market condition to start the production of cassava based bioethanol. For the whole chain analysis the bioethanol production starts at the field level. Cassava is planted, fertilized and harvested, then transported to the refinery directly or to the cassava chip drying facility. Finally the root or the dried chip enters the refining process and is converted to bioethanol. The refineries are at isolated locations and have to rely on fossil energy supplies for electricity and steam demand. There are different cases shown in the *Figure 7*. The factory 1 uses coal and grid electricity for the refining process, the technology applied is not high efficient. This results in total GHG emissions of 111 g CO_{2eq}/MJ for a low input agriculture case (national average) and in 116 g CO_{2eq}/MJ for a high input agriculture (more fertilizer, higher yields). With these values more GHG are emitted as compared to the use of fossil gasoline. This option would not be regarded as sustainable. The factory 2 is more efficient, but still uses coal and grid electricity. Part of the necessary steam is produced by biogas produced from the wastewater in the process. These results of a total GHG emission between 57 and 62 g CO_{2eq}/MJ are near to saving requirements also displayed in *Figure 7* and would just be sustainable. The third factory is attached to a sugar mill complex and receives steam and electricity produced from bagasse. As this is renewable energy, the overall GHG balance is much better with a level of 20 to 25 g CO_{2eq}/MJ.

The individual contributions of the different processing steps on the final result are assessed as well. Agriculture and transport shows a slightly higher contribution compared to the chip production and the final transport to the refinery. Both efforts together represent only minor contributions to the final GHG emissions. The main impact on final GHG emissions is through the refining process. Employment of biogas can reduce the total figure. But more efforts are necessary to reduce GHG emissions in processing, especially in the refining step.

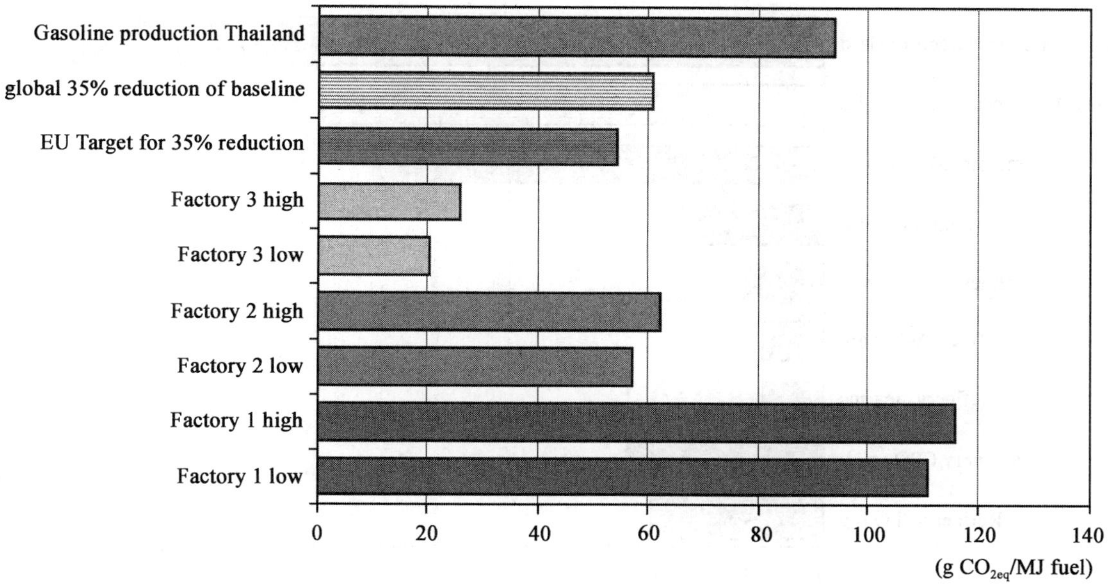

Figure 7 Results for greenhouse gas balance, cassava based bioethanol

4.5 Biodiesel from oil palm sector

Biodiesel is produced in Thailand from different resources related to palm oil production. Major supplier of input material for the biodiesel process is palm oil, either as crude palm oil or as refined palm oil (in competition to edible oil production). Out of roughly 1.2 million t of CPO produced in 2007 a share of 400 000 t is assigned to biodiesel production and represents 72% of present total biodiesel production. Another source is stearine, which is a byproduct of the CPO refining process with a market share of 24%. Some biodiesel producers also use waste cooking oil as feedstock material, which allocates some 4% of the total market. Total biodiesel production at the end of 2008 was at 1.2 MLPD. This amount of biodiesel produced relates to a total market share of roughly 2.3% of total diesel consumption (B2 is mandatory and B5 is available). Besides this market oriented biodiesel production there are numerous small scale biodiesel and community projects active producing biodiesel for local consumption. The results of the LCA and the GEMIS calculation are shown in *Figure 8*. The base case from CPO to biodiesel on large scale ends up with 20 g CO_{2eq}/MJ biodiesel. This in comparison to fossil diesel of 93 g CO_{2eq}/MJ enables substantial GHG savings and is slightly lower than the base case for molasses bioethanol. The small scale operation is not that effective and needs more specific energy and materials. Therefore this value rises up to 24 g CO_{2eq}/MJ). Stearine and waste cooking oil are byproducts and do not bear the chain emissions from agriculture and palm oil processing. Thus, the resulting GHG emissions are quite low with 7 g CO_{2eq}/MJ. The small scale processing of waste cooking oil again has higher values because of the less efficient material and energy use. As assessed above, the present mix for biodiesel production in Thailand comes up with a final figure of 17 g CO_{2eq}/MJ. This in fact is the lowest figure for all biofuel activities. In this analysis no GHG emissions for land use change have been considered, as most of the oil palm plantations are established and expansion of plantation area does not encroach into forests or displaces other food production area. However, these assumptions would need further clarification. Also the management of byproducts (wastewater, empty fruit bunch, fibers) has not been analyzed fully, but this could have large positive or negative effects on total GHG emissions (depending on the definition of the system boundary).

Looking at the contribution to GHG emissions, the plantation causes the highest share on the total.

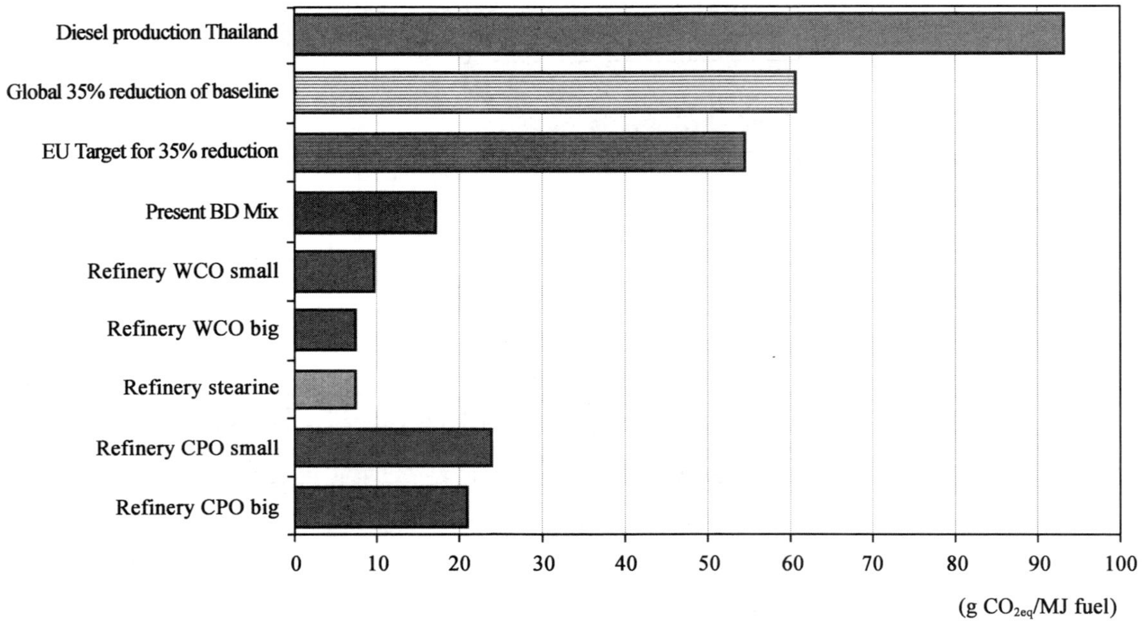

Figure 8 Results for greenhouse gas balance, palm oil based biodiesel

Milling and transport are minor contributors, also because in oil palm mills the energy for milling is generated by using the byproduct fiber for heat and power. Energy demand for refining is less than compared for the bioethanol systems, so even fossil energy demand would not increase the overall GHG emissions substantially.

5 Conclusions

Somehow contrary to the prevailing opinion, the highest potential for increased power production from biomass resources exists for the case of bagasse in the sugar industry. For the full utilization of the potential, however, the efficiency question at factory level is the main issue. Increased export levels are only possible if the efficiency of the power production is increased (shift from backpressure technology to high pressure condensing-extracting turbines) and the downstream sugar processing sees some improvements. Economic prospects are good also in cases where major investments are necessary. The field based resources cane tops, trash and leaves have a very large potential, but the resources need to be collected and transported. These efforts reduce the economic viability slightly and the net GHG savings as well.

Rice husk would be fully utilized mainly for energy purposes in Thailand in the near future. The net energy and GHG savings are slightly lower compared to bagasse but still favorable. Also the economic analysis shows, that projects are less attractive but still viable. Rice straw is a field based residue and faces a whole set of technical and economic difficulties for a widespread employment. Without specific measures and incentives no substantial contribution to power production is expected.

The actual contribution to net GHG savings is leveled between bagasse and rice husk at some 500 000 t/a (*Figure 9*). But due to the recent developments rice husk would overtake bagasse as the main contributor of GHG savings (see *Figure 9*). Larger potential are still unused for the field based residues. Therefore, proposed measures should concentrate on tapping the full potential for bagasse and developing strategies to reach part of the theoretical potential in the field based residues. But this would need more time for development of technologies needed and more efforts (research and development plus financial incentives) to make these options more attractive.

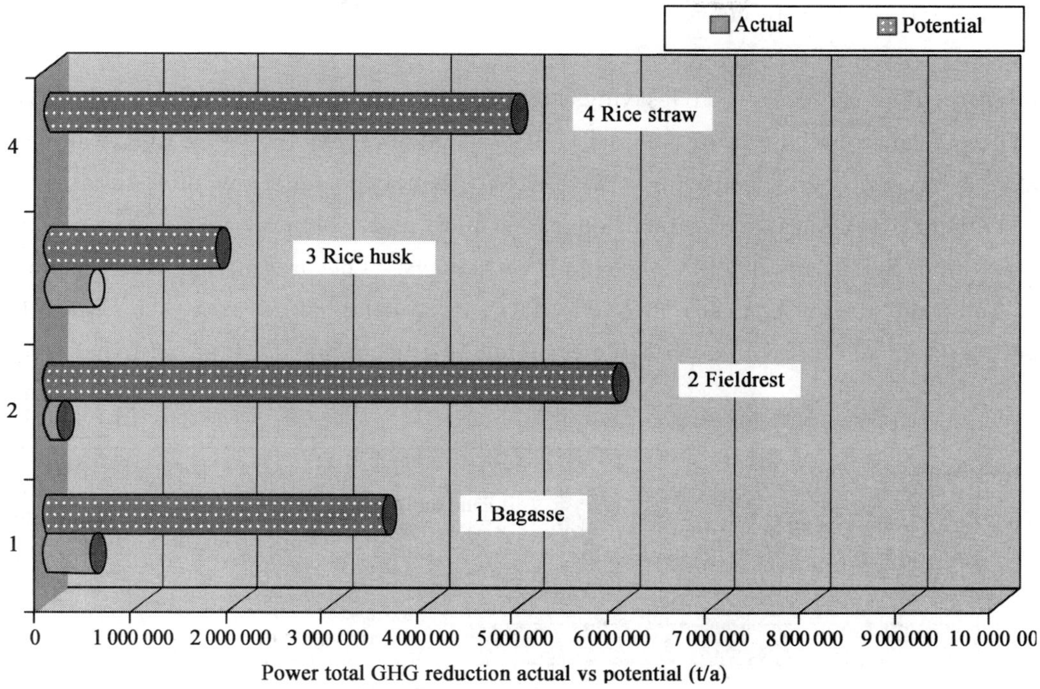

Figure 9 Total and actual potential for national GHG savings, power sector

The present use of sugar based byproducts results in tolerable levels of GHG emissions associated with the whole production chain of bioethanol. This holds only true if for the high energy demand in refining the opportunities are met by relying on sugar mill based energy supplies, which is regarded as nearly GHG free. In cases, where refining would utilize only fossil fuels (which are not existent right now), the net energy and GHG gains are too low to encourage a further production. Theoretical opportunities are also for converting sugar (sugar raw juice) into bioethanol, which would increase the total potential. Production costs for the sugar based bioethanol routes are below the reference price and offer financial viability in most circumstances.

Although use of cassava for bioethanol would offer high potentials the prospects are not that promising as in most cases fossil fuels are employed for the production chain and especially in the refining process. Results from the case studies do not reduce GHG emissions and show no limited financial viability at present conditions. Here, emphasis is needed for further process improvements and full use of other byproducts for energy generation and cost reduction.

At the present level of investigation biodiesel from palm oil offers good chances for GHG net reduction and primary energy savings. However, there is a restriction on the total available land area for the production of palm oil, which may influence the overall set targets for biodiesel consumption in the future. The use of byproducts and residues may reduce the stress on land area and would also lower production cost. Presently, biodiesel could only be produced above market prices for fossil diesel and therefore needs substantial subsidies.

Highest GHG savings with more than 10 000 kg CO_{2eq}/hm^2 are possible for the biodiesel production from palm oil mainly due to the high level of final product per area and the complete conversion into biodiesel. Second best is the bioethanol production from sugar at a level of some 7 000 kg GHG savings per hm^2. The fossil option reduces the net savings to 3 500 kg only. Latter value is to be expected for the cassava options as well, but only in cases where process improvements are made compared to actual case studies. Molasses is a byproduct with a low productivity per hm^2, thus leaving only limited potentials in the area consideration.

The potential for molasses based GHG savings for bioethanol is limited as other usages also have to be satisfied. The present contribution to national GHG savings of some 480 000 t CO_{2eq} can not be enlarged substantially. Largest potentials exist for cassava based ethanol and sugar raw juice (if all the production is converted to ethanol). Biodiesel from palm oil has a quite high potential also considering that the total acreage is less than half than that for sugar and for cassava. The present net GHG savings amount to nearly 1 million t and are the biggest individual GHG saver in the biofuel area (see *Figure 10*). But also here competing usage and difficulties with the economic justification could hinder further large increases.

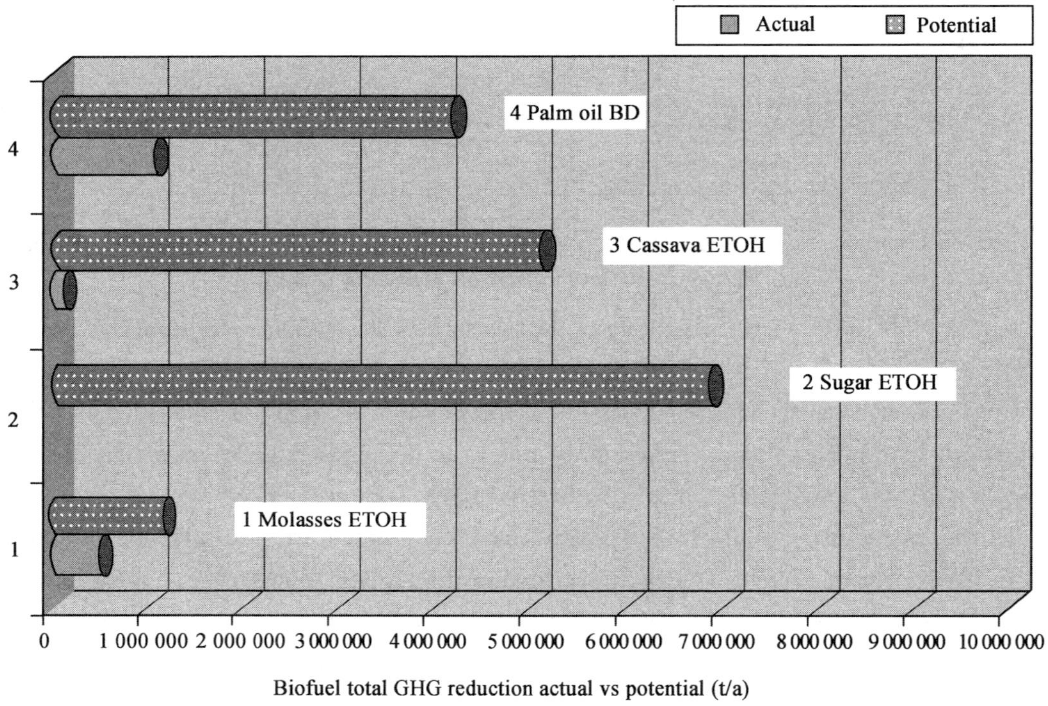

Figure 10 Total and actual potential for national GHG savings, biofuel sector

Acknowledgements

The author gratefully acknowledges the contribution of various colleagues from JGSEE and from the industrial sector during interviews and field trips. Most of the research work was conducted using research grants from the Thailand Research Fund, made available under the Energy Policy Project Phase 2 on behalf of the Ministry of Energy, Energy Policy and Planning Office. Part of the initial funding was also provided under the CIM-program of GTZ, Germany.

References

Amranand P. , (2007): *Renewable Energy Policy in Thailand*. RE Asia 2007, Bangkok, June 6.

DEDE, (2008): *Thailand Energy Situation* 2007. Department of Alternative Energy Development and Efficiency, Ministry of Energy, available as booklet or at www. dede. go. th.

DEDE, (2009): *Update on bioethanol and biodiesel*. Department of Alternative Energy Development and Efficiency, Ministry of Energy, available at www. dede. go. th

EPPO, (2008): *Development strategy on alternative energy for* 2008-2022. Energy Policy and Planning Office, Ministry of Energy, different documents at www. eppo. go. th.

EPPO, (2009): *State of IPP, SPP and VSPP*. Energy Policy and Planning Office, Ministry of Energy, different documents, latest version September 2009 under www. eppo. go. th.

EU, (2009): *Directive* 2009/28/*EG, Promotion of the use of energy from renewable sources*. Brussels, 23 April 2009.

GEMIS: *Global Emission Model for Integrated Systems*. Öko-Institute Darmstadt, free software, description, examples and handbook under www. gemis. de.

IPCC, (2006): *IPCC Guidelines for National Greenhouse Gas Inventories Volume* 2 *Energy*. Intergovernmental Panel on Climate Change, IGES, Japan.

Siemers W., (2009): *Development of tools to estimate alternative strategies and options in using biomass for energy*. Final report, Thailand Research Fund, Bangkok.

NaOH 固态碱处理法增强秸秆厌氧消化产气率研究

朱保宁，李秀金*，庞云芝，刘研萍

（北京化工大学，北京 100029）

【摘要】本文讨论了使用不同浓度的 NaOH 对秸秆样品进行化学预处理，并将之用做厌氧消化的原料来测试其沼气产量。结果表明，当 NaOH 的添加量为秸秆干重的 6%时，预处理的效果较好。此时沼气的累计产气率可达 850 mL/g VS，秸秆充分降解的时间缩短为 35 d，相较于未经预处理的秸秆样品，产气潜力提高了一倍。该项创新技术为以作物秸秆或其他能源作物为原料的沼气工程提供了有效的盈利方案。

【关键词】秸秆预处理；NaOH；沼气产量

IMPROVING THE ANAEROBIC BIOGAS YIELD OF RICE STRAW THROUGH LOW MOISTURE NaOH ALKALINE PRETREATMENT

ZHU Baoning, LI Xiujin*, PANG Yunzhi, LIU Yanping

Beijing University of Chemical Technology

Contact: Prof. LI Xiujin, Beijing University of Chemical Technology, PO Box 107, 15# Beisanhuan East Road, Chaoyang District, 100029 Beijing, China, Tel.: +86-13661070453, E-Mail: xjli@mail.buct.edu.cn

EXECUTIVE SUMMARY

This paper discusses the use of NaOH to pre-treat rice straw which is used as feedstock for anaerobic biogas production. Various NaOH proportions added to the straw for pre-processing were evaluated in order to identify a cost-effective way to enhance the biogas yield within a given time period. The results showed that a NaOH portion of 6% was the optimal ratio. The cumulative biogas yields achieved was 850 mL/g and the digestion time (HRT) could be reduced to 35 days. The patented solid-state NaOH pretreatment can double the biogas potential of rice straw and would be a cost-effective solution for commercial scale anaerobic digestion plants which use crop straw and other energy crops as feedstock

Key words: Straw pre-treatment; NaOH; biogas yield

1 Introduction

Modern technologies applied in agriculture lead to increasing cereal production accompanied with a much larger portion of crop straw. As one of the major cereal crops in China, rice is reported to become the overwhelming food for most families. At present, most of the rice straw is burnt in open-air area rather than being ensilaged because the rice straw doesn't have good mouth-feeling for livestock. Crop straw mainly consists of lignocelluloses fractions which include cellulose, hemi-cellulose and lignin. In rice straw, a particularly higher portion of ash contents has also been observed. Rice straw could be good substrate for biofuel production because of the high carbon contents. However, cellulose compounds are usually imbedded into lignin matrix which protects cellulose from enzyme attack. Pretreatments have to be applied to release cellulose or hemi-cellulose from lignin. Steam explosion is the most commonly used method for pretreatment of lignocelluloses materials (McMillan, 1994). Cullis et al. (2004) indicated that both increased chip size and increased moisture content resulted in improved solids recovery and increased hemicelluloses-derived sugar recovery as well as minimized condensation of lignin. The advantages of biological pretreatment include low energy requirements and mild environmental conditions, while a disadvantage is the long time period required for lignin degradation. Acid pretreatment can utilize either dilute or concentrated acids to improve cellulose hydrolysis. Pre-hydrolysis with dilute acid at temperatures higher than 121℃ is very effective for increasing the enzymatic digestibility of cellulose (Grohmann et. al, 1986). In spite of benefits, acid pretreatment presents potential problems such as an acid waste stream, the formation of compounds such as organic acids which are toxic to anaerobic bacteria during digestion (Ingram et al. , 1997) and the need for corrosion-resistant equipment (Brown, 2003). The mechanism of alkali pretreatment is believed to be saponification of intermolecular ester bonds crosslinking xylan hemicelluloses and other components such as lignin and hemicelluloses. After alkali pretreatment, the porosity of the material is increased due to the extensive swelling facilitated by removal of the crosslinks (Tarkow and Feist, 1969). Pretreatment of corn stover by Varga et al. (2002) with 10% sodium hydroxide (NaOH) for 60 minutes under pressure at 121℃ in the autoclave decreased the lignin fraction by more than 95% and increased the enzymatic conversion more than four times to 79. 4% as compared to untreated stover. In addition, by using dilute NaOH (0. 5% w/w) and increasing the reaction time to 90 minutes, 80. 1% enzymatic conversion was achieved (Varga et al. , 2002). Varga et al. (2002) soaked corn stover in NaOH for 24 hours prior to H_2SO_4 pretreatment and achieved nearly theoretical maximum enzymatic conversion (95. 7%). NaOH was the most effective at removing the lignin (29% removed).

In China, the ethanol production from crop straw is always been limited because of not only the technology shortage of cellulose enzyme efficiency, but also the poor quality of raw materials which contain a large amount of soil and ash. Compared to alcohol production, anaerobic digestion of biomass does not have to run in clean circumstance. The digestion of crop straw results in the production of biogas, which mainly consists of methane and carbon dioxide. In this way, energy from crop straw could be released and be used as a substitute for fossil fuels, thereby serving as a CO_2-neutral energy source. The digestate, which contains large amount of anaerobic microorganisms and residue of digestion, could be used as high quality soil amendments. From this standpoint, the anaerobic digestion of rice straw could be an environmental-friendly technology for the sustainable development in China's rural areas. In this paper, the biogas potential of alkaline pretreated rice straw is evaluated and several industrial-scale applications are introduced.

2 Materials and methods

2.1 Materials

Rice straw was collected right after harvesting from a farm in SHUNYI County, Beijing and exposed in open-area for 5 days to be dried on site. After arrival at the lab, raw rice straw was chopped to 2 mm particles by a blade trommel mill (GPA-15, WUXI QINGGONG machinery) and sealed in Ziploc bags as the materials for alkaline treatment. The moisture content (MC) and volatile solids content (VS) of chopped straw are 8% and 67. 7%, respectively. Sodium hydroxide was used to pretreat the straw samples. Anaerobic sludge was collected from a mesophilic digester in GAOBEIDIAN municipal waste water treatment plant and used as the inoculum for lab-scale digestion. The sludge was sealed in a gas tight bucket and stored in a refrigerator for no more than 1 month to keep the activity. Before the experiment, seed sludge was put into a 35±1℃ incubator for 3 days to wake up the bacteria and get rid of the substrate in the sludge. The total solids (TS) and volatile solids (VS) of activated sludge are 5.6% and 2.1%, respectively.

2.2 Experimental design and data analysis

2.2.1 Semi-dry NaOH pretreatment of rice straw

A novel pretreatment protocol was used in this study to reduce the water consumption and pretreatment duration. Certain amount of NaOH was prepared prior to pretreatment to reach the NaOH/straw ratio (by dry weight) to 4%, 6%, 8% and 10%. Tap water was measured to reach the water/straw ratio to 8:10. NaOH was then dissolved into water and the solution was mixed with rice straw samples for pretreatment. Straw samples were covered by plastic film and left in ambient temperature for 3 days to help the saponification reaction. After the 3-day pretreatment, rice samples were air-dried in 105℃ oven.

2.2.2 Evaluation of biogas potential and composition

Batch anaerobic digestion was conducted to evaluate the biogas yield of pretreated and raw rice straw. The experiments were carried out in triplicates for 50 days. 1.5 L conical beakers were used as the reactors. A rubber tube connected the headspace of each reactor with water-replacement equipment which was used to measure the biogas production volume. Biogas composition was analyzed on a Gas Chromatography which equipped with TCD detector. The operational parameters were 100℃ (injector), 120℃ (oven), 100℃ (detector).

3 Results and discussion

3.1 Characteristics of NaOH treated rice straw

The volatile solids of pretreated and untreated straw samples were measured. The ratio of VS/TS and dry matter loss would represent the degree of chemical degradation. According to *Table 1*, higher portion of NaOH led to a litter bit higher VS lost. However, the data of dry matter loss showed opposite trends. The results may indicate that, with the increasing of NaOH portion, more and more ash contents were converted into vapors. Total carbon (TC), cellulose, hemicellulose and lignin concentration were decreased after the pretreatment. If this phenomenon is not enough to indicate the degradation of straw components, the changes of cold water/hot water/benzene-ethanol extractives definitely represent the release of water-soluble compounds.

Table 1 Characteristics of rice straw samples

Parameters	Untreated	4% NaOH treated	6% NaOH treated	8% NaOH treated	10% NaOH treated
VS/TS	0.74	0.71	0.71	0.70	0.70
Dry matter loss (%)	—	5.5	5.3	3.5	1.2
TC(% of DM)	41.5	N/A	37.2	N/A	N/A
Cellulose(% of DM)	33.4	N/A	29.9	N/A	N/A
Hemicellulose(% of DM)	28.2	N/A	19.1	N/A	N/A
Lignin(% of DM)	7.4	N/A	5.7	N/A	N/A
Cold water extractives (%)	15.6	N/A	35.0	N/A	N/A
Hot water extractives (%)	17.8	N/A	39.6	N/A	N/A
Benzene-ethanol extractives (%)	10.4	N/A	5.1	N/A	N/A

3.2 Daily biogas yield and Cumulative biogas yield

Figure 1 to Figure 5 shows the daily and cumulative biogas yields of rice straw samples. For the straw samples pretreated with 4% NaOH (*Figure 1*), the trial of organic loading rate (OLR) at 35 g TS/L showed quick biogas production 58 mL/(g VS · d) in the 11st day. Other trials showed biogas production peaks at least 2 days later. These trends can also be found from cumulative biogas yield curves. After the 50 d experiments, cumulative biogas yield of all four samples arrived at almost a same point of 650 mL/g VS, indicating that a 50 d experiment period is enough to completely degrade the straws.

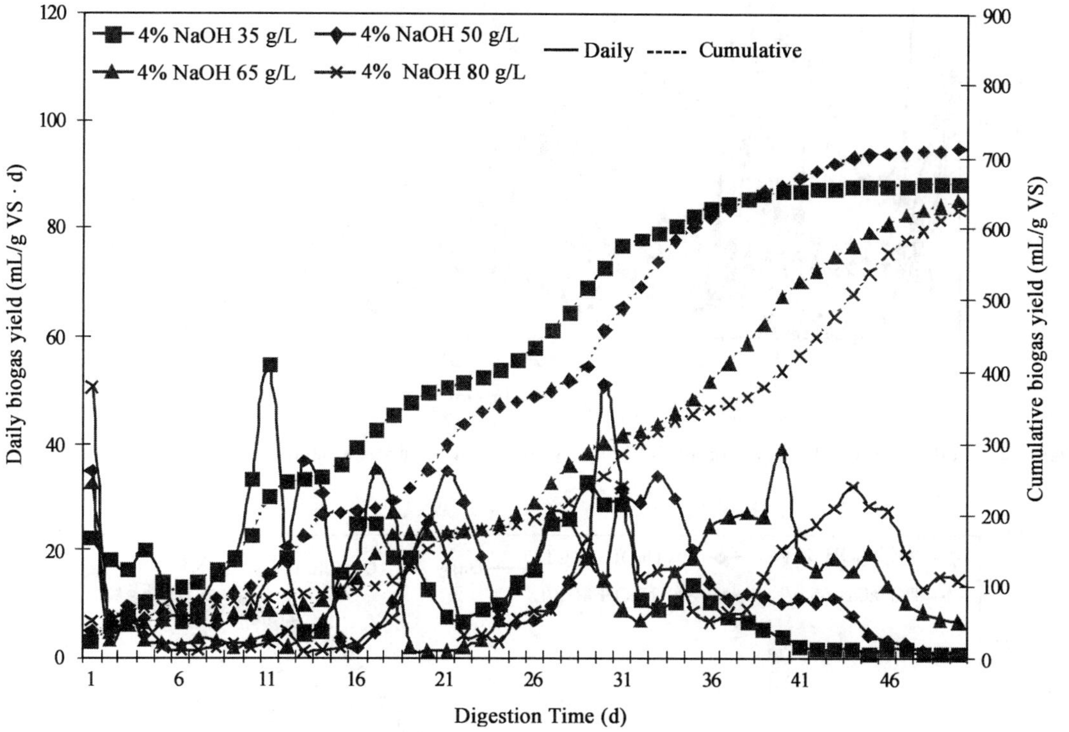

Figure 1 Daily biogas yields and cumulative biogas yields of the rice straw samples pretreated with 4% NaOH

For the straw samples pretreated with 6% NaOH (*Figure 2*), the trial of organic loading rate (OLR) at 35 g TS/L showed quick biogas production 70 mL/(g VS · d) in the 3rd day. Other trials showed biogas production peaks at least 10 days later. Ultimately the cumulative biogas yield of all four samples arrived at 680-850 mL/g VS, indicating that the OLR of 35 g TS/L and 50 g TS/L are the optimal choice for rice straw digestion. Compared to the cumulative biogas yields of 4% NaOH pretreated samples, 6% NaOH pretreated samples showed significantly better biogas production potential.

For the straw samples pretreated with 8% NaOH (*Figure 3*), the trial of organic loading rate (OLR) at 80 g TS/L showed quick biogas production 45 mL/(g VS · d) in the 2nd day. Other trials showed biogas production peaks at least 10 days later. The cumulative biogas yields at higher OLR were higher than the results of straw samples pretreated with 4% or 6% NaOH. It may represent that higher portion of NaOH are good for the digestion at high OLRs.

For the straw samples pretreated with 10% NaOH (*Figure 4*), it can be found that the cumulative biogas yields of all four trials are similar as the results from 8% NaOH pretreated samples. From the cost-effective standpoint, the NaOH portion of 10% may not be preferable.

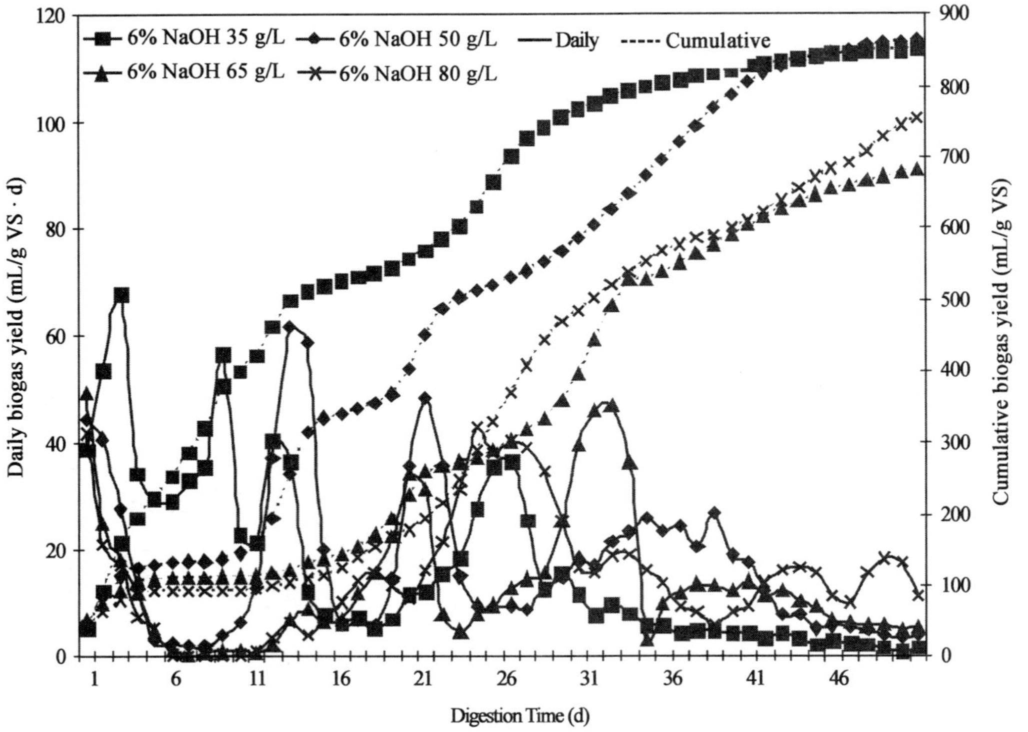

Figure 2 Daily biogas yields and cumulative biogas yields of the rice straw samples pretreated with 6% NaOH

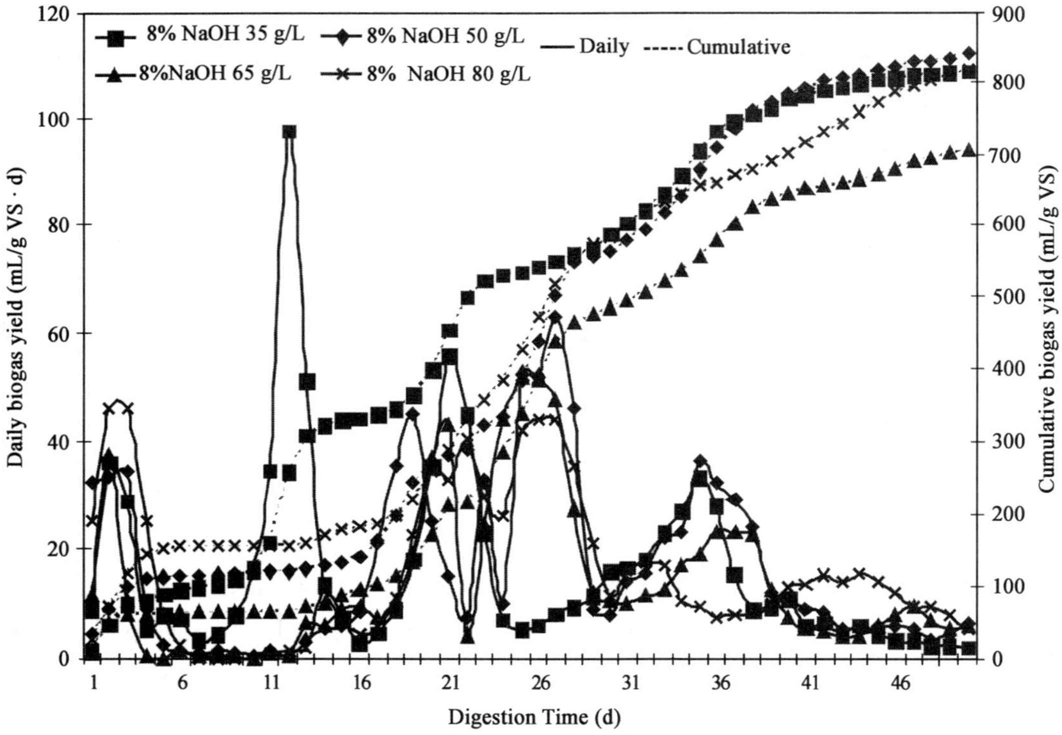

Figure 3 Daily biogas yields and cumulative biogas yields of the rice straw samples pretreated with 8% NaOH

The daily biogas yields and cumulative biogas yields of untreated straw samples showed comparatively low results, indicating that the alkaline pretreatment on the rice straw samples did help enhancing the biogas production potential.

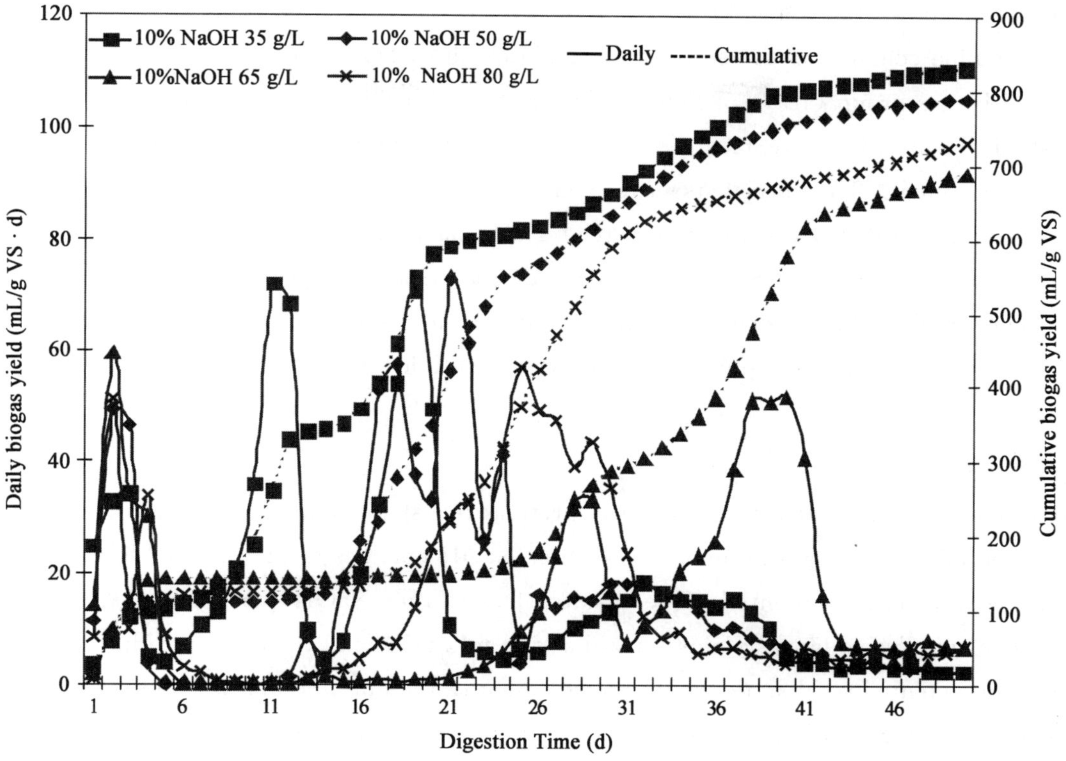

Figure 4 Daily biogas yields and cumulative biogas yields of the rice straw samples pretreated with 10% NaOH

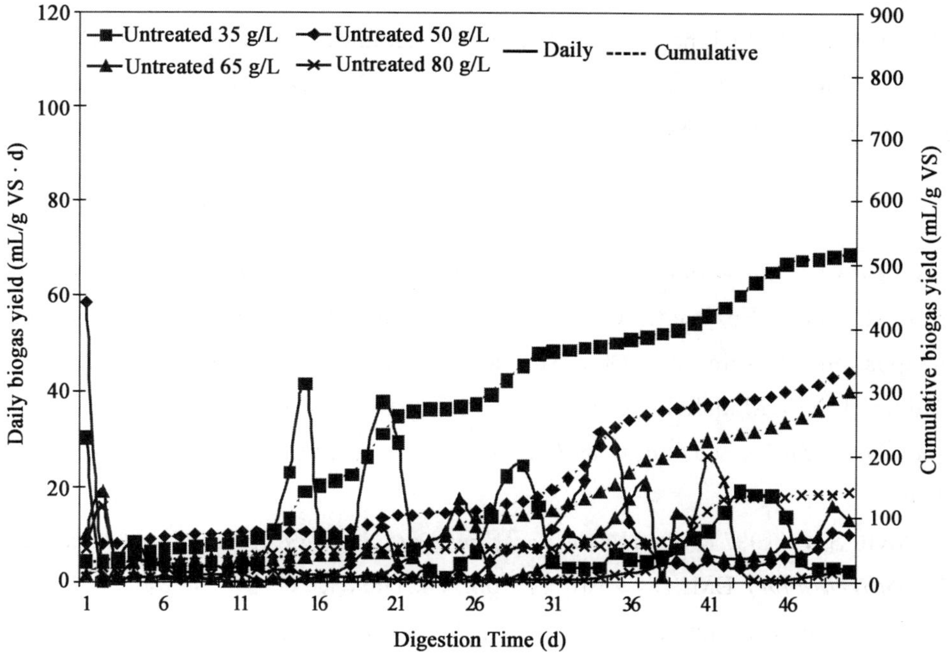

Figure 5 Daily biogas yields and cumulative biogas yields of the untreated rice straw samples

3.3 Biogas composition

The biogas compositions of all trials were analyzed every 3 days. From the beginning of experiments, concentrations of CO_2 were decreasing fast. On the first biogas production peak, the CO_2 counts for at least 70% of biogas. But after 10 days of digestion, the methane contents increased to 55%-70% and the

CO_2 contents went down to around 40%.

3.4 Introducing the commercial scale applications of rice straw digestion

The industrial biogas group (IBG) of Beijing University of Chemical Technology was established in year 2001. Most of the members are faculties and postgraduate students in the environmental science and engineering department. The specialists have plenty experiences on the designing of anaerobic digesters and biogas plants. Until now, seven commercial scale anaerobic digestion plants were built under the guidance from IBG in provinces Shandong, Heilongjiang, Guangxi, and Beijing, providing fuel gas to 100-500 farmer families. The feedstock of these plants include NaOH pretreated corn stover, rice straw, dairy manure, etc. Raw biogas is purified to remove the sulfide and water.

4 Conclusions

The study discussed in this paper used NaOH to pretreat rice straw samples which were subsequently used as feedstock for anaerobic biogas production. Various NaOH portions were evaluated to get a cost-effective way to enhance the biogas potential. The results showed that NaOH portion of 6% was the optimal parameter. The cumulative biogas yields arrived at 850 mL/g VS and the digestion time could be reduced to 35 days. The patented solid-state NaOH pretreatment can double the biogas potential of rice straw and could be a cost-effective solution for commercial scale anaerobic digestion plants which use crop straw and energy crops as feedstock.

Acknowledgements

This study was funded by China National High Technology Research and Development Program 2008AA062401/02. The authors would like to thank Mr. LUO Qingming, Mr. GAO Zhijian and Mrs. YANG Dongyan for sharing their experiences.

References

McMillan J. D., (1994): *Conversion of hemicellulose hydrolysates to ethanol*. Himmel M. E., Baker J. O., Overend R., eds. Enzymatic Conversion of Biomass for Fuels Production, American Chemical Society Symposium. Washington, DC, American Chemical Society, pp. 411-437.

Cullis I. F., Saddler J. N., Mansfield S. D., (2004): *Effect of initial moisture content and chip size on the bioconversion efficiency of softwood lignocellulosics*. Biotechnology and Bioengineering, 85 (4), pp. 413-421.

Ingram G. C., Doyle S., (1997): *Dual role for fimbriata in regulating floral homeotic genes and cell division in Antirrhinum*, EMBO J, 16(21), pp. 6521-6534.

Grohmann K., Torget R., Himmel M., (1986): *Dilute acid pretreatment of biomass at high solids concentrations*. Aspen Bibliography, pp. 135-151.

Brown B., Lee K. L., Nesic S., (2003): *Corrosion in Multiphase Flow Containing Small Amounts of* H_2S. CORROSION, pp. 125-131.

Ingram L. O., Lai X., Moniruzzaman M., Wood B. E., York S. W., (1997): *Fuel Ethanol Production from Lignocellulose Using Genetically Engineered Bacteria*. Fuels and Chemicals from Biomass, Chapter 3, pp. 57-73.

Tarkow, H. , Feist, W. C. (1969): *A mechanism for improving digestibility of lignocellulosic materials with dilute alkali and liquid ammonia*. Robert F. Gould (Ed.) Cellulases and their applications, Advances in Chemistry Series, No. 95, Amer. Chem. Soc. , Washington, D. C.

Varga, E. , Szengyel Z. , (2002): *Chemical pretreatments of corn stover for enhancing enzymatic digestibility*. Applied Biochemistry and Biotechnology, 98-100(1): 73-87.

HYDROGEN AND METHANE PRODUCTION FROM CORN STRAW VIA A TWO-STAGE ANAEROBIC FERMENTATION

YANG Zhiman, MENG Xianwu, XU Xiaohui, GUO Rongbo

Qingdao Institute of BioEnergy and Bioprocess Technology, Chinese Academy of Sciences

Contact: Prof. GUO Rongbo, Qingdao Institute of BioEnergy and Bioprocess Technology, Chinese Academy of Sciences, No. 189 Songling Road, Laoshan District, Qingdao 266101, PR China, Tel.: +86-532-8066-2708, E-Mail: guorb@qibebt.ac.cn

EXECUTIVE SUMMARY

The goal of this paper was to develop a two-stage process to produce hydrogen and methane from corn straw. The biogas production and biodegradability efficiency of straw were compared between one-and two-stage fermentation processes. The two-stage process generated the low level of 7.5 mL H_2/g volatile solid (VS) due to the existence of hydrogen-consuming bacteria. The methane yield of 290 mL CH_4/g VS was 22% higher than the level in the one-stage process. Additionally, biodegradability efficiency increased from 27% in the one-stage process to 42.9% in the two-stage process, which can be contributed to the fact that the improved hydrolysis and biodegradability of corn straw occurred in the hydrogen production stage.

Key words: Two-stage anaerobic fermentation; corn straw; hydrogen; methane

玉米秸秆两相发酵制备氢气和甲烷研究

杨智满，孟宪武，许晓晖，郭荣波*

（中国科学院青岛生物能源与过程研究所，青岛 266101）

【摘要】本研究的目的是建立玉米秸秆两相厌氧发酵联产氢气和甲烷技术，通过以沼气产量和秸秆降解效率为技术指标考察两相发酵和单相发酵技术的优劣。在两相发酵过程中，由于发酵过程中存在氢营养型细菌，从而导致第一相产氢阶段只产生了少量的氢气（7.5 mL H_2/g VS）。第二相发酵的甲烷产量是 290 mL CH_4/g VS，比单相发酵的甲烷产量提高了 22%。另外，由于第一相产氢阶段显著提高了秸秆的水解效率和生物降解性，从而引起秸秆的降解效率从单相发酵的 27%提高到两相发酵的 42.9%。

【关键词】两相厌氧发酵；玉米秸秆；氢；甲烷

1 引言

1.1 研究背景

中国是世界农业大国，农作物秸秆年总产量超过 7×10^8 t，其中玉米秸秆产量占 35%左右。玉米秸秆除少量作为饲料、部分还田和生活用能之外，绝大部分被焚烧或者处于弃之无用状态，造成了极大的资源浪费和环境污染问题。因而，研究人员急需开发一种环境友好的秸秆利用途径。玉米秸秆经厌氧发酵后可以生

* 通信作者：郭荣波，男，博士，研究员，博士生导师，中国科学院青岛生物能源与过程研究所，山东省青岛市崂山区松岭路 189 号，guorb@qibebt.ac.cn，0532-80662708

产清洁的可再生能源——沼气。

厌氧发酵是多个种属的厌氧细菌参与的复杂过程，主要包括四个阶段：水解、产酸发酵、产氢产乙酸和产甲烷阶段[1]。传统的厌氧发酵过程主要是在一个反应器内完成上述的四个阶段，通常检测不到氢气，以甲烷和 CO_2 为终产物[2]。然而，厌氧发酵的产甲烷阶段的最佳 pH 是在 6.8～7.2，而水解和产酸发酵阶段的最佳 pH 在 5.5～6.5[1]。另外，单相厌氧发酵还具有生物质水解速率慢、系统启动时间长、发酵周期长等缺点[3]。因此，将厌氧发酵过程中的产酸发酵和产甲烷两个阶段分开，并分别对这两个过程进行优化，将是一个理想的能量回收途径，同时可实现秸秆生物质的高效利用。两相厌氧发酵联产氢气和甲烷技术(简称为两相厌氧发酵)已经广泛用于处理各种类型的废弃生物质[4,5]。两相厌氧发酵主要由产酸发酵和产甲烷两个阶段组成。在产酸发酵阶段，有机物多聚体(例如：碳水化合物、蛋白质、脂等)被转化成为有机酸和氢气，然后产生的有机酸在产甲烷阶段被转化成为甲烷[6]。与传统的单相厌氧发酵相比，两相厌氧发酵具有很多优势，例如：提高甲烷产量[4]，提高底物的降解率[7]和提高厌氧发酵过程的稳定性[1]。然而，到目前为止，秸秆厌氧发酵产沼气的技术主要是单相厌氧发酵工艺，而有关秸秆两相厌氧发酵制备生物气的技术则很少涉及。

玉米秸秆中纤维素和半纤维素占 70%，木质素占 15%～20%。秸秆生物质的复杂结构组成，导致很难被降解。然而，由于传统的预处理方法(物理、化学处理，或者二者组合)虽然提高了秸秆生物质的糖化率，但是同时又产生了大量的对菌体有毒性的物质，导致预处理后秸秆的发酵受抑制[8]。此外，预处理过程中需要大量的化学试剂，无疑增加了预处理成本，引起环境污染。因而，开发一种低成本适于秸秆规模化厌氧发酵产沼气技术具有重要的现实意义。本研究以粉碎后的秸秆为研究对象，通过建立两相厌氧发酵技术，来规避物化处理技术的弊端。

1.2 研究目的

本研究以经粉碎处理后的玉米秸秆生物质为研究对象，针对玉米秸秆发酵制备沼气水解速度慢、产气效率低等瓶颈性问题，运用两相厌氧发酵技术，通过第一相的秸秆水解酸化产氢技术，提高玉米秸秆的溶解度和可降解性，揭示秸秆生物质水解、酸化和产甲烷过程中的物质转化规律，寻求发酵法生产甲烷的最大能力，建立适于秸秆生物质两相厌氧发酵制备甲烷的工艺，从而使整个系统的能源转化率都获得突破性提高。

1.3 研究内容

本研究以取自秸秆厌氧发酵过程中的发酵液为接种物，以粉碎后的玉米秸秆为原料，研究两相厌氧发酵制备沼气的过程特性。通过分析累积产气量、秸秆溶解效率、中间代谢产物分布和秸秆的降解效率，比较两相厌氧发酵和传统单相厌氧发酵工艺的优缺点。

2 研究方法

2.1 秸秆

实验所用玉米秸秆取自青岛市崂山区农村地区。玉米秸秆主要成分的质量分数是：纤维素 35.3%、总纤维素 63.4%、木质素 13.3%、水分 9.3%、灰分 10.4%、总固体(Total solid, TS)90.7%、挥发性固体(Volatile solid, VS)80.3%。使用前用秸秆粉碎机粉碎处理，其粒径分布如表 1 所示。

Table 1 Distribution of particle size of crushed corn straw

表 1 玉米秸秆粉碎后的粒径分布

粒径尺寸(目)	>80	80>粒径>60	60>粒径>40	<40
分布(%)	19.49	17.55	38.19	24.77

2.2 接种物

接种物取自本团队中试规模(连续稳定运行半年以上)的秸秆沼气单相发酵过程中的发酵液。接种物的总固体含量、挥发性固体含量和 pH 分别为 20 g/L、13 g/L 和 pH 7.5。为了灭活甲烷菌，两相发酵第一相的接种物先经热处理，条件为 95℃，30 min[3]。两相发酵第二相和单相发酵的接种物未经任何处理。

2.3 实验步骤

发酵方式为批次发酵，批次试验在 300 mL 的厌氧瓶中进行，其工作体积为 100 mL。每个试验组做三个重复。加入接种物和发酵底物后，用高纯氮气排掉厌氧瓶中残留的氧气，再用丁基橡胶塞封口，最后放到摇床上振荡发酵(37℃ ，180 r)。

两相发酵中的第一相，厌氧瓶中加入 40 mL 热处理后的接种物和 2.0 g 粉碎的秸秆，用 10 N 的 HCl 将发酵液的 pH 调到 6.0，工作体积 100 mL。两相发酵中的第二相，厌氧瓶中加入 40 mL 未处理的接种物和 60 mL 第一相发酵液，初始 pH 调到 7.0。单相发酵作为对照组，厌氧瓶中加入 40 mL 未处理的接种物，1.2 g 秸秆，用自来水调至 100 mL。初始 pH 用 10 N NaOH 调至 7.0。

2.4 分析方法

TS 和 VS 的测定参照美国《水和废水标准检验方法》进行[3]。纤维素、综纤维素、酸不溶木质素含量的测定参照 Van-soest 的方法进行[10]。SCOD (Soluble chemical oxygen demand)的测定：发酵液或者预处理液经过 10 000 rpm 5 min 离心后取上清。然后用兰州连华环保科技有限公司的 5B-3(B) COD 测定仪测定 SCOD。发酵液的 pH 值使用精密 pH 计测定(Starious，德国)。

采用排水法测定气体体积，气相色谱(SP6890，山东鲁南瑞虹化工仪器有限公司)测定产生的氢气和甲烷，色谱柱为 2 m 长的不锈钢管，5A 分子筛为填料，氮气作为载气。进样口、柱温和 TCD(热导检测器)温度分别为 100℃、50℃ 和 100℃。挥发性脂肪酸(VFA)的检测：使用美国瓦里安 450GC 气相色谱测定，色谱柱为 30 m×0.25 mm×0.25 μm 毛细管柱，氮气作为载气。进样口和氢火焰检测器温度分别为 220℃ 和 240℃。柱温采用程序升温，首先在 80℃保持 1 min，然后以 25℃/min 速度上升到 120℃，保持 0 min，然后再以 30℃/min 速度上升到 135℃保持 0 min，再以 15℃/min 上升到 180℃保持 1 min。

3 结果与讨论

3.1 两相发酵第一相发酵特点

3.1.1 氢气生产

两相发酵第一相累积产氢量的变化如图 1 所示。接种物预处理和未处理实验组，氢气产量变化的趋势相似，即在 1 d 后，氢气产量达到最大；然后，随着发酵时间的延长，氢气产量表现出下降趋势，提示发酵过程中存在氢气消耗现象。氢气消耗现象的出现与发酵液中同型乙酸菌和丙酸菌的存在密切相关，其中丙酸菌利用 H_2 和碳水化合物产生丙酸、乙酸和少量的戊酸，而同型乙酸菌利用 CO_2 和 H_2 生成乙酸[11,12]。热处理后的接种物发酵秸秆后，仍存在氢气消耗现象，提示丙酸菌或者/和同型乙酸菌可能共同参与了秸秆发酵过程。此外，两个实验组的氢气产量不存在显著区别，提示接种物热处理与否对氢气产量影响不大。然而，

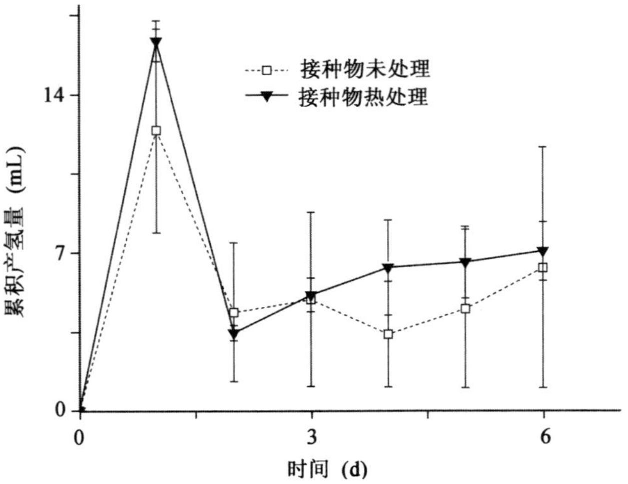

Figure 1 Profile of hydrogen production in the hydrogen stage of two-stage process

图 1 两相发酵第一相累积产氢量的变化

Wang 等[13]和 Mu 等[14]通过比较厌氧消化污泥接种物预处理方法对葡萄糖氢气产量的影响时发现，热处理是最有效的方式。我们的结果与前人的研究结果不同之处可能是接种物和底物不同造成的。

3.1.2 秸秆溶解率

SCOD 是表征生物质降解效果的重要指标。在两相发酵过程中，第一相中秸秆溶解程度的变化过程，如图 2 所示。接种物预处理和未处理实验组，SCOD 浓度的变化趋势相似，即：随着发酵时间的推移，SCOD 浓度逐渐增加；第 6 d 时，SCOD 浓度缓慢增加。值得注意的是，接种物热处理组的秸秆溶解率显著高于接种物未处理实验组；发酵 3 d 后，两组 SCOD 浓度的差异极其显著。这些结果表明热处理的接种物更有利于秸秆降解，同时也表明第一相发酵可明显促进底物的水解，这与 Pakarinen 等[15]的研究结果一致。

3.1.3 VFA 的分布

在发酵产氢过程中，氢气的产生总是伴随着挥发性脂肪酸（VFA）的生成。VFA 的生成反映了产氢微生物代谢途径的变化。因此，VFA 的变化分析对于优化产氢条件具有重要意义。秸秆第一相发酵过程中，发酵液中的 VFA 主要由乙酸、丙酸和丁酸组成。VFA 分布的变化如图 3 所示。接种物预处理和未处理实验组，乙酸和丙酸的产量和氢气产量变化的趋势相反，即：随着第一相发酵的进行，呈现逐出逐渐增加的趋势。提示丙酸菌或/和同型乙酸菌可能共同参与了秸秆发酵过程，热处理并不能有效地灭活丙酸菌和同型乙酸菌。在碳水化合物发酵中，Hussy 等[16]认为丁酸的产量和氢气生产呈正相关。而本研究中随着丁酸产量的逐渐增加，氢气产量下降，进一步证实了丙酸菌或/和同型乙酸菌可能共同参与了秸秆发酵过程。另外，从图 2 和图 3 可知，对于接种物预处理和未处理实验组而言，总 VFA 浓度的变化趋势与 SCOD 浓度变化一致，表明 SCOD 浓度的增加与总 VFA 浓度相关。虽然发酵液中可能存在可溶性碳水化合物，但是由于同步糖化发酵作用，发酵液中可溶性碳水化合物浓度相当低[17,18]。因此，SCOD 主要取决于发酵液中的总 VFA 产量。

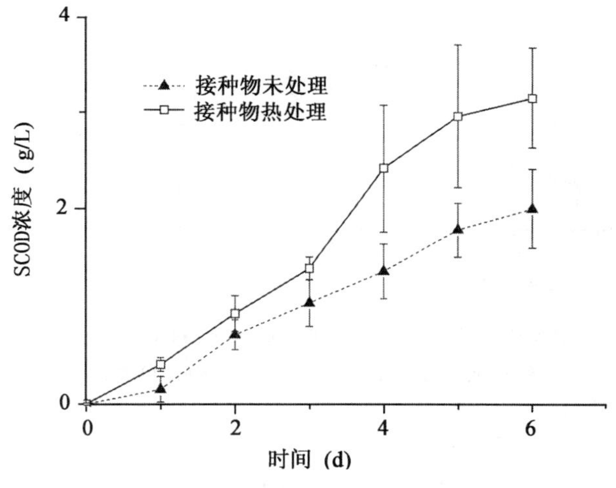

Figure 2 Profile of SCOD in the hydrogen stage of two-stage process

图 2 两相发酵第一相发酵液中 SCOD 浓度的变化

3.2 两相发酵第二相发酵特点

从图 4 可知，两相发酵中的第二相发酵产甲烷的效率显著高于单相发酵。第二相发酵的甲烷产量和甲烷生成速率分别是 290 mL CH_4/g VS 和 11.3 mL CH_4/d，产生的生物气中甲烷含量在 50%左右（V/V），VS 降解率为 42.9%。而当粉碎后的秸秆直接进行单相发酵产甲烷时，甲烷产量和甲烷产生速率则只有 237 mL CH_4/g VS 和 7.4 mL CH_4/d，甲烷含量在 50%左右，VS 降解率在 27%左右。因此，第二相发酵显著提高了甲烷产量和甲烷产生速率，分别提高了 22%和 53%，并显著提高了秸秆的降解效率。此外，就两相发酵第二相和单相发酵而言，发酵 10 d 后，基本上检测不到 VFA 的存在，表明秸秆发酵过程中更多的 VFA 被转化成为甲烷。本研究中两相厌氧发酵时的秸秆甲烷产量为 290 mL CH_4/g VS，高于 Zhong 等报道的单相发酵时的 280 mL 沼气/g TS(NaOH 预处理)和 240 mL 沼气/g TS(生物预处理)[9],显著高于 Lu 等报道的两相发酵的 163 mL CH_4/g TS[20]和未处理的 211 mL CH_4/g VS[21],另外,也显著高于 Zheng 等报道单相发酵的 182～197 mL CH_4/g TS(NaOH 预处理)[22]。上述比较性的结果表明本研究建立的两相厌氧发酵技术优于传统的单相厌氧发酵技术。据报道,当使用酿酒厂、厨房和橄榄厂等的废弃物,以及青贮牧草[7,15,23,24]作为发酵物料时,两相发酵也优于单相发酵。

上述研究结果表明本研究建立的秸秆两相厌氧发酵在技术上具有可行性。两相厌氧发酵获得了较高的甲烷产量,是由于第一相的产氢阶段显著改善了秸秆的水解过程和生物降解性[15]。本研究中,在一相产氢阶段的 pH 6.0 是秸秆水解和酸化的最佳 pH。pH 6.0 有利于产酸产氢菌的生长和繁殖,因而,产氢菌群间接提高了秸秆发酵的甲烷产量[6]。

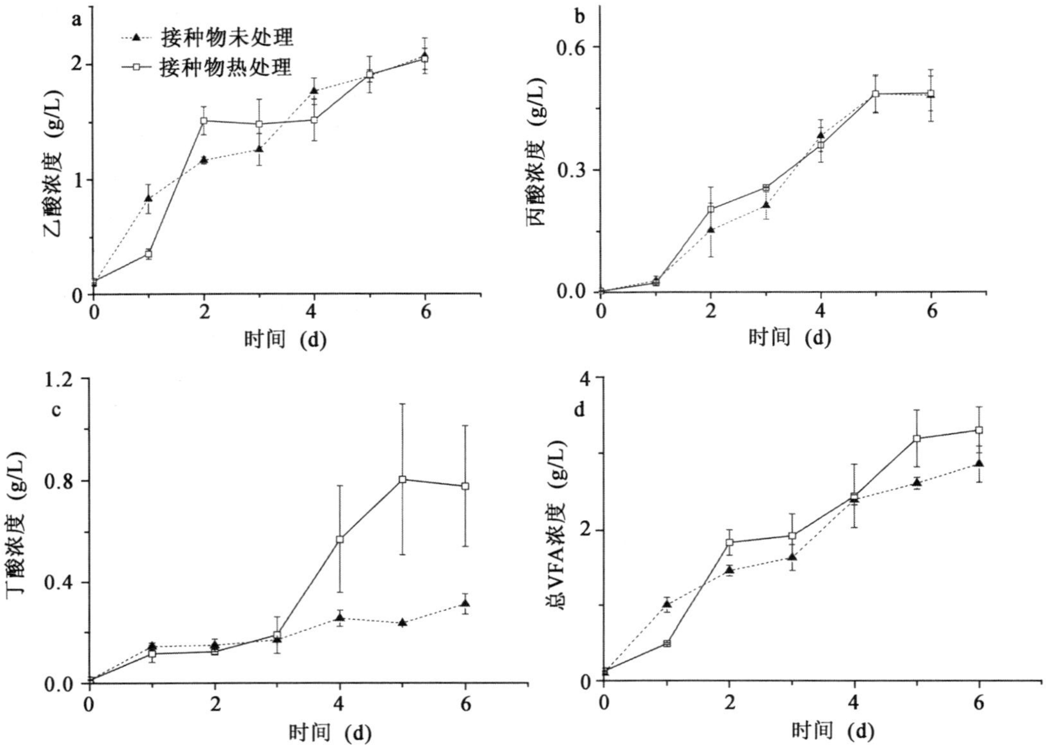

Figure 3 Profile of VFA in the hydrogen stage of two-stage process

图 3 产氢相两相发酵第一相发酵液中 VFA 分布的变化

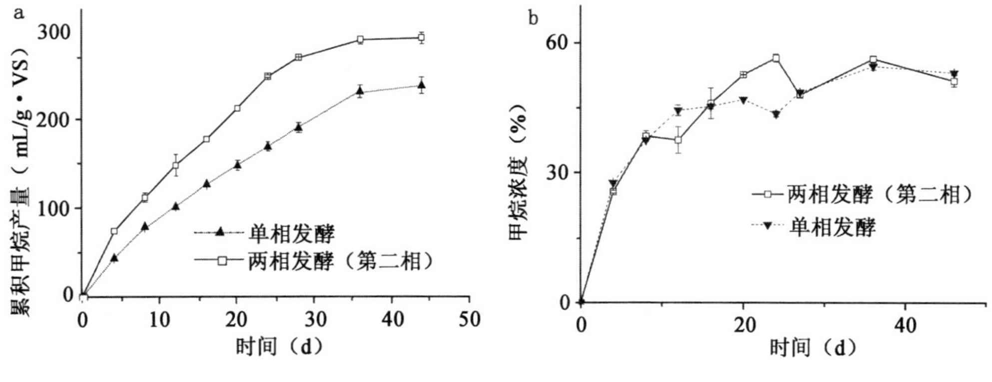

Figure 4 Profiles of methane production (a) and methane content (b) in one-stage process and methane stage of two-stage process

图 4 两相发酵(第二相)和单相发酵累积甲烷产量(a)和甲烷浓度(b)变化

4 结论

本研究成功建立了秸秆两相厌氧发酵联产氢气和甲烷技术，并显著提高了甲烷产量。第一相产氢阶段的发酵类型为丁酸型发酵，氢气产量是 7.5 mL H_2/g VS。接种物预处理显著提高了秸秆的水解效率。第二相甲烷产量是 290 mL CH_4/g VS，比单相发酵的甲烷产量高 22%；秸秆的降解效率从单相发酵的 27%提高到两相发酵的 42.9%，上述结果表明秸秆两相发酵优于单相发酵。

项目及基金支持

感谢以下基金对本研究的支持：十一五国家科技支撑计划重大项目“特大型养牛场热电肥联产沼气工程”(2008BADC4B06)；科技部 863 计划课题“多原料混合发酵”(2010AA101604)；山东省科技发展计划项目“黄河三角洲秸秆沼气干式发酵”(2008GG20007003)。

参考文献

[1] Ward A. J., Hobbs P. J., Holliman P. J., Jones D. L. Optimisation of the anaerobic digestion of agricultural resources [J]. Bioresource Technology, 2008, 99:7928-7940.

[2] Pang Y. Z., Liu Y. P., Li X. J., Wang K. S., Yuan H. R. Improving Biodegradability and Biogas Production of Corn Stover through Sodium Hydroxide Solid State Pretreatment [J]. Energy & Fuels, 2008, 22:2761-2766.

[3] Yang Z., Guo R., Fan X., Luo S. Hydrogen and methane production from lipid-extracted microalgal biomass residues [J]. International Journal of Hydrogen Energy, 2011, 36:3465-3470.

[4] Wang Y., Zhang Y., Meng L., Wang J., Zhang W. Hydrogen-methane production from swine manure: Effect of pretreatment and VFAs accumulation on gas yield [J]. Biomass and Bioenergy, 2009, 33:1131-1138.

[5] Wang X., Zhao Y. C. A bench scale study of fermentative hydrogen and methane production from food waste in integrated two-stage process [J]. International Journal of Hydrogen Energy, 2009, 34:245-254.

[6] Ueno Y., Fukui H., Goto M. Operation of a Two-Stage Fermentation Process Producing Hydrogen and Methane from Organic Waste [J]. Environmental Science & Technology, 2007, 41:1413-1419.

[7] Liu D., Liu D., Zeng R. J., Angelidaki I. Hydrogen and methane production from household solid waste in the two-stage fermentation process [J]. Water Research, 2006, 40:2230-2236.

[8] BIinod P., Sindhu R., Singhania R. R., Vikrams S., Devi L., Nagalakshmi S., Kurien N., Sukumaran R. K., Pandey A. Bioethanol production from rice straw: An overview [J]. Bioresource Technology, 2010, 101:4767-4774.

[9] Apha. Standard methods for the examination of waste and wastewater [M]. American Public Health Association, Washington, DC, USA, 1999.

[10] Van-Sest P., Rrobertson J. Methods for dietary fibre, neutral detergent fibre, and nonstrach polysaccarides in relation to animal nutrition [J]. Dair Sei, 1991, 74:3579-3583.

[11] Selembo P. A., Perez J. M., Llloyd W. A., Logan B. E. Enhanced hydrogen and 1,3-propanediol production from glycerol by fermentation using mixed cultures [J]. Biotechnology and Bioengineering, 2009, 104:1098-1106.

[12] Hussy I., Hawkes F. R., Dinsdale R., Hawkes D. L. Continuous fermentative hydrogen production from a wheat starch co-product by mixed microflora [J]. Biotechnology and Bioengineering, 2003,

84:619-626.

[13] Wang J., Wan W. Comparison of different pretreatment methods for enriching hydrogen-producing bacteria from digested sludge [J]. International Journal of Hydrogen Energy, 2008, 33:2934-2941.

[14] Mu Y., Yu H. Q., Wang G. Evaluation of three methods for enriching H_2-producing cultures from anaerobic sludge [J]. Enzyme and Microbial Technology, 2007, 40:947-953.

[15] Pakarinen O. M., Tahti H. P., Rintala J. A. One-stage H_2 and CH_4 and two-stage H_2+CH_4 production from grass silage and from solid and liquid fractions of NaOH pre-treated grass silage [J]. Biomass and Bioenergy, 2009, 33:1419-1427.

[16] Hussy I., Hawkes F. R., Dnsdale R., Hawkes D. L. Continuous fermentative hydrogen production from sucrose and sugarbeet [J]. International Journal of Hydrogen Energy, 2005, 30:471-483.

[17] Olofsson K., Ertilson M., Liden G. A short review on SSF-an interesting process option for ethanol production from lignocellulosic feedstocks [J]. Biotechnology for Biofuels, 2008, 1:1-14.

[18] Lay J. J. Biohydrogen generation by mesophilic anaerobic fermentation of microcrystalline cellulose [J]. Biotechnology and Bioengineering, 2001, 74:280-287.

[19] Zhong W., Zhang Z., Qiao W., Fu P., Liu M. Comparison of chemical and biological pretreatment of corn straw for biogas production by anaerobic digestion [J]. Renewable Energy, 2011, 36:1875-1879.

[20] Lu Y., Lai Q., Zhang C., Zhao H., Ma K., Zhao X., Chen H., Liu D., Xing X. H. Characteristics of hydrogen and methane production from cornstalks by an augmented two-or three-stage anaerobic fermentation process [J]. Bioresource Technology, 2009, 100:2889-2895.

[21] Raposo F., Banks C. J., Siegert I., Heavean S., Boria A R. Influence of inoculum to substrate ratio on the biochemical methane potential of maize in batch tests [J]. Process Biochemistry, 2006, 41:1444-1450.

[22] Zheng M., Li X., Li L., Yang X., He Y. Enhancing anaerobic biogasification of corn stover through wet state NaOH pretreatment [J]. Bioresource Technology, 2009, 100:5140-5145.

[23] Blonskaja V. Menert A., Vilu R. Use of two-stage anaerobic treatment for distillery waste [J]. Advances in Environmental Research, 2003, 7:671-678.

[24] Koutrouli E. C., Kalfas H., Gavala H. N., Skiadas I. V., Stamatelatous K., Lyberatos G. Hydrogen and methane production through two-stage mesophilic anaerobic digestion of olive pulp [J]. Bioresource Technology, 2009, 100:3718-3723.

利用碳酸钠调节改善餐厨垃圾厌氧消化性能

S. Y. Xu, J. W. C. Wong

（香港浸会大学，香港）

【摘要】餐厨垃圾是城市固体废弃物（MSW）最主要的组成部分之一，使用合适的方法替代填埋处理餐厨垃圾能延长垃圾填埋场的使用寿命。厌氧消化能产生含甲烷约60%的沼气，是从餐厨垃圾回收能源和资源的有效途径；生成的沼渣沼液还可以作为土壤改良剂。餐厨垃圾厌氧发酵的主要问题是水解酸化过程中产生的乳酸会导致pH值下降，抑制细菌生长。因此本研究的目的是在餐厨垃圾固液两相发酵（HASL）系统的水解酸化阶段投加碳酸钠控制pH值的效果，尤其是控制乳酸的效果。

本研究采用的改进型固液两相发酵（HASL）系统是由一个滤床（LB）酸化反应器和一个上流式厌氧污泥床（UASB）产甲烷反应器组成，它们的有效容积分别为4.6 L和10.0 L。1.0 kg餐厨垃圾，0.2 kg厌氧污泥和0.1 kg木屑混合后投加到LB反应器内，渗滤液产生并被收集到反应器下端容器内。大约50%的渗滤液加入UASB反应器，剩下50%渗滤液用相同体积的水稀释后直接或者投加碳酸钠将pH值调节到6.0后回流到LB反应器。两个反应器都保持35℃恒温，实验时间持续30 d。在有碳酸钠调节的系统中，渗滤液的pH值在第三天从3.9降低到3.3之后慢慢增加到6.0～7.0并保持稳定直到实验结束。而在对照系统中，渗滤液的pH值逐渐下降，到实验结束降低到2.7。在有碳酸钠调节的系统中，高pH值明显阻止了水解酸化阶段的乳酸生成，渗滤液中乳酸浓度最高达8.1 g/L，占总化学需氧量（COD）的26%；而在参照系统中，渗滤液中的乳酸浓度达到21.5 g/L，占总化学需氧量（COD）的56%。LB反应器中微生物活性通过分析渗滤液中的酶活性监测。与对照系统相比，有碳酸钠调节的系统产生的渗滤液中含有更高的α-淀粉酶、羧甲基纤维素酶（CMCase）和滤纸纤维素酶（FPase）活性，从而显示了更高的生物活性。酶活性的提高可以归结于相对较低的乳酸浓度。在参照系统和碳酸钠调节的系统中产生的渗滤液COD累积分别为197.9 g和350.7 g。通过pH值调节，LB反应器中的餐厨垃圾水解比例明显增加，提高了77%的COD。在产甲烷的UASB反应器中，两组反应器的有机负荷（OLR）相对很低，范围在0.1～1.7 kg COD/(m^3·d)。在有碳酸钠调节的系统中UASB反应器的COD去除率明显更高，比对照系统中甲烷产量增加了62%，表明由于其渗滤液中乳酸比例较低，更有利于甲烷的生成。综上所述，投加碳酸钠调节pH值有效地防止了水解酸化阶段的乳酸生成，增强了LB反应器的生物活性，产生了更多的COD，增加了UASB反应器的甲烷产量。

【关键词】碳酸钠；UASB沼气发酵罐；餐厨垃圾；沼气

SODIUM CARBONATE AMENDMENT ENHANCED THE PERFORMANCE OF ANAEROBIC DIGESTION OF FOOD WASTE*

S. Y. Xu, J. W. C. Wong

Hong Kong Baptist University

Contact: Prof. Jonathan W. C. Wong, Sino-Forest Applied Research Centre for Pearl River Delta Environment, Department of Biology, Hong Kong Baptist University, Hong Kong SAR, P. R. China, Tel: +852-3411-7056, E-mail: jwcwong@hkbu.edu.hk

* This paper was initially presented at the Orbit 2009 CHINA. 本文已发表于Orbit 2009 CHINA。

EXECUTIVE SUMMARY

Food waste is the single-largest component of the municipal solid waste (MSW) and diverting them from landfilling using other appropriate treatment methods would increase life span our landfills. Anaerobic digestion is an energy and resource conservation technology for food waste treatment due to the production of biogas containing appr. 60% methane; and the digestate which can be applied as a soil conditioner to land. The major concern associated with anaerobic digestion of food waste is the generation of lactic acid during hydrolysis-acidogenesis stage resulting in pH based growth inhibition on microbes. Therefore, the objective of the present study was to investigate the effect of sodium carbonate on controlling the pH during the hydrolysis-acidogenesis stage of food waste anaerobic digestion in a modified hybrid anaerobic solid-liquid (HASL) system with the focus on the controlling the generation of lactic acid.

The modified HASL system adopted in this study consisted of a leach bed (LB) acidogenic reactor and an upflow anaerobic sludge blanket (UASB) methanogenic reactor with the working volume of 4.6 L and 10.0 L, respectively. Synthetic food waste, 1.0 kg, was mixed with 0.2 kg of anaerobic sludge and 0.1 kg of saw dust and loaded in the LB reactor. Leaching naturally occurred, and the leachate was collected in the lower chamber of the reactor. About 50% of the LB leachate was fed to UASB, and the remaining 50% was diluted by same volume of water and recycled back to LB reactor either without treatment (control) or after adjusting the pH to 6.0 with sodium carbonate (Na_2CO_3 treatment). Reactors were incubated at 35℃ and the experiment was continued for 30 days. In the Na_2CO_3 treatment, the pH of the leachate decreased from 3.9 to 3.3 on day 3 and increased slowly thereafter to 6.0-7.0 and maintained till the end of the experiment. While in the control treatment, the pH of leachate gradually declined until the end of the experiment to 2.7. Higher pH in Na_2CO_3 treatment significantly prevented the generation of lactic acid during the hydrolysis and acidogenesis. Maximum concentration of lactic acid in the leachate of Na_2CO_3 treatment was 8.1 g/L and constituted about 26% of total chemical oxygen demand (COD); while in the control, lactic acid was accounting for 56% of total COD and reaching a maximum of 21.5 g/L. The microbial activities of LB reactors were monitored by analyzing the enzyme activities in the leachate. Compared to the control treatment, leachate from Na_2CO_3 treatment exhibited higher activities of α-amylase, carboxyl methyl cellulose (CMCase) and filter paper cellulose (FPase) indicating higher microbial activity. This increase in enzyme activities may be attributed to the comparatively lower concentration of lactic acid. The cumulative COD of leachates in control and Na_2CO_3 treatments were 197.9 g and 350.7 g, respectively. Obviously, the rate of hydrolysis in LB reactors was increased by pH adjustment as indicated by 77% increase in total COD production. In the methanogenic UASB reactors, the OLRs were in relatively low levels, ranging from 0.1-1.7 kg COD/(m^3 · d) for both treatments. The COD removal efficiency of leachate from Na_2CO_3 treatment was significantly higher than that from the control treatment in methanogenic reactor. Correspondingly, the methane yield in this treatment was 62% higher than that in the control treatment, implying feedstock with lower concentration of lactic acid was more favourable to the subsequent methanogenesis. To conclude, pH adjustment with sodium carbonate effectively prevented the generation of lactic acid during the hydrolysis-acidogenesis stage, enhanced the microbial activity in LB reactor resulted in higher COD production and enhanced the methane yield in the UASB reactor.

Key words: Sodium carbonate; UASB biogas reactor; food waste; biogas

1 Introduction

Food waste is the single-largest component of the waste stream by weight, constituting about 30%-55% in different countries. According to Hong Kong Environmental Protection Department (HKEPD), food wastes constituted about 35% of the municipal solid waste (MSW) in 2006. In the past years, the food wastes were mostly disposed of by land filling, causing groundwater contamination through leachates; and global warming through greenhouse gases emission. Anaerobic digestion, a potential waste treatment technology, not only provides a means for pollution prevention, but also allows for energy, biomass and nutrient recovery, converting a disposal problem into a profit centre. Food wastes are reported to produce 200 to 500 L CH_4/kg VS (Forster-Carneiro et al., 2008). Recent studies have examined the use of anaerobic digestion in food waste (Lim et al., 2008; Stabnikova et al., 2008b), but some characteristics of food waste, especially low pH, offer unique challenges to hinder adopting anaerobic digestion for putrescible waste treatment (Lee et al., 1999; Zhang et al., 2005). Hydrolysis, an initial process of anaerobic digestion, is rate-limiting, especially when the substrate is in particulate form (Myint et al., 2007). Therefore, enhancing the hydrolysis efficiency of organic solid wastes is believed to increase the yield of methane gas per unit biomass.

The effectiveness of an anaerobic digestion system depends on the concentration of volatile fatty acids (VFAs). Food waste has been regarded as readily biodegradable because of its high volatile fraction (Appr. 90% of total solids) (Kim et al., 2008). Due to the fast decomposition of the easily degradable organic matters in the food waste, high concentration of VFAs may accumulate, resulting in low pH, which would inhibit the microbial activities (Ying and Yang, 2004). The intense acidification during the food waste acidogenesis is inhibitory to the methanogenesis in a one stage anaerobic digester. Furthermore, as the growth rate of acidogenic and methanogenic bacteria were quite different, physical separation of these two stages enables us to maintain appropriate environment, leading to a high overall reaction rate and biogas yield (Bouallagui et al., 2004). The volumetric gas production can potentially be increased through operating the digesters at a high solid concentration, such as a leach bed system (Weiland, 2003). Therefore, the two-phase hybrid anaerobic solid-liquid (HASL) system is an attractive system due to its simple design, easy process control and low investment cost, and in fact it has a wide application in kitchen wastes, grass, alfalfa silage, and agro-residues (Nordberg et al., 2007; Stabnikova et al., 2008b; Wang et al., 2006).

Among the organic acids produced, lactic acid was reported as an inhibitory substance to both acidogenesis and methanogenesis (Narendranath et al., 2001; Chen et al., 2007; Zhang et al., 2007a; Lü et al., 2008). Kim et al. (2003) found that the lactic acid was the earliest intermediate but as the reaction proceeds, short chain VFAs were produced as final end products with a decrease in lactic acid. Besides, the toxicity of fatty acid includes not only pH based growth inhibition but also anion-specific effects on metabolism that also affect growth (Warnecke and Gill, 2005; Lim et al., 2008; Lü et al., 2008). Thus, controlling the pH during the initial degradation phase is the primary issue to be resolved.

Recycling the leachate within hydrolysis reactor could accelerate degradation of organic solid wastes, owing to more efficient dispersion of inoculums, nutrients and degradation products, as well as improving total extracellular enzyme activities (Cirne et al., 2007; Stabnikova et al., 2008a). Therefore, the modified HASL system with internal leachate recirculation was chosen for food waste treatment. Recycle stream with/without pH adjustment to 6.0 by Na_2CO_3 into leach bed (LB) reactor was compared in this study to investigate the effect of pH adjustment on the hydrolysis of food waste, as well as exploring the way of improving the methane yield of food waste. Besides pH, chemical oxygen demand (COD) and VFAs, the changes in extracellular enzyme activities in LB reactors were investigated.

2 Methodology

2.1 Food waste

Components and their ratio in the synthetic food waste mix are presented in *Table 1*. A representative portion was collected from the well mixed food waste for the analysis of total and volatile solids, which were 39.5% and 97.1%, respectively. Saw dust, at 100 g/kg food waste, was mixed to increase the porosity of LB reactor. Anaerobic microbial sludge, used as inoculums for the acidogenic reactor and methanogenic reactor, were collected from the Shatin wastewater treatment plant, Hong Kong.

Table 1 Components and their contribution in synthetic food waste mix used in the study

Component	TS (%)	VS/TS (%)	Composition in the mix (% of wet weight)
Bread	72.4	97.3	35
Boiled rice	27.5	99.3	25
Cabbage	4.9	88.3	25
Boiled pork	30.2	97.9	15
Sludge	5.2	86.1	—
Saw dust	100	90.2	—

2.2 Experimental set-up

Four identical sets of two-phase HASL systems, each set consisting of an acidogenic LB reactor and a modified upflow anaerobic sludge blanket (UASB) methanogenic reactor with effluent recirculation, were used in the present study. The schematic diagram of the system is presented in *Figure 1*. The total and working volumes of LB reactor were 6.4 L and 4.6 L, respectively. In the upper chamber of LB reactor, a sand bed containing 1 kg of acid washed and oven dried sand was placed above nylon screen to facilitate the filtering of the leachate. The working volume of the UASB reactor is 10 L with inner diameter of 200 mm and a height of 1.4 m. Prior to the experiment, the UASB reactors had been operated for more than two months with synthetic wastewater under OLR of 1.0 kg COD/(m^3 · d). The chemical composition of synthetic wastewater was as follows (mg/L): glucose 30.96 g, sodium acetate 19.23 g, NH_4Cl 4.30 g, K_2HPO_4 1.60 g.

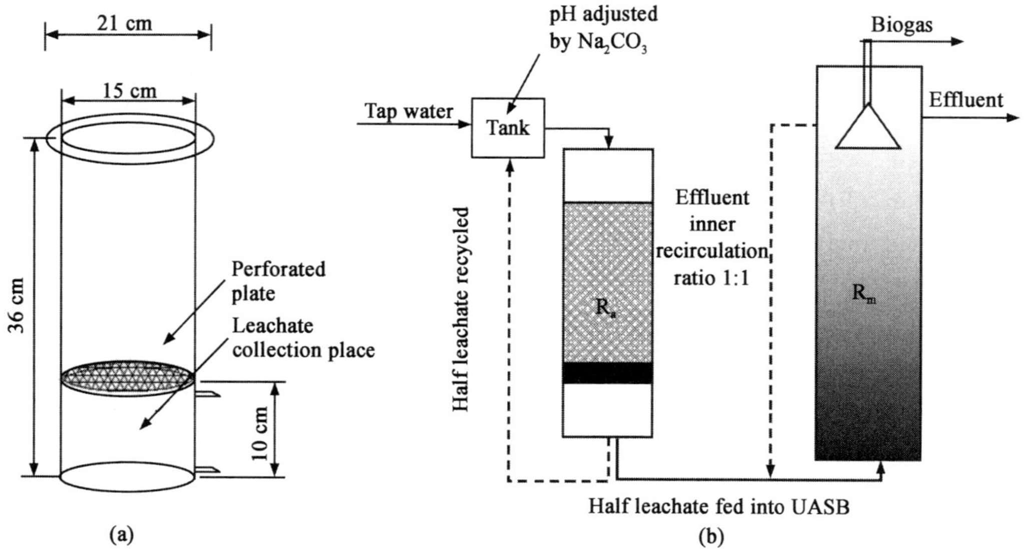

Figure 1 Schematic of the experimental setup: (a) LB reactor, (b) HASL system

Initially, the LB reactor was packed with a mixture of 1.0 kg food waste, 0.2 kg anaerobic sludge and 0.1 kg saw dust, and then 1.0 L of tap water was added. Leaching occurred naturally and the leachate was collected in the lower chamber of the reactors daily. About 50% of the LB leachate was fed to the UASB, and the remaining 50% was diluted by same volume of tap water and then recycled back to LB reactor either without treatment (control) or after adjusting the pH to 6.0 with sodium carbonate (Na_2CO_3 treatment). Each treatment had two replicates. Reactors were incubated at 35℃ and the experiment was conducted for 30 days. Leachate samples collected from LB reactor were analysed for pH, COD, ammonia (NH_4^+-N), total nitrogen (TN), enzyme activities and VFAs concentrations. The concentrations of COD and VFAs in the effluent from UASB reactor were also measured. Methane was collected after absorbing CO_2 in the biogas by 5.0 N NaOH solutions, and the volume of methane was measured by weighing the displaced water.

2.3 Chemical analysis

Total solids (TS) and volatile solids (VS) of the food waste mix and the components were determined by drying at 105℃ for 24 h and igniting at 550℃ for 16 h, respectively. The pH of the leachate was measured directly with a pH electrode (Orion 920, Thermo). The concentrations of COD, NH_4^+-N and TN in leachate were determined according to the Standard Methods (APHA, 2005). For the analysis of VFAs, leachate samples were filtered through 0.45 μm cellulose acetate membrane and analyzed with HPLC (Agilent 1100, USA) with acetonitrile and 0.02 M KH_2PO_4 as mobile phase. A sample volume of 20 μL was separated using RPC18 column and the eluents were monitored at 215 nm with a diode array detector. Concentrations were determined using a standard curve obtained by injecting standard solutions of lactic, acetic, propionic, butyric, iso-butyric, valeric and iso-valeric acids. The sum of these seven fatty acids are reported as the concentration of total fatty acids (TFA).

The α-amylase was assayed using the Bernfeld method (1955). The assay system contained 0.5 mL of enzyme solution and 0.5 mL of 1.0% soluble starch solution (pH 6.9 in phosphate buffer, 0.05 mol/L), incubating at 37℃ for 90 min. To assay the carboxy methyl cellulose (CMCase), leachate samples were incubated with carboxy methyl cellulose in citrate buffer (pH 4.8, 0.05 mol/L), pre-warmed to the incubation temperature of 50℃ (Nakamura and Kitamura, 1988). For the filter paperase activity (FPase), the samples were incubated with filter paper strips (Ghose, 1987) and the assay procedure was similar to CMCase. The glucose released in the assays of α-amylase, CMCase and FPase, was quantified by spectrophotometer at 540 nm (DR/4000, HACH, USA) using glucose for preparing the standard curve. One unit of enzyme activity was defined as the amount of enzyme that releases 1 μg of glucose per min. Protease activity was assayed by first pre-warming the mixed solution of 1 mL sample with 1 mL of 2% of casein in sodium acetate (pH 7, 0.2 mol/L) for 30 min (40℃). The reaction was then terminated by the addition of 2 mL of 0.4 mol/L trichloroacetic acid solution. After the mixture was centrifuged at 3 000 g, 1 mL supernatant, 5 mL 0.4 mol/L Na_2CO_3 and 1 mL foline-phenol reagent were added (Lowry et al., 1951). The absorbance of the mixture was read at 660 nm. The activity of protease was expressed as the amount of enzyme that releases 1 μg of tyrosine per min.

3 Results and discussion

3.1 Hydrolysis and acidogenesis

3.1.1 pH, volume and COD of leachate from LB reactor

The pH, volume, COD concentration and cumulative COD profiles of leachate were analyzed over time to examine the operation of LB reactors and investigate the influence of pH adjustment on the efficiency of hydrolysis and acidification processes, and the results are presented in *Figure 2*. pH is one of the major conditions affecting the product formation in anaerobic acidogenesis (Horiuchi et al., 2002; Zhang et al., 2005). When the pH was adjusted by Na_2CO_3, the pH of leachate from Na_2CO_3 treatment increased gradually to 5 in 20 days and kept on increasing to 6.9 by day 30. Without pH adjustment in control reactor, the pH of leachate decreased to 2.7 in 30 days and remained the same till the end of the experiment. The initial pH of both Na_2CO_3 treatment and control reactors decreased from 3.9 to 3.3 in the first 3 days, which might be attributed to the large proportion of easily degradable organic matter. Due to the acidogenic bacteria, fast decomposition of carbohydrate resulted in high concentrations of short chain fatty acids, which decreased the pH rapidly. Correspondingly, the COD concentrations were the highest on day 3, i.e. 39.1 g/L and 43.8 g/L in the control and Na_2CO_3 treatment, respectively (*Figure 2c*). This trend agrees with the results of Wang et al. (2006), who reported a positive correlation between low leachate pH and high VFA concentration in the acidogenic reactor.

With a similar system, Dogan et al. (2009) found that the initial pH of LB reactor was 3.0 and the system tended to buffer itself towards a high pH (4.0 after 40 days) during acidogenesis stage, which was different from our result that the control reactor without pH adjustment turned out to be over acidified with pH lower than 3. This difference may be related to the inoculums and the leachate volume of the LB reactor. Their inoculums was acidogenic culture, which might be able to withstand low pH conditions during the hydrolysis and acidogenesis stagesm, which requires further study to prove this assumption. On the other hand, as presented in *Figure 2b*, although the initial volume of leachate collected from the control was similar to Na_2CO_3 treatment, it decreased gradually to lower than Na_2CO_3 treatment by day 10. It implies that the pH of system might also affect the leachability of LB reactor. In a recent study on the anaerobic biogasification of undiluted dairy manure in LB reactor, the leaching was rather poor and the fine particulate materials contained in the manure as well as its low porosity were reported as the main parameters leading to this phenomenon (Demirer and Chen, 2008). Due to the poor leachability, a large amount of VFAs produced during acidogenesis could not be leached out and was retained in the biomass, consequently affecting the activities of acidogenic bacteria. On the contrary, an increase in the leaching capacity might have led to the increased amount of water added in Na_2CO_3 treatment. Another reason for the irreversible decrease of pH in the might be related to the ratio of microorganism to feed, as too much food could result in higher concentrations of VFAs impeding the restoration of microbial balance.

Chemical oxygen demand of leachate from Na_2CO_3 treatment decreased gradually below 8.0 g/L after 26 d of operation and almost constant at this level for the rest of the experimental period. Meanwhile, COD of control reactor declined steeply after day 3 and began to increase gradually from day 7. Till day 18, the value of COD from the control was even higher than that of Na_2CO_3 treatment. In fact, the variations observed in the COD concentration might not only be related with the hydrolysis/acidogenesis processes but also with the dilution rates in the LB reactors due to the addition of water. The cumulative COD of leachate from the Na_2CO_3 treatment and control were 350.72 g and 197.85 g, respectively (*Figure 2d*). To conclude, the hydrolysis and acidogenesis of food waste in LB reactor was significantly improved by 77% due to pH adjustment.

3.1.2 Total nitrogen and ammonia of leachate from LB reactor

In the LB reactors, the TN concentrations of leachate were close to 1 000 mg/L, while the highest

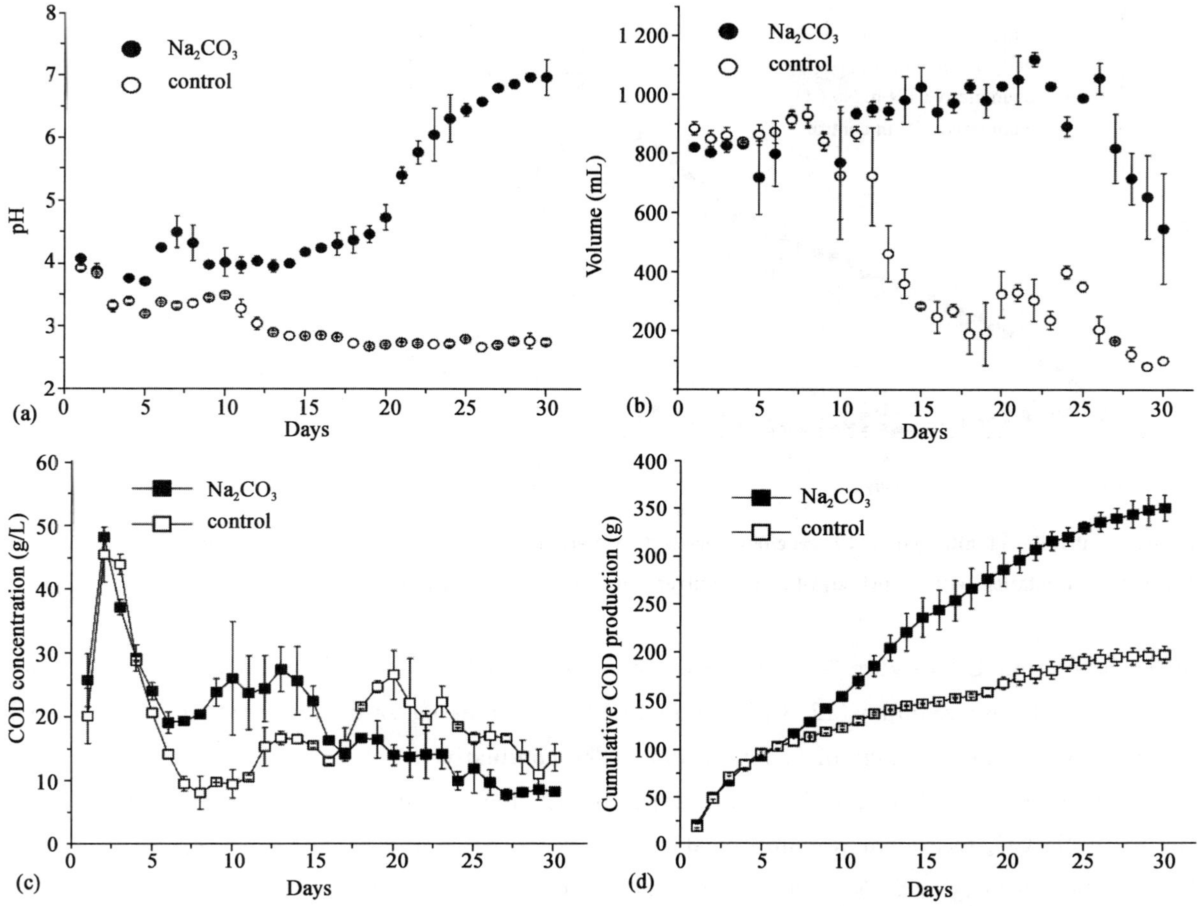

Figure 2 Characteristics of leachate from LB reactors with and without pH adjustment: (a) pH, (b) volume, (c) COD concentration, (d) Cumulative COD production. Leachate recirculation without pH adjustment (control) and with pH adjustment to 6.0 (Na_2CO_3)

concentration of ammonia was less than 200 mg/L. Although the total COD leached out from feedstock was enhanced significantly by pH adjustment, the cumulative TN of Na_2CO_3 treatment and the control were 8.49 g and 5.48 g, respectively. The effect of pH changes on the degradation of organic nitrogen was not as significant as on the degradation of carbohydrate. Similar results were also reported by Lü et al. (2008).

The initial concentrations of ammoniacal-N from both LB reactors were relatively high, in the range of 139.2-145.3 mg/L (*Figure 3*), which might be due to the degradation of organic nitrogen by methanogenic bacteria in the inoculum. Subsequently, the concentration of ammonia declined sharply during the initial 5 days, which supports the assumptions. Similar to the variations of pH and COD, the differences in ammoniacal-N between Na_2CO_3 treatment and control started from day 3. The ammoniacal-N concentration in the leachate from Na_2CO_3 treatment was slightly higher than control until day 25; and a steep increase was observed thereafter. Increased pH of Na_2CO_3 treatment on day 25 might be more favourable for the growth and activities of denitrifying bacteria, thus producing more ammonia.

3.1.3 Changes in the concentrations of organic acids

The hydrolysis of complex organic polymers to simple organic compounds by hydrolytic microorganisms and their conversion to VFAs by acidogenic microbes take place in LB reactors. Organic acid concentrations

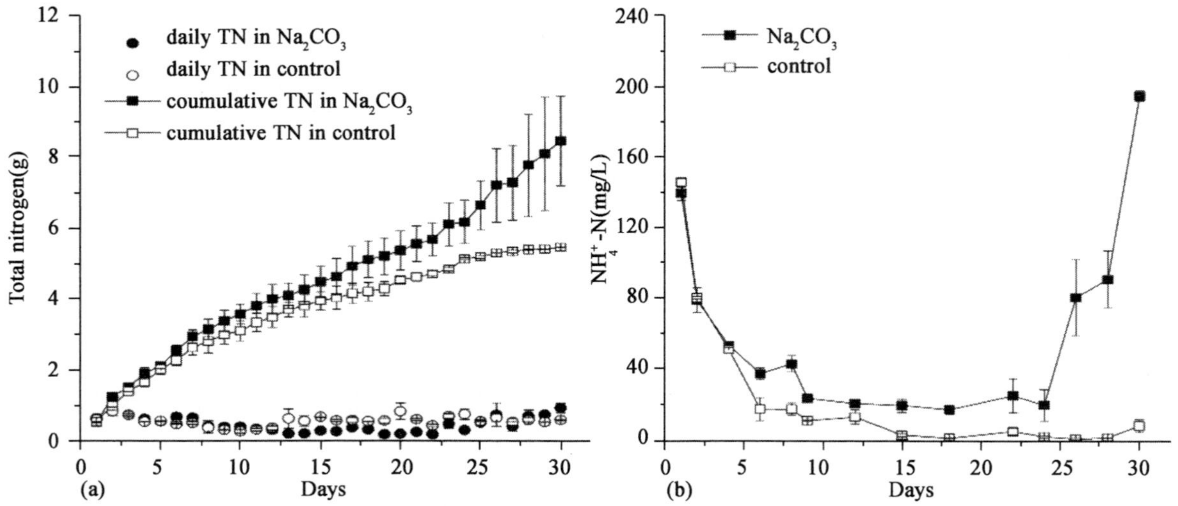

Figure 3 Profiles of nitrogen in LB reactors during digestion: (a) total nitrogen (TN); (b) ammonia (NH_4^+-N). Leachate recirculation without pH adjustment (control) and with pH adjustment to 6.0 (Na_2CO_3)

in the leachate from LB reactors are presented in *Figure 4*. The Total Fatty Acid (TFA) of Na_2CO_3 treatment reached the maximum (12.8 g/L) on day 10 and decreased gradually thereafter, while the TFA in the control varied in the range of 6.2-19.8 g/L. The cumulative TFA generation in Na_2CO_3 treatment and the control were 94.5 g and 77.8 g, respectively. Thus, due to the pH adjustment, the TFA production of Na_2CO_3 treatment was improved by about 25% as compared with control.

The composition of fatty acids in leachate from the two treatments differed from each other (*Figure 4* (b) and (c)). Lactic acid was the main fraction of fatty acid in the control with a maximum of 21.5 g/L on day 20, which was even higher than the total concentration of other six VFAs. However, the maximum concentration of lactic acid in leachate from Na_2CO_3 treatment was only 8.1 g/L on day 10, where acetic and butyric acids were the second and third abundant species of VFAs. Lactic acid in the leachate of Na_2CO_3 treatment contributed about 26% of total COD; while in the control, lactic acid was accounting for 56% of total COD. It was reported that lactic acid was the main fermentation products of hydrolysis-acidogenesis in the two-phase anaerobic digestion process for kitchen wastes (Chen et al., 2007; Zhang et al., 2007a). Lactic acid concentration could constitute approximately 50% of the COD in the hydrolysis-acidification liquid and lactate can inhibit the fermentative microorganisms (Narendranath et al., 2001). Lü et al. (2008) found that lactate substantially reduce the carbohydrate hydrolysis. Furthermore, the toxicity includes not only pH based growth inhibition but also anion-specific effects on metabolism (Warnecke and Gill, 2005). Therefore, high concentration of lactic acid might hinder the hydrolysis and acidogenesis of food waste in the control without pH adjustment. Akao et al. (2007) found that pH 5.5 facilitated the lactate fermentation when compared with 6.0 and 6.5. Despite very low pH in the control reactor, lactate concentration was higher which might be due to the low leachability or the prevalence of more specific acid tolerant microbes. For example *Lactobacillus salivarius* could survive at about pH 3.5 and maintain a high population, competitively inhibiting the growth of other microbes and thereby contribute to the food waste degradation (Shin et al., 2002). Besides, acetic acid was higher than butyric acid in the initial days of digestion and then the butyric acid turned into the predominant species of VFAs in the latter days. This might be a result of the transition of microbial diversity in the reactors as observed previously (Lee et al., 2009).

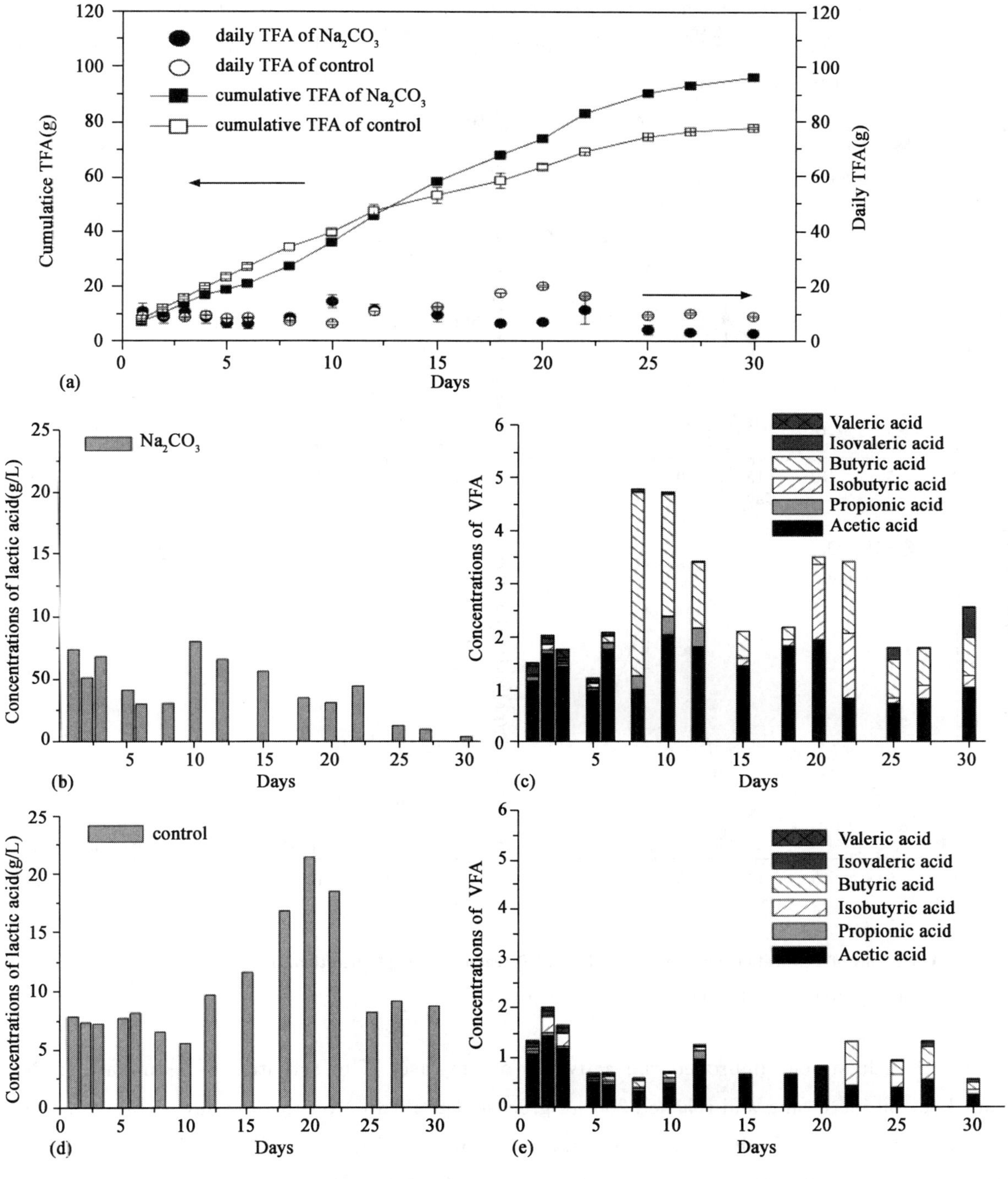

Figure 4 The concentrations of lactic acid and VFAs in the leachate from LB reactors: (a) daily & cumulative TFA in leachate; (b) lactic acid concentration of Na_2CO_3 treatment; (c) VFAs concentrations of Na_2CO_3 treatment; (d) lactic acid concentrations of control; (e) VFAs concentrations of control. Leachate recirculation without pH adjustment (control) and with pH adjustment to 6.0 (Na_2CO_3)

3.1.4 Changes in enzyme activities

The microbial activities of LB reactors were investigated through analyzing four kinds of enzyme activities in the leachate, i.e. α-amylase, CMCase, FPase and protease, and the results are presented in *Figure 5*. Maximum activities of the enzymes occurred at different times during the study period. The initial activities of starch-hydrolyzing enzyme were much higher than cellulose-hydrolyzing and protein hydrolyzing enzymes. In fact, enzyme activities were not only related to inoculum but also the feedstock compositions (Parawira et al., 2005). High α-amylase activities in LB reactors might be resulted from high proportion of starch in food waste.

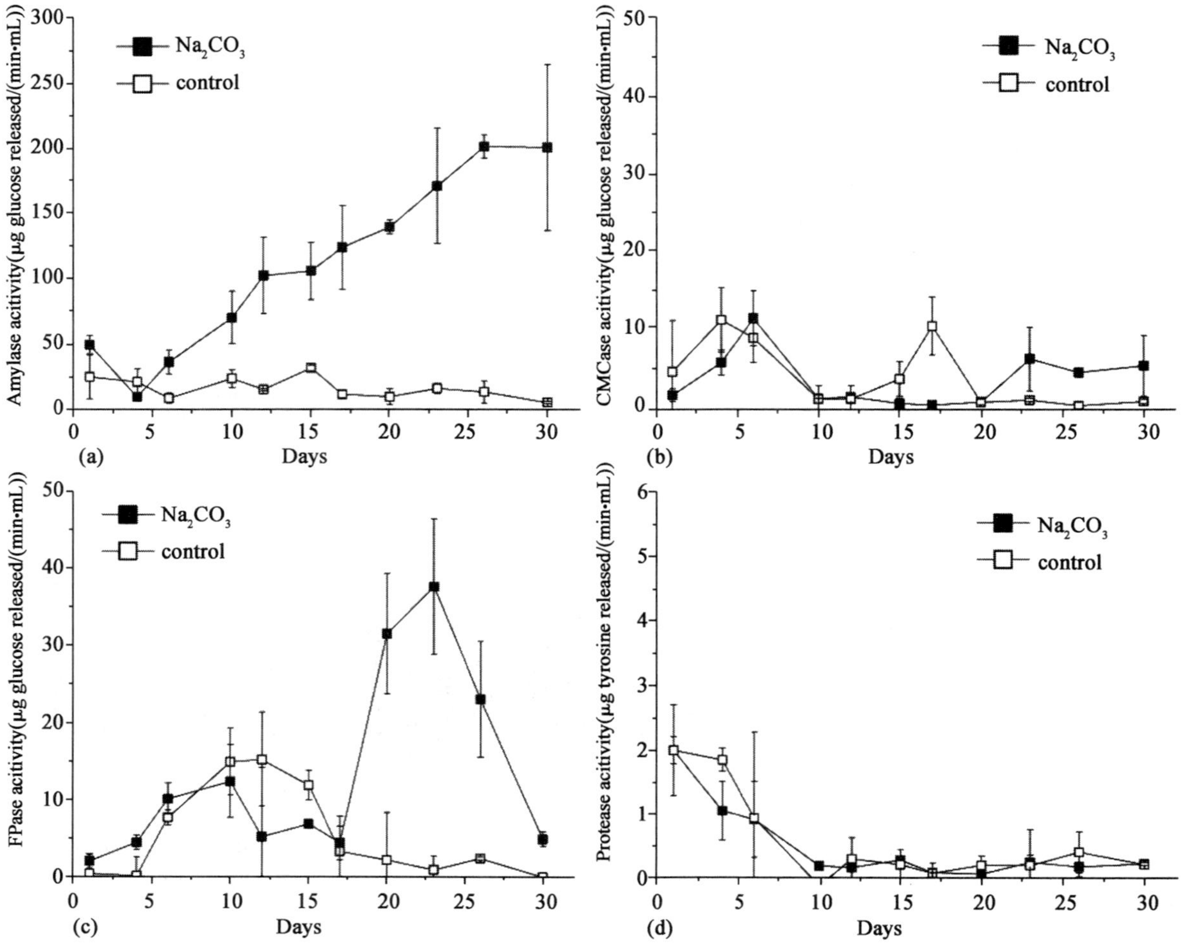

Figure 5 Variations of extracellular enzyme activities in LB reactors through digestion. Leachate recirculation without pH adjustment (control) and with pH adjustment to 6.0 (Na_2CO_3)

Obviously, pH adjustment improved the activities of amylase. The activities of α-amylase in Na_2CO_3 treatment kept on increasing and reached 200 μg glucose/(min · mL) on the 30th day, while it was around 13 μg glucose/(min · mL) in the control. The initial enzyme activities of CMCase and FPase in LB reactors were 1.67 and 2.00 μg-glucose/(min · mL), respectively. Both the CMCase and FPase increased gradually and the activities of these two cellulases in Na_2CO_3 treatment were slightly higher than the control in the last 10 days. Protease levels were low in both acidogenic reactors throughout the study, ranging from 0.2 to 2.0 μg tyrosine/(min · mL). To conclude, the trend of fluctuation in enzyme activities was corresponding to the variations of COD and TN in leachate from LB reactors. The higher enzyme activities in Na_2CO_3 treatment may be partially attributed to the comparatively lower concentration of lactic acid.

3.2 The performance of HASL system

The OLRs of UASB reactors were in relatively low levels, ranging from 0.1-1.7 kg COD/(m^3 · d) (*Figure 6*). Due to the low OLR of reactors, the removal efficiencies of influent COD were high (>90%)

and the COD removal efficiency of leachate from the control was lower than Na_2CO_3 treatment, which might be attributed to higher proportion of lactic acid in leachate from the control. Lactic acid was reported to be inhibitory to the methanogens (Zhang et al., 2007a). The acidogenesis products with lower concentration of lactic acid were more favourable to the subsequent methanogenesis. Chen et al. (2007) evaluated the influence of lactic acid on the methanogenesis in a two-phase anaerobic digestion of kitchen wastes. The maximum organic loading rate was lower in the digester fed with lactic acid than that fed with glucose, where VFA and COD removal were deteriorated compared to that fed with glucose. Lactate was easily converted into propionic acid and acetic acid, and resulted in the accumulation of propionic acid in the methanogenic reactor (Skiadas et al., 2000; Ren et al., 2002). Since the methanogenesis of propionate was slower compared with acetate and butyrate as propionate was an undesirable intermediate product in the methanogenic process (Demirel and Yenigun, 2002; Gallert and Winter, 2008). Therefore, in addition to enhancing the ratio of TFA/COD, reducing the proportion of lactic acid in TFA should be another important aspect to improve the biogas production in two-phase anaerobic digestion of food waste. As results presented above, owing to pH adjustment by Na_2CO_3, the proportion of lactic acid in acidogenesis products in Na_2CO_3 treatment was reduced, which might facilitate the methane production.

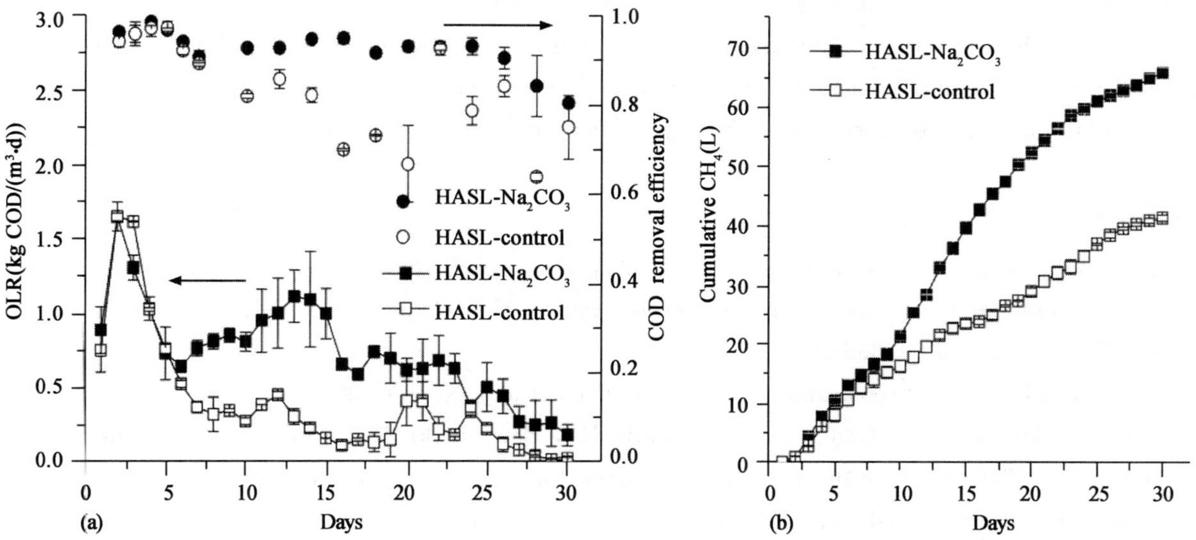

Figure 6 Performance of methanogenesis reactors: (a) OLR and COD removal efficiency, (b) cumulative methane production. Leachate recirculation without pH adjustment (control) and with pH adjustment to 6.0 (Na_2CO_3) in the LB reactors

The total organic removal efficiencies in the HASL system are presented in *Table 2*. After 30 days digestion, TS and VS removal efficiencies of Na_2CO_3 treatment were 86.15% and 87.39%, respectively, which was enhanced by 41% and 36% when compared to the control. Correspondingly, the methane yield of Na_2CO_3 treatment was 0.19 L CH_4/g VS_{add}, 62% higher than control (0.12 L CH_4/g VS_{add}). Therefore, both the hydrolysis/acidogenesis and the methanogenesis of food waste were strengthened by pH adjustment, evidenced by cumulative COD production and methane yield. The higher methane production will be beneficial to energy recovery of food waste.

Table 2 Organic removal efficiency of food waste in HASL system

	Na_2CO_3 treatment			Control		
	Before digested(g)	After digested (g)	Removal efficiency (%)	Before digested (g)	After digested (g)	Removal efficiency (%)
TS	403.33±0.47	55.85±6.8	86.15	403.33±0.47	157.155±15.6	61.04
VS	398.03±0.89	50.18±8.00	87.39	398.03±0.89	141.59±12.70	64.42
Protein	111.71±7.79	54.92±12.22	50.84	111.71±7.79	77.43±7.15	30.68
Methane yield	0.19 L CH_4/g VS_{add} [a]			0.12 L CH_4/g VS_{add} [a]		
	0.22 L CH_4/g VS_{rem} [b]			0.18 L CH_4/g VS_{rem} [b]		

Note: [a] VS_{add}, VS added into the LB reactor.
[b] VS_{rem}, VS removed in the LB reactor.

4 Conclusions

Hydrolysis and acidogenesis rate of food waste in LB reactor was enhanced by pH adjustment, showing 77% of cumulative COD in leachate higher than reactor without pH adjustment. This improvement was also evidenced by the increase in methane yield, from 0.12 to 0.19 L CH_4/g VS_{add}. With pH adjustment using Na_2CO_3, the activity of amylase improved significantly, while cellulose and protease showed marginal increase. The high concentration of lactic acid generated in the LB reactor without pH adjustment might be the inhibitory substances to further hydrolysis/acidogenesis of food waste.

References

Akao S., Tsuno H., Horie T., Mori S., (2007): *Effects of ph and temperature on products and bacterial community in l-lactate batch fermentation of garbage under unsterile condition.* Water Research, 41(12), pp. 2636-2642.

Bernfeld P. ,(1955): *Amylases α and β.* Academic Press Inc., New York.

Bouallagui H., Torrijos M., Godon J. J., Moletta R., Ben Cheikh R., Touhami Y., Delgenes J. P., Hamdi M., (2004): *Two-phases anaerobic digestion of fruit and vegetable wastes: Bioreactors performance.* Biochemical Engineering Journal, 21(2), pp. 193-197.

Chen L., Jiang W. Z., Kitamura Y., Li B. M., (2007): *Enhancement of hydrolysis and acidification of solid organic waste by a rotational drum fermentation system with methanogenic leachate recirculation.* Bioresource Technology, 98(11), pp. 2194-2200.

Cirne D. G., Lehtomäki A., Björnsson L., Blackall L. L., (2007): *Hydrolysis and microbial community analyses in two-stage anaerobic digestion of energy crops.* Journal of Applied Microbiology, 103(3), pp. 516-527.

Demirer G. N., Chen S., (2008): *Anaerobic biogasification of undiluted dairy manure in leaching bed reactors.* Waste Management, 28(1), pp. 112-119.

Demirel B., Yenigun O., (2002): *The effects of change in volatile fatty acid (VFA) composition on methanogenic upflow filter reactor (UFAF) performance.* Environmental Technology, 23(10), pp. 1179-1187.

Dogan E., Dunaev T., Erguder T. H., Demirer G. N., (2009): *Performance of leaching bed reactor converting the organic fraction of municipal solid waste to organic acids and alcohols.* Chemosphere, 74(6), pp. 797-803.

Forster-Carneiro T., Perez M., Romero L. I., (2008): *Influence of total solid and inoculum contents on performance of anaerobic reactors treating food waste*. Bioresource Technology, 99(15), pp. 6994-7002.

Gallert C., Winter J., (2008): *Propionic acid accumulation and degradation during restart of a full-scale anaerobic biowaste digester*. Bioresource Technology, 99(1), pp. 170-178.

Ghose T. K., (1987): *Measurement of celllulase activities*. Pure and Applied Chemistry, 59(2), pp. 257-268.

Guerrero L., Omil F., Mendez R., Lema J. M., (1999): *Anaerobic hydrolysis and acidogenesis of wastewaters from food industries with high content of organic solids and protein*. Water Research, 33(15), pp. 3281-3290.

Horiuchi J. I., Shimizu T., Tada K., Kanno T., Kobayashi M., (2002): *Selective production of organic acids in anaerobic acid reactor by pH control*. Bioresource Technology, 82(3), pp. 209-213.

Kim J. K., Han G. H., Oh B. R., Chun Y. N., Eom C. Y., Kim S. W., (2008): *Volumetric scale-up of a three stage fermentation system for food waste treatment*. Bioresource Technology, 99(10), pp. 4394-4399.

Kim M., Gomec C. Y., Ahn, Y., Speece R. E., (2003): *Hydrolysis and acidogenesis of particulate organic material in mesophilic and thermophilic anaerobic digestion*. Environmental Technology, 24 (9), pp. 1183-1190.

Lee C., Kim J., Hwang K., O'flaherty V., Hwang S., (2009): *Quantitative analysis of methanogenic community dynamics in three anaerobic batch digesters treating different wastewaters*. Water Research, 43(1), pp. 157-165.

Lee J. P., Lee J. S., Park S. C., (1999): *Two-phase methanization of food wastes in pilot scale*. Applied Biochemistry and Biotechnology, 79(1-3), pp. 585-593.

Lim S. J., Kim B. J., Jeong C. M., Choi J. D. R., Ahn Y. H., Chang H. N., (2008): *Anaerobic organic acid production of food waste in once-a-day feeding and drawing-off bioreactor*. Bioresource Technology, 99(16), pp. 7866-7874.

Lowry O. H., Rosebrough N., Farr A. L., Randall R. J., (1951): *Protein measurements using foline phenol reagent*. Journal of Biological Chemistry, 193(1), pp. 265-275.

Lü F., He P. J., Shao L. M., Lee D. J., (2008): *Lactate inhibits hydrolysis of polysaccharide-rich particulate organic waste*. Bioresource Technology, 99(7), pp. 2476-2482.

Moletta R., Verrier D., Albagnac G., (1986): *Dynamic modelling of anaerobic digestion*. Water Research, 20(4), pp. 427-434.

Myint M., Nirmalakhandan N., Speece R. E., (2007): *Anaerobic fermentation of cattle manure: Modeling of hydrolysis and acidogenesis*. Water Research, 41(2), pp. 323-332.

Nakamura K., Kitamura K., (1988): *Cellulases of cellulomonas uda*. Methods in Enzymology, 160(1), pp. 211-216.

Narendranath N. V., Thomas K. C., Ingledew W. M., (2001): *Acetic acid and lactic acid inhibition of growth of Saccharomyces cerevisiae by different mechanisms*. Journal of the American Society of Brewing Chemists, 59(4), pp. 187-194.

Nordberg A., Jarvis A., Stenberg B., Mathisen B., Svensson B. H., (2007): *Anaerobic digestion of alfalfa silage with recirculation of process liquid*. Bioresource Technology, 98(1), pp. 104-111.

Parawira W., Murto M., Read J. S., Mattiasson B., (2005): *Profile of hydrolases and biogas production during two-stage mesophilic anaerobic digestion of solid potato waste*. Process

Biochemistry, 40(9), pp. 2945-2952.

Shin J. Y., Yang S. Y., Song M. D., (2002): *Isolation and identification of lactic acid bacteria for direct-fed microbials*. Konkuk Journal of Natural Science (Kor.) 13, pp. 37-43.

Skiadas I. V., Gavala H. N., Lyberatos G., (2000): *Modelling of the periodic anaerobic baffled reactor (PABR) based on the retaining factor concept*. Water Research, 34(15), pp. 3725-3736.

Stabnikova O., Liu X. Y., Wang J. Y., (2008a): *Anaerobic digestion of food waste in a hybrid anaerobic solid-liquid system with leachate recirculation in an acidogenic reactor*. Biochemical Engineering Journal, 41(2), pp. 198-201.

Stabnikova O., Liu X. Y., Wang J. Y., (2008b): *Digestion of frozen/thawed food waste in the hybrid anaerobic solid-liquid system*. Waste Management, 28(9), pp. 1654-1659.

Wang J. Y., Liu X. Y., Kao J. C. M., Stabnikova L., (2006): *Digestion of pre-treated food waste in a hybrid anaerobic solid-liquid (HASL) system*. Journal of Chemical Technology and Biotechnology, 81(3), pp. 345-351.

Warnecke T., Gill R. T., (2005): *Organic acid toxicity, tolerance, and production in* Escherichia coli *biorefining applications*. Microbial Cell Factories, 4(1), pp. 25-32.

Weiland P., (2003): *Production and energetic use of biogas from energy crops and wastes in Germany*. Applied Biochemistry and Biotechnology, 109(1-3), pp. 263-274.

Wu M., Sun K., Zhang Y., (2006): *Influence of temperature fluctuation on thermophilic anaerobic digestion of municipal organic solid waste*. Journal of Zhejiang University Science B, 7(3), pp. 180-185.

Ying Z., Yang S. T., (2004): *Effect of ph on metabolic pathway shift in fermentation of xylose by Clostridium tyrobutyricum*. Journal of Biotechnology, 110(2), pp. 143-157.

Zhang B., Cai W. M., He P. J., (2007a): *Influence of lactic acid on the two-phase anaerobic digestion of kitchen wastes*. Journal of Environmental Sciences-China, 19(2), pp. 244-249.

Zhang B., He P. J., Lu F., Shao L. M., Wang P., (2007b): *Extracellular enzyme activities during regulated hydrolysis of high-solid organic wastes*. Water Research, 41(19), pp. 4468-4478.

STUDY ON DRY FERMENTATION OF STRAW TO PRODUCE BIOGAS

CHEN Zhiyuan[1,2], YAO Jiangang[2]

1. *Zhejiang Province Key Laboratory of Biomass Fuel*, 2. *Hangzhou Energy and Environment Engineering Co., Ltd.*

Contact: CHEN Zhiyuan, Hangzhou Energy and Environment Engineering Co., Ltd., No. 118 East Fengqi Road, Hangzhou, PR China, Mobile: +86-15158879600, E-Mail: chenzhiyuan2004@163.com

EXECUTIVE SUMMARY

The batch fermentation process was adopted in this experiment to study the biogas production rate of corn stalk, rice straw, tobacco stalk and cassava. The results indicate that the batch fermentation process is suitable for corn stalk, rice straw, tobacco stalk and cassava which are all qualified feedstock for anaerobic fermentation. The biogas production rates related to TS (dry substance) were 413 mL/g, 336 mL/g, 333 mL/g and 222 mL/g respectively, and the biogas production rates related to VS (volatile solids) were 470 mL/g, 387 mL/g, 426 mL/g and 241 mL/g.

Key words: Straw; dry fermentation; gas production rate

秸秆厌氧干发酵产沼气的研究*

陈智远[1,2,**]，姚建刚[1]

（1. 浙江省生物质燃料利用技术研究重点实验室，杭州 310020；2. 杭州能源环境工程有限公司，杭州 310020）

【摘要】本试验以玉米秸、稻草、烟叶秆、木薯秆为代表的秸秆作为原料，在温度38℃，采用批量发酵工艺进行高浓度厌氧发酵产气研究。试验结果表明，玉米秸、稻草、烟叶秆及木薯秆采用批量发酵方式产沼气是可行的，其TS产气率分别为413 mL/g、336 mL/g、333 mL/g、222 mL/g，而VS产气率分别为470 mL/g、387 mL/g、426 mL/g、241 mL/g，是厌氧沼气发酵的好原料。

【关键词】秸秆；干发酵；产气率

1 引言

农业固体废弃物是指在整个农业生产过程中被丢弃的有机类物质，主要包括农业生产和加工过程中产生的植物残余类废弃物、动物残余类废弃物和农村城镇生活垃圾等[1]。据孙永明[1]等报道，我国每年产生固体废弃物高达几十亿吨，而每年产生农作物秸秆总量约 7×10^8 t，除去用于造纸、饲料及造肥还田外，还有一大部分未充分利用，大量的剩余秸秆的随地堆弃和任意焚烧，造成了大气污染、土壤污染、火灾事故、堵塞交通等大量社会、经济和生态问题[2,3]；与此同时，秸秆作为沼气干发酵是以固体有机废弃物为原料（总固体含量在20%以上），利用厌氧菌将其分解为CH_4、CO_2、H_2S等气体的发酵工艺[4]。与湿发酵相比，主要优点是可以适应各种来源的固体有机废弃物、运行费用低并提高容积产气率、需水量少或不需水，节约用水、产生沼液少后续处理费用低等[5]。本文对玉米秸、稻草、烟叶秆及木薯秆的高浓度厌氧发酵产气潜力进行研究。

---

* 本文已收录于《农业工程技术（新能源产业）》，2009(10)：24-26.

** 陈智远，男，本科，助理工程师，杭州能源环境工程有限公司，浙江杭州江干区凤起东路118号7楼 310020，chenzhiyuan2004@163.com，15158879600

2 材料与方法

2.1 材料与试验装置

玉米秸和稻草取自杭州郊区某农场，烟叶秆与木薯秆分别取自云南昆明郊区某卷烟厂和某农场，经切碎后2～3 cm待用。污泥则取自杭州市种猪试验场的沼气站。原料的TS与VS见表1。厌氧装置采用自制的1.5 L发酵装置，采用排水法计量气体，试验装置见图1。

Table 1 TS and VS of Feedstock

表1 原料的TS与VS

项目	玉米秸	稻草	烟叶秆	木薯秆	污泥
TS(%)	84.42	86.33	87.96	23.90	11.64
VS(%)	73.96	75.02	68.68	22.00	7.32

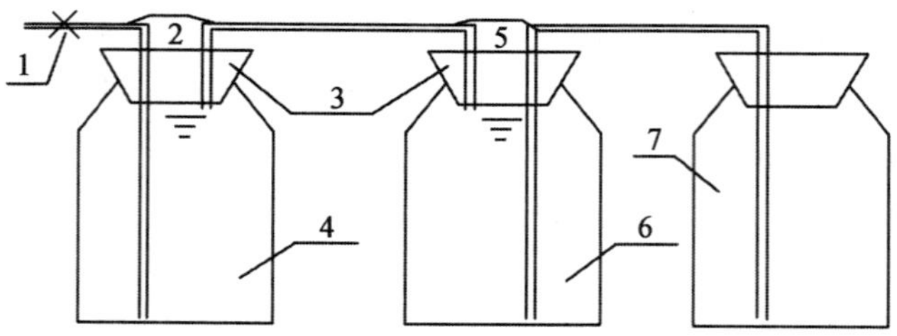

1. 止水夹 2. 胶管 3. 盖子 4. 发酵瓶 5. 胶管 6. 集气瓶 7. 集水瓶

Figure 1 Schematic diagram of reactor

图1 反应装置示意图

2.2 试验设计

试验设4个试验组和1个空白组，每组3个平行，在38℃的恒温间内发酵。将100 g玉米秸、稻草、烟叶秆分别和800 g污泥混合均匀后加入发酵瓶中，将100 g木薯秆与600 g污泥混合均匀后也加入发酵瓶中，空白则将1 000 g污泥加入发酵瓶中。

2.3 分析项目及方法

TS测定是将待测混合物置于已烘干、称重的硬质玻璃杯中，(105±2)℃烘干至恒重，称重计算，而VS测定是将待测混合物置于已烘干、称重的坩埚中，(550±10)℃灼烧至恒重，称重计算[6]。pH值采用精密试纸法。

每天定时测定发酵产气量，即测定集水瓶中水的体积量为日产气量。利用沼气分析仪(武汉四方沼气分析仪)及根据沼气燃烧的火焰颜色参照CH_4含量标准卡联合检测CH_4浓度[7]。

3 结果与讨论

3.1 发酵前后的相关测定及分析

从图2可以看出，各试验组发酵前后的TS及VS均有所下降，这说明原料被消耗并生产沼气。图中数据表明玉米秸、稻草、烟叶秆及木薯秆的TS减少率分别为24.68%、17.71%、29.84%、14.72%，而VS减少率为28.53%、31.67%、50.32%、23.04%。发酵前后，在TS和VS减少率方面，玉米秸和稻草的都大于木薯秆的，两者的干物质产气量也大于木薯秆，表明秸秆原料的利用程度(TS、VS降解率)和产气量存在一定关联。但烟叶秆的减少率大于玉米秸和稻草，干物质产气量小于玉米秸，基本跟稻草持平，这可能与烟叶秆被烘烤过，而玉米秸和稻草是风干的有关，也可能是与各发酵原料秸秆不同相关。

玉米秸、稻草、烟叶秆及木薯秆发酵后的pH值分别为8.6、8.2、8.5、8.1，与发酵前的pH值(各组都大约为7.3)相比均略微增大,表明各处理组均未出现酸累积抑制甲烷生成的现象[8]。

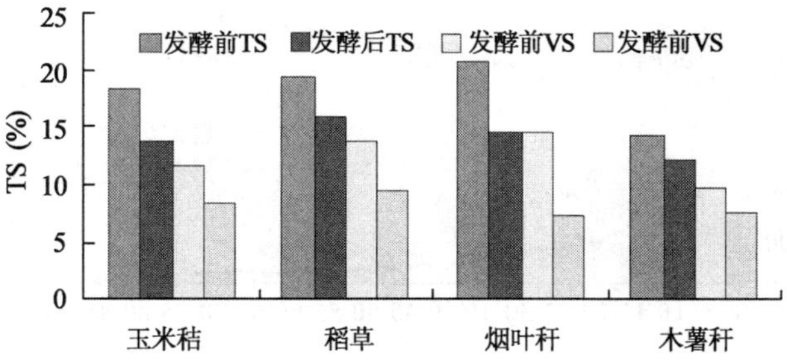

Figure 2　TS, VS content before and after fermentation

图2　发酵前后的TS与VS

3.2　产气量的统计及分析

试验持续90 d,各试验组的甲烷含量逐渐升高,都在启动发酵的第1 d迅速产气,并且从第2 d所产的沼气能够连续燃烧,火焰呈晴蓝或云水蓝色,参照比色卡,甲烷含量较高(55%～70%),而空白组则在发酵第1 d就能点燃;从图3可知,各组所产沼气中平均甲烷含量相差不大,都达到60%以上,表明各试验组厌氧发酵正常。

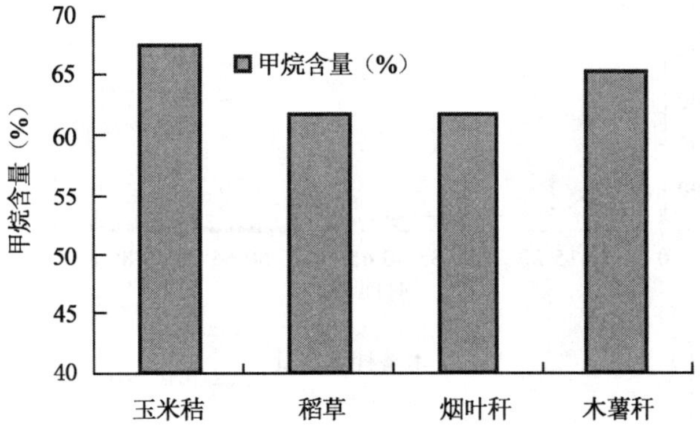

Figure 3　Methane content

图3　各组甲烷含量

图4是厌氧发酵过程中各秸秆的日产气量及累计产气量的变化曲线。从图4曲线上看,玉米秸秆组和稻草组都为两产气高峰(一大一小),两产气高峰都处于发酵前15 d;而烟叶秆和木薯秆都在发酵第2 d达到产期高峰,并且只有一个高峰,这可能与所使用发酵原料含有半纤维素、纤维素、木质素等比例不同相关。

玉米秸、稻草、烟叶秆及木薯秆在发酵启动15 d的累计产气量占总产气量分别为60.9%、42.4%、51.8%和73%;在发酵30 d时累计产气量占总产气量的78.1%、65.8%、69.1%和83.2%;而在发酵60 d时累计产气量占总产气量的92.6%、85.9%、90.6%和95.5%;表明四种秸秆在发酵进行60 d时,原料中的能够发酵的物质大部分转化为沼气。1 000 g污泥空白在整个试验过程中累计产气量为8 974 mL;按比例去除污泥空白的产气量,玉米秸、稻草、烟叶秆及木薯秆的净累计产气量分别为34 782 mL、29 050 mL、29 263 mL和5 305 mL。玉米秸、稻草、烟叶秆及木薯秆的TS产气率分别为413 mL/g、336 mL/g、333 mL/g、222 mL/g,而VS产气率分别为470 mL/g、387 mL/g、426 mL/g、241 mL/g。

4　结论

经试验测定,在温度38℃、没有经过预处理高浓度发酵的条件下,玉米秸、稻草、烟叶秆及木薯秆的TS

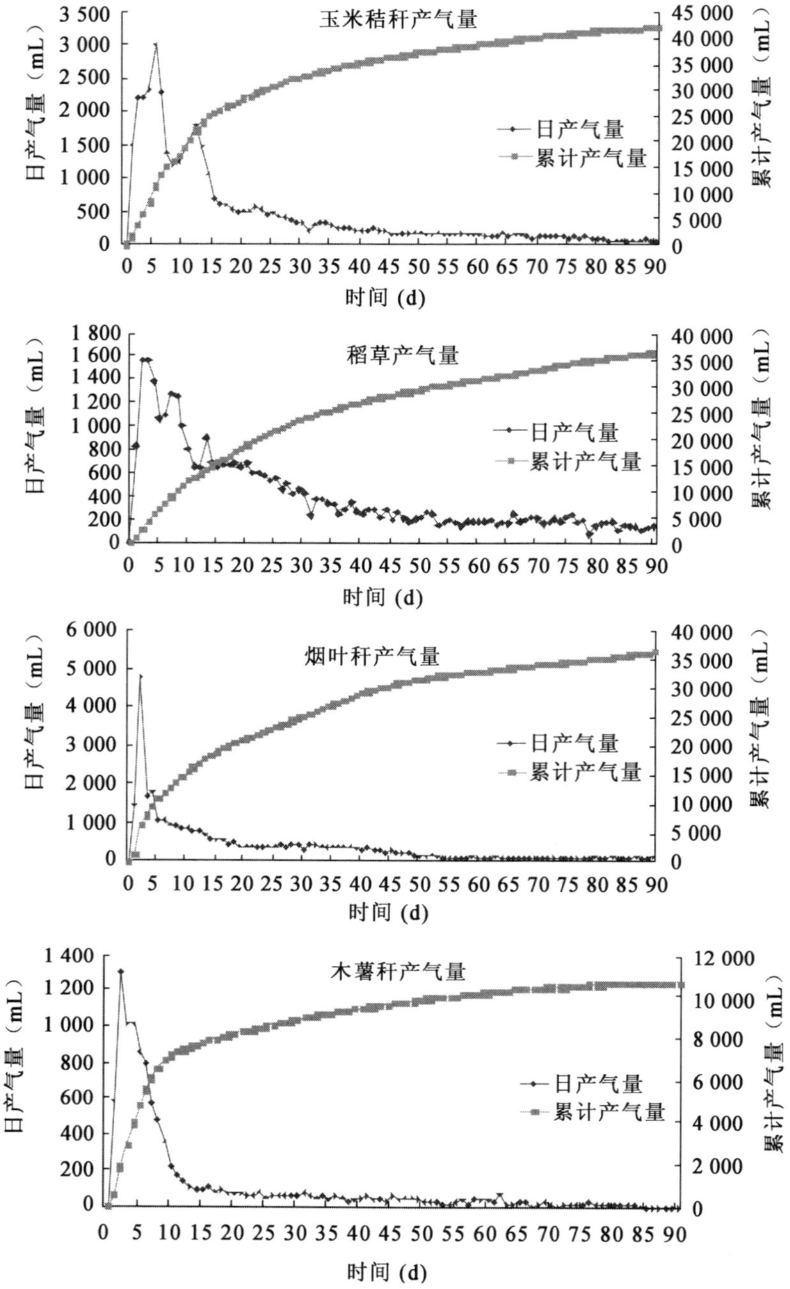

Figure 4　Daily gas production and accumulated gas production of straw

图 4　秸秆日产气量及累计产气量

产气率分别为 413 mL/g、336 mL/g、333 mL/g、222 mL/g，而 VS 产气率分别为 470 mL/g、387 mL/g、426 mL/g、241 mL/g；玉米秸、稻草、烟叶秆和木薯秆都高浓度沼气发酵的好原料，采用批量发酵方式产沼气是可行的；发酵前后玉米秸和稻草的都的 TS 与 VS 减少率大于木薯秆的，两者的干物质产气量也大于木薯秆，表明秸秆原料降解程度和产气量存在一定关联；本试验是在使用畜禽场沼液沼渣作为接种污泥的情况下进行的，为解决发酵原料不足的问题提供试验的指导。

项目及基金支持

“十一五”国家科技支撑计划“新型高效规模化沼气工程”重大项目课题(2008BADC4B10)

参考文献

[1] 孙永明，李国学，张夫道，等. 中国农业废弃物资源化现状与发展战略[J]. 农业工程学报，2005，21(8)：169-173.

[2] 边炳鑫，赵由才. 农业固体废弃物的处理综合利用[M]. 北京：化学工业出版社，2005：250-305.

[3] 董佑福，侯方安. 重新认识秸秆发展秸秆循环经济[J]. 当代农机，2007，(9)：14-15.

[4] 李想，赵立欣，韩捷，等. 农业废弃物资源化利用新方向——沼气干发酵技术[J]. 中国沼气，2006，24(4)：23-27.

[5] 曲静霞，姜洋，何光设，等. 农业废弃物干法厌氧发酵技术的研究[J]. 可再生能源，2004，(2)：40-41.

[6] 中科院成都生物研究所. 沼气发酵常规分析方法[M]. 北京：北京科技出版社，1981：1-90.

[7] 江蕴华，余晓华. 利用火焰颜色判断沼气中甲烷含量[J]. 中国沼气，1983，(3)：28.

[8] Chyi Y. T.，Dague R. R. Effects of particulate size in anaerobic acidogenesis using cellulose as a sole carbon source[J]. Water Environmental Research，1994，66：670-678.

集装箱式高浓度有机垃圾两步厌氧发酵装置研究

王彤[1],李俊涛[2]

(1. 德国 BEB 生物质能有限公司,柏林 12555;2. 成都德通环境工程有限公司,成都 610041)

【摘要】餐厨垃圾中含有大量的有机物,故具有很高的生物降解性,是厌氧发酵和生物降解生产沼气的理想原料。厌氧处理后,餐厨垃圾转化为可再生能源——沼气和生物肥料。德国柏林生物质能有限公司(BEB Berlin)和成都德通联合研发出了一种经济的选择性收集有机废弃物生物处理系统。收集和处理系统,分别在两个不同的装置中(10 英尺装置,40 英尺装置)。本系统可以根据单独的或工厂化优化设计确定规模。湿法废弃物处理单元有从单个家庭使用到集中处理系统的不同规模,并且可根据工程项目所在地点,优化沼气工程设计,使其与收集系统相连。整个系统设计紧凑、空间需求相对较低,处理能力可根据模块数量进行叠加。在东京/日本,设计并建成了安装在一个 40 英尺集装箱的 5 个反应器、总容积 14 m^3,日处理 400 kg 有机废弃物的沼气工程点,该工程点已稳定运行约 10 年之久。

【关键词】集装箱式沼气发酵装置;餐厨生物可降解垃圾

TWO-PHASE ANAEROBIC DIGESTION OF HIGH STRENGTH ORGANIC WASTES WITH A CONTAINER TYPE BIOGAS PLANT

TONG Boitin[1], LI Juntao[2]

1. *BEB BioEnergy Berlin GmbH*; 2. *Chengdu DeTong Environmental Engineering Co., Ltd*

Contact: Tong Boitin, BEB BioEnergy Berlin GmbH, Köpenicker Str. 325, 12555 Berlin, Germany, E-Mail: t. boitin@web. de

EXECUTIVE SUMMARY

Kitchen waste contains a large number of organic substances, so it has very high biodegradability and is a kind of optimal materials that is easy to be fermented by anaerobe digestion and biodegraded to produce biogas. After anaerobic treatment, kitchen wastes will be transformed to renewable energy resource biogas and bio-fertilizer. BEB Berlin and Chengdu DeTong have developed a modular system for the selective collection and economic biological treatment of organic wastes. The collection and treatment system are separated into two containers (10-ft containers and 40-ft container), allowing individual and optimal plant sizes. Different scales from single household use to central treatment systems are available for the wet-waste treatment unit and, depending on the site of location, an optimal biogas plant is connected to the collection system. The compact design of the whole system requires relatively little space, and the treatment capacity can be modular increased. The realised plant was designed with a capacity of 400 kg organic waste per day, installed in a 40ft-Container with 5 reactors of 14 m^3 total in Tokyo/Japan. The container biogas plant has been operated optimally about 10 years in Tokyo.

Key words: Container type biogas plant; kitchen bio-waste

1 Introduction

The model of 'Sustainable Development', which was declared by the UN Environment Development Conference 1992 in Rio de Janeiro ('Global Summit') and was concretised during the following conferences in Berlin (1995), Kyoto (1997) and Buenos Aires (1998), shows that concern about the environmental issues is becoming a very important topic not only in industrialised countries like Japan and Germany but in the entire world. So, on one hand modern waste management is a society duty-on the other hand it offers a great potential for cost reductions in the fields of waste management, energy consumption and agricultural applications.

The BEB two-phase anaerobic digestion process for the high strength organic wastes such as kitchen waste, as developed by the BEB BioEnergy Berlin GmbH and Chengdu DeTong Environmental Engineering Co., Ltd, is a new alternative for the processing of easily biodegradable wastes into biogas and compost, without environmental burden. The organic wastes are produced in large quantities in large-scale kitchen, tourist centres, food industries and markets. The solid contents of this waste is between 10% and 20% and is easy digestible.

This organic wastes treatment with anaerobic process to produce biogas is a very good way, and can reduce environmental pollution and keep ecological balance. But in single stage anaerobic digestion, there is a problem of acids accumulation and the associated pH drops with the loading rate increasing. It could be seen that with two phase anaerobic process, there was no acidification problem of fermenting liquid and providing rapid-rate and efficient substrate conversion. Therefore, an optimal biological way to treat these wastes is using a two-phase anaerobic digestion process.

The solid matter, after being crushed, is digested in a hydrolyser by a population of acid-forming bacteria under semi-anaerobic condition. The effluent from the hydrolyser is treated in a methanizer. Biodegradation achieved in two weeks was around 75%. The process was stable, even when the pH was at the lowest levels (around 4-5 in hydrolyser). The process has been developed through laboratory and technical stages for various kinds of raw materials from Germany and China. The realised plant was designed with a capacity of 400 kg organic waste per day, installed in a 40ft-Container with 5 reactors of 14 m^3 total in Tokyo/Japan. The container biogas plant has been operated optimally about 10 years in Tokyo.

1.1 40ft-Container-biogas plant system

The BEB 40ft-container Biogas system is a plant for the hygienic and energetic biological treatment of kitchen wastes and swill from e.g. hotels, restaurants and food industry. The collected and homogenised wet wastes will be pumped from the external wet waste treatment unit (kitchen shooter/satellite unit) into the biogas plant or will be transported via tank truck from decentralised collection units to the biogas plant system.

The following chapters give a brief introduction of the complete treatment system.

1.2 10ft-Container pre-treatment system

The 10ft-container per-treatment unit is intended for the disposal of wet refuse. Many kinds of organic wastes, food residues and kitchen waste can be treated and pre-stored in this air-proof and odourless system.

Figure 1 Wastes storage and compressor in 10ft container

Figure 2 Pre-treatment in 10ft container

With this system, the occurring wet refuse is treated without addition of any chemicals or disinfectants. The occurring food residues are directly filled into the input station by hand (small scale) or with a hydraulic lift-and-tip system for garbage cans (big scale). Here, the standardised receptacle is placed onto the lift system and afterwards the charging of the WWTU (Wet waste Treatment Unit) evacuation of the can runs automatically.

After feeding the delivery system, the organic material of the constituents is roughly comminute within specialised comminuting equipment. Afterwards, the wet refuse directly passes through an inlet slide valve to the main tank of the system. With a combined comminuting and pumping unit (cutter blade pump) the wet refuse is regularly circulated to achieve homogeneity conveyance.

1.3 40ft-container biogas plant

The completely equipped 40ft-container with biogas system is developed for the treatment of 400 kg wet wastes per day. The waste material is pumped into a tank and reactor system where the biotechnological decomposition of the organic material-the anaerobic treatment-takes place. Within the first stage, the complex organic wastes will be hydrolysed in a semi-aerobic milieu. The hydrolysed organic waste material is fed by means of a screw pump into the biogas reactors. The biomass will be decomposed by bacteria under exclusion of air at a thermophilic temperature level. A biogas storage tank acts as buffer system. A gas storage tank is helpful for the continuous flow of gas due to the possible changing demand of the added biogas utilisation system.

The process is automatically controlled by a PCL-system that guarantees few staff attention for regular operation.

Figure 3 40ft-container-biogas

Figure 4 Inside view of 40ft container biogas plant

1.4 Biogas utilization

A co-generation (CHP-combined heat and power) unit is the mainly used system for biogas utilisation nowadays. In that combined process, the biogas is used in a combustion engine which is connected with an electric generator. Hence, electrical and heat energy are produced. The modular design of CHP units and the bright range of capacities from 3 kW_{el} up to some MWel make CHPs suitable for any purpose of biogas transformation into electricity and thermal power.

The temperature level of the thermal power fits optimal to the heat demand of the biogas process-so the needed electricity and thermal power for the biogas process is produced out of the biogas itself. CHP units with a capacity of up to 50 kW_{el} are equipped with soundproof hoods-so noise emission is very low and installation in a basement boiler room is possible. Bigger engines are placed in separated soundproof buildings or containers.

Figure 5 Gas storage in 40ft container biogas plant in Tokyo

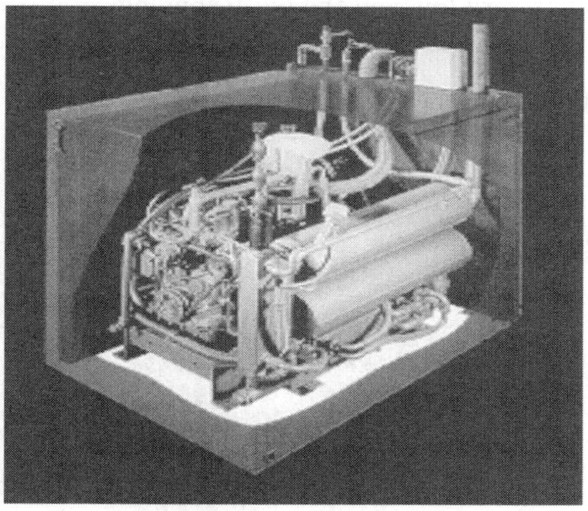

Figure 6 CHP unit

1.5 Biofilter

A biofilter cleans the exhaust gas from the hydrolysis step and the air from the delivery area (if built). Because of a few H_2S-content of this gas it has a strong and unpleasant smell. If there is no waste air cleaning unit installed at the biogas plant site, a biofilter is necessary to avoid odour. The biofilter is realised as open bed or integrated in closed tanks (see *Figure 7* and *Figure 8*) in relation to the throughput. The biological gas treatment method that is realised in this system works with aerobic bacteria. In an air conditioned milieu and a specific humidity these bacteria exist on a compost-like culture medium.

Figure 7 Biofilter on the top of 40ft container biogas plant

Figure 8 View of BioFilter

1.6 Automatic control system

The process is automatically controlled by a PLC-system that guarantees little staff attention for regular operation. The control system includes an electro-pneumatically interface which makes the free control of all valves and sliders possible. Pumps and other electrical units will be triggered by control contactors. This control system allows the complete automatically operation of the biogas plant. All process data and parameters of the system can be checked by process visualisation and can be changed system. Maintenance effort is very low and the bio-technological process runs for itself if it has been started up once perfectly by ECB specialists. The control room as well as a small laboratory are installed inside of a 40'-standard container for the bigger plant systems.

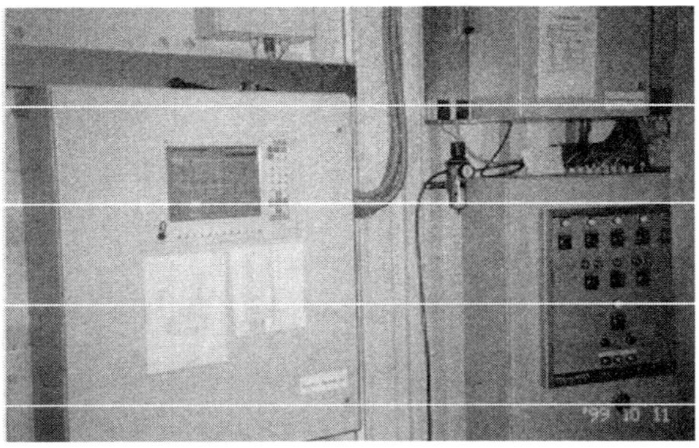

Figure 9 Controlling system for 40ft container biogas plant

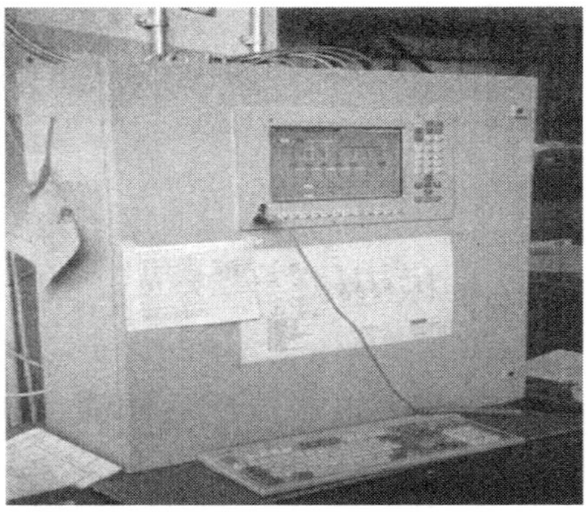

Figure 10 Display of the controlling system

2 Materials and methods

2.1 Input materials

The input material fed to the digesters is collected from kitchen and markets in Tokyo. The total solid content of this kind of organic waste is between 8% and 15% and VS in TS around 85%. Therefore this waste with high organic matter is a very good material for anaerobic treatment with the BEB two-step anaerobic process. The components of the waste have big differences in different days and seasons, which will be adjusted by process arrangement controlled by the automatic computer control. *Table 1* below shows the analytical average value of the organic waste. The wastes were crushed in order to obtain particles less than 20 mm through a Waste Disposal System.

Table 1 Characteristic of the Components of the Organic Waste

Components	Average value	Components	Average value
TS (%)	8-15	N-NH_4 (mg/L)	100-200
VS (%)	7-14	COD (g/L)	150-200
Total C (g/kg)	45-65	BOD_5 (g/L)	95-125
Total N (g/kg)	2.5-6.0	VFA (g/L)	0.8-1.5
Total P (g/kg)	0.3-0.5	pH	3.5-6.0

2.2 Anaerobic reactors

The pilot plant is a mobile container biogas plant, which consists of a hydrolyser (4 m^3) and four methanizers with volume of 4×2.5 m^3. The hydrolyser is a semi-anaerobic reactor which is fed with a mixture consisting of organic waste and methanized effluent. It is operated at ambient temperature. The four methanizers are operated at the 55℃ and semi-continuously fed with effluent from the hydrolyser stage. A sequential recycling of the effluent allows both a homogenization of the liquid content and maximal biomass retention.

The following graphic and photo shows the inside of a 40ft-biogas container:

Figure 11 Inside of the 40ft-biogas container

2.3 Research conditions

Table 2 The research conditions for the container biogas plant

Research conditions	Hydrolyzer	Methanizers
Input TS (%)	8-15	5-10
Temperature (℃)	20-40	55
pH-Value	3.5-6.0	7.0-7.8
CH_4 in biogas (%)	<20	60-70
H_2S in biogas (mg/L)	0.1-0.6	<0.1
Agitation	semi-continuously with agitator and circulation pump	continuously with agitator in CSTR Reactor

3 Results and discussion

Within the BEB two-phase anaerobic treatment systems, hydrolysis of polymers and acidification of easily degradable compounds take place in the first phase (hydrolysis phase), yielding volatile fatty acids. In the second phase (methanization phase), what mainly takes place is acetogenesis and methanogenesis. In this approach the crushed solid organic matter loaded semi-continuously, is broken down in the first stage by acid-forming and hydrolytic bacteria. The soluble organic matter and VFA are pumped into the methane Completely Stirred Tank Reactor (CSTR), containing a high concentration of methanogenic bacteria and producing biogas.

3.1 Hydrolysis phase

The Hydrolyzer is a semi-anaerobic stirred reactor and loaded with a mixture of organic waste and methanized effluent from sedimenter. It is operated at a temperature of 30-40℃. In this step the high molecular components of the organic waste like proteins, carbohydrates and fats are broken down by hydrolytic and acid-forming bacteria, and yielding soluble small organic matters such as glucose and volatile fatty acids. The hydrolytic reactor was loaded with a mixture of organic wastes and effluents of methanizer with ratio of 1:1 by volume. The characteristics of the substrate (see *Table 1*) presented that the C:N:P ratio was well balanced, being around 124 : 9 : 1, therefore no other nutrients were added to the hydrolyser.

The main operating parameters of the hydrolyser are:

- Hydraulic retention time (HRT): 3-5 d
- TS-influent: 8%-15%
- pH in reactor: 3.5-6.0
- TS-loading rate: 25-35 kg TS/ (m^3 · d)
- VS-removal rate: ca. 20%

3.2 Methanization phase

After a hydraulic retention time of approx. 3-5 days in hydrolyser, the effluent feeds the methanizer, a stirred continuous anaerobic reactor, operating at a temperature of 55℃. The methanizer contains a high concentration of methanogenic bacteria, which can only use the small soluble organic matter such as volatile fatty acids from hydrolyser. During the retention time of approx. 10-15 d, methanogenic bacteria transform the volatile fatty acids directly into biogas. The biogas with 60%-70% methane content contains very small amounts of hydrosulphide (H_2S), usually less than 0.1%.

- TS-Loading Rate and VS-Removal Rate

The average VS removal rate of 70% in methanizer was achieved at a loading rate from 5 to 15 kg TS/(m^3 · d). The VS removal rate can be as high as 80% in some cases. Even under a shock loading rate of 15 kg TS/(m^3 · d), the VS removal rate achieved more than 65% with 10 days of hydraulic retention time.

- Biogas Production and Loading Rate

There is a very strong dependence between the loading rate and the daily gas production in anaerobic treatment processes. Such a relationship can be observed that biogas production yield was about 0.75 m^3/kg VS removed. The CH_4 contents of this biogas reached 65%-75% in some cases. Biogas production in the container biogas plant reached around 40 m^3/d.

- pH, Alkalinity and VFA

pH, alkalinity and concentration of volatile fatty acids are operating parameters for an effective control of the anaerobic treatment process. pH values vary in range 7.4 to 7.8 in methane reactor. There is no low pH problem in this period for the methane reactor. This mainly originates from the characteristics of the two-phase process. The alkalinity in the anaerobic reactor was maintained above 5 500-7 000 mg/L as $CaCO_3$ and it generally averaged about 6 500 mg/L without any adjustment. Average volatile fatty acid concentrations in the methane reactor reached about 3 500 mg/L. VFA concentration of 4 000 to 6 000 mg/L in the methane reactor could easily be tolerated in the case of high organic loading.

The main operating parameters of the methanizer are:

- Hydraulic retention time (HRT): 10-15 d
- pH in reactor: 7.0-8.0
- TS-loading rate: 10-16 kg TS/(m^3 · d)
- VS-removal rate: ca. 80%
- CH_4-production rate: 0.4-0.5 m^3 CH_4/ (kg VS · d)
- CH_4-content in biogas: 60%-70%

3.3 Two-Phase Anaerobic Treatment System

It is evident that biogas production rate is quite high (about 40 m^3/d) with 70% methane content using two-phase system. The maximal loading rates was 10 kg TS/(m^3 • d) and the average VS removal rate was 85%. Obviously, the phases separation improved the experiment results, especially the methane yield.
The average VFA concentration reached separately 6 g/L in the methanizer and 8 g/L in the hydrolyser. As can be seen, the two-phase system could take higher VFA concentration, a maximum of about 11 000 ppm in the hydrolyser and a maximum of about 8 000 ppm in methanizer. No problems arose even when the pH was at the lowest level (a value of 4.5 for the hydrolyser and value of 7.6 for the methanizer) The low pH values and high VFA concentration from hydrolyser do not affect the methanizer pH, which was always kept around 7.4-7.8 as a result of the high value of alkalinity. Nevertheless, the alkalinity value in methanizer should be maintained above 5 g/L as $CaCO_3$.
Leachate ammonia nitrogen (NH_4^+-N) concentration was checked to ensure the right nutrient balance. This is more convenient than just verifying substrate composition. The results showed that it was fairly constant during all the experiment, with values around 2 000 mg/L.

3.4 Effect of recycling on system performance

General speaking, this organic wastes have a high TS concentration (about 8%-15%) and a low pH values (about 3.5-4.5). To control the optimal pH in the two-phase system, recycling of effluent to the influent stream has been done by BEB process. So recirculation rate was claimed to be a significant parameter in the operation of a two-phase digestion. This has four major consequences:
The base TS influent, the TS-concentration is diluted to effective influent TS, as well as to avoid feeding block.
Recycling of the methanized effluent which had high pH value mostly over 7.5, to maintain the suitable pH in hydrolyser.
High alkalinity in the effluent augmented the alkalinity in the influent, at the same time, enhanced system stability and recycled the alkalinity.
Because the methanizer is Completely Stirred Tank Reactor (CSTR), it has a disadvantage, that is, biomass can be lost following the effluent. So biomass recycling can overcome this problem.
The whole control system is controlled by an automatic control system (PLC). Pumps, stirrers and other electrical units will be triggered by control contactors and additional hand control facilities. Measured data like temperatures, pressure, flow rate or pH value are shown on the screen and stored in the internal data storage for process documentation.
This control system allows a complete automatic operation of the container biogas plant. All process data and parameters of the system can by checked by process visualisation and can be changed by using a special code. A manual operation of the units is possible in case of blackout failure of the control system.

4 Conclusions

According to the results obtained, it can be concluded that the two-phase system to treat high concentration and easily degradable solid organic wastes is highly effective for VS removal and bio-energy recovery. This two-phase system is highly stable and able to withstand high peak loading. Moreover, the two-phase operation permits high conversion efficiencies (more than 85% of volatile matter was converted to methane in this process), and high methane yields with 0.5 m^3 CH_4/kg VS can be achieved. Therefore, this BEB two-phase anaerobic treatment process has been identified to be a good and high efficient way to easily treat degradable organic wastes.

References

Constant M., Naveau H., Ferrero G., Nyns E., (1989): *Biogas End-use in European Community*. Elsevier. New York, USA.

Yamamoto Y., Katsumata K., (1997): *Technology to create high concentrations of methane in digestion gas*. Pro. 8th international conf. on anaerobic digestion, 3, pp. 488-491.

Mao T., Show K. Y., (2006): *Performance of high-rate sludge digesters fed with sonicated sludge*. Water Science & Technology, 54(9), pp. 27-33.

Zhang B., He Z. G., Zhang L. L., et al., (2003): *Semi-continuous anaerobic digestion of food waste using a hybrid anaerobic solid-liquid bioreactor*. Water Science & Technology, 48 (4), pp. 169-174.

EFFECT OF TEMPERATURE ON ANAEROBIC DIGESTION OF LIVESTOCK MANURE

CHEN Zhiyuan[1,2], CAI Changda[1,2], SHI Dongwei[2]

1. *Zhejiang Province Key Laboratory of Biomass Fuel*; 2. *Hangzhou Energy and Environment Engineering Co., Ltd.*

Contact: CHEN Zhiyuan, Hangzhou Energy and Environmental Engineering Co. Ltd., 6F, No. 118 East Fengqi Road, Hangzhou, PR China, Mobile: 15158879600, E-Mail: chenzhiyuan2004@163.com

EXECUTIVE SUMMARY

This paper analyses the temperature effects on the anaerobic digestion process of pig manure, chicken manure and cow dung. The pH value, biogas production capability, COD and NH_4^+-N in the liquid supernatant have been tested at 35℃, 25℃, and ambient temperature. The optimum temperature and biogas production potential of the mentioned manure were confirmed, and the results did provide important parameters for improving the fermentation performance and biogas production of pig manure, chicken manure and cow dung, which could guide a more effective treatment of livestock manure.

Key words: Temperature; anaerobic fermentation/digestion; pig manure; chicken manure; cow dung

不同温度对畜禽粪便厌氧发酵的影响

陈智远[1,2,*]，蔡昌达[1,2]，石东伟[2]

（1. 浙江省生物质燃料利用技术研究重点实验室，杭州 310020；2. 杭州能源环境工程有限公司，杭州 310020）

【摘要】 针对猪、鸡、牛 3 种粪便以及 35℃、25℃和常温 3 种温度条件，研究了不同温度对不同粪便厌氧发酵过程的 pH 值、产沼气性能、上清液 COD 及 NH_4^+-N 的影响。通过以上试验，确定了各种粪便厌氧发酵的适宜温度及各温度下的沼气产气潜力，为提高猪粪、鸡粪和牛粪便厌氧发酵性能及沼气产量提供了重要参数，对畜禽粪便的有效处理具有一定的指导意义。

【关键词】 温度；厌氧发酵；猪粪；鸡粪；牛粪

1 引言

在畜牧业迅速发展的同时，也产生了大量的畜禽粪污（粪和尿）。畜禽养殖废弃物（粪、尿和冲栏废水）产量大而且集中，远远超过了自然分解能力，如不科学处理，势必严重污染环境和水源，而且传播疾病，也影响畜牧业自身的正常和健康发展。因此，畜禽粪污的资源化处理和利用对降低环境污染有着重要的意义。厌氧消化是处理畜禽粪便的有效途径，不仅能解决燃料短缺问题，而且实现了畜禽养殖废弃物无害化处理及多层次资源化利用。温度是影响厌氧消化的重要因素之一，于晓章等[1]研究了温度对厌氧嗜热菌群产甲烷能力的影响，Kettunen R. H. 等[2]研究了低温对甲烷菌的影响。本试验针对猪、鸡、牛粪便分别在 35℃、25℃和常温条件下厌氧发酵产生沼气性能，拟研究在各温度条件下对厌氧发酵过程的 pH 值、产沼气性能、上清液 COD（化学耗氧量）及 NH_4^+-N（氨氮）的影响。

* 本文已收录于《贵州农业科学》，2009，37（12）：148-151.

通讯作者：陈智远，男，本科，助理工程师，杭州能源环境工程有限公司，浙江杭州江干区凤起东路 118 号 7 楼 310020，chenzhiyuan2004@163.com，15158879600

2　材料与方法

2.1　试验设计与装置

猪粪、鸡粪和牛粪分别取自成都郊区某养殖场的新鲜粪便，TS（干物质）分别为 23.28%、30.90%和 17.79%。接种污泥取自成都郊区某沼气站，并在适宜温度下放置直至不产气，TS 为 13.57%。将取回的新鲜猪、鸡和牛粪便除去较大杂物，人工适当搅拌，按一定比例和活性污泥（均按 TS 计）分别进行混合：猪粪：污泥＝4：6，鸡粪：污泥＝3：7，牛粪：污泥＝4：6，并加水稀释到 TS 为 8%的发酵液。将各处理粪便的发酵液分别装入预先组装好的发酵罐内，密封。再分别将发酵罐按设计放在桌面上（裸露在空气中）和恒温水浴锅中，恒温水浴锅温度分别保证在（35±1）℃、（25±1）℃，进行厌氧发酵，每日观察发酵情况。试验采用单相厌氧发酵工艺，试验装置如图 1 所示。

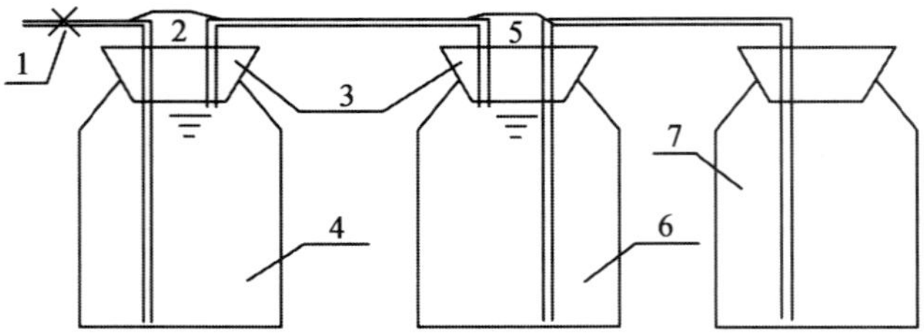

1. 止水夹；2. 胶管；3. 胶塞；4. 发酵罐；5. 玻璃管；6. 集气瓶；7. 集水瓶

Figure 1　Schematic diagram of experimental setup

图 1　试验装置示意图

2.2　测定指标与方法

pH 值，使用酸度计每 2 d 测量 1 次；沼气产量，利用排水法测定沼气日产量和最终的产气总量；COD，采用快速催化消解法，每 7 d 测定 1 次厌氧发酵物的上清液 COD 值；NH_4^+-N，采用纳氏试剂比色法，每 7 d 测定 1 次厌氧发酵物上清液的氨氮值。

3　结果与分析

发酵液调配好后，测定其 TS，猪粪、鸡粪、牛粪发酵液 TS 分别为 7.23%、6.10%、7.40%。试验历时将近 2 个月，图 2 是试验期间的水温变化，水温在前 20 d 左右，升高较为迅速，之后基本维持在 20～24℃。

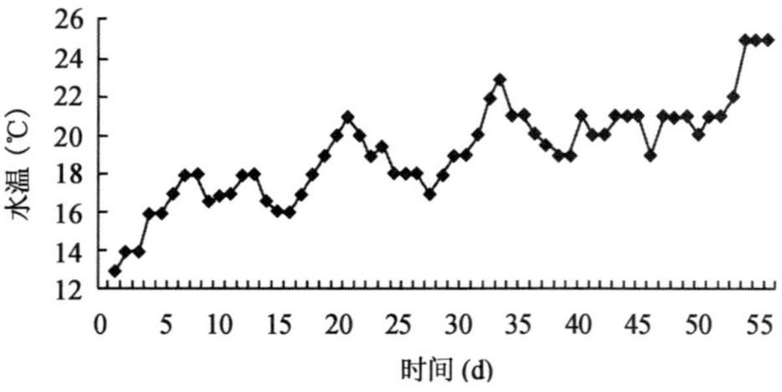

Figure 2　Water temperature changes at ambient temperature

图 2　常温条件下水温变化

3.1　pH 值的变化

3.1.1　猪粪发酵液

初始 pH 值为 7.41（图 3），随着厌氧发酵进行，在 35℃、25℃和常温 3 种温度条件下 pH 值都开始逐渐降低，并分别在发酵第 4 d、8 d 和 18 d pH 值下降到最低点，为 6.54、6.45 和 6.49，这是由于发酵初期细菌对高分子化合物进行水解形成挥发性脂肪酸等中间产物，使发酵体系 pH 值偏低[3,4]；随着厌氧发酵的进行，产甲烷菌群的数量增多，对产酸简短的产物利用率提高，料液中有机酸被消耗，同时料液中的氨化作用（蛋白质分解和脱氨作用，产生缓冲剂——氨[5])使料液的 pH 值逐渐升高到 7～7.5；在发酵第 46 d 各温度组 pH 值又缓慢下降，直至试验结束，这是由于发酵微生物在发酵后期外部环境(食物缺乏、pH 值上升等)不宜，即发酵微生物将进入衰亡期而大量死亡留下大量体内营养物质，发酵微生物少，发酵作用产生酸类使 pH 值又下降。各温度组在厌氧发酵过程中 pH 值在各时间点略有不同，但变化趋势基本相同。

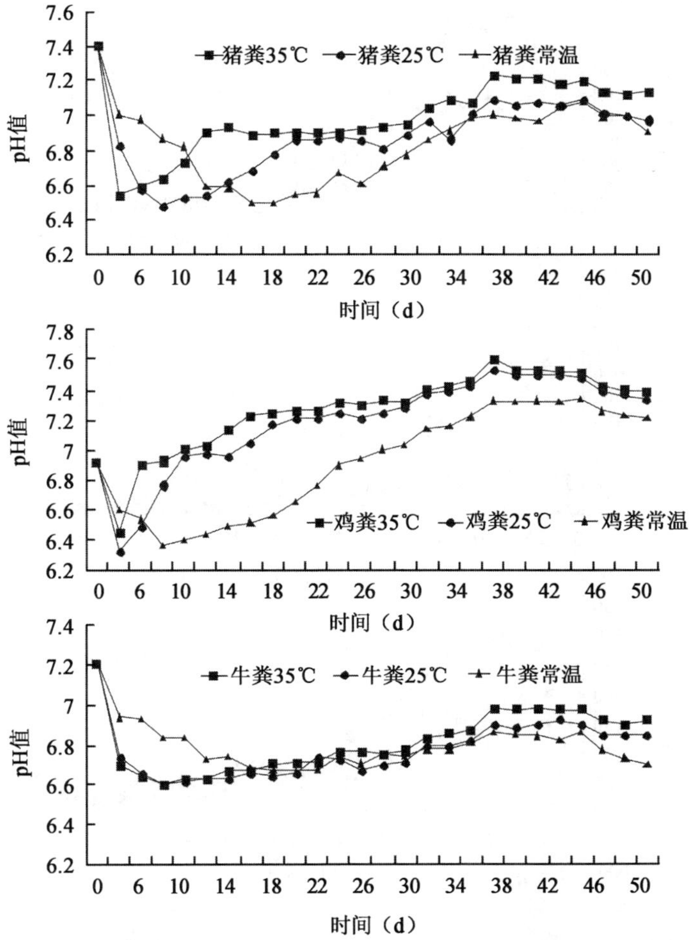

Figure 3 The effect of temperature on substrate's pH

图 3 温度对料液 pH 值的影响

3.1.2 鸡粪发酵液

初始 pH 值为 6.91，随着厌氧发酵进行，在 35℃、25℃和常温 3 种温度条件下 pH 值都开始逐渐降低，并分别在发酵第 4 d、6 d 和 8 d 下降到最低点，为 6.47、6.35 和 6.32，随后 pH 值上升到 7～7.6；发酵第 38 d 左右各温度组 pH 值又缓慢下降，直至试验结束。图 3 显示，35℃、25℃条件下的 pH 变化很相近，可能是一方面所选用的接种物即污泥中微生物都是中温厌氧发酵后的底物，所以发酵瓶中微生物菌群以适宜中温发酵的微生物群落为主。一般认为，不同温度的厌氧发酵是由不同温度类型的微生物进行的[6]，当环境温度改变时，微生物活性降低，此时产酸菌就占据了相对优势，生长较旺。另一方面，在一定温度范围内，温度高、酶的活性也高的缘故。

3.1.3 牛粪发酵液

初始 pH 值为 7.21，随着厌氧发酵进行，在 35℃、25℃和常温 3 种温度条件下 pH 值都开始逐渐降低，并分别在发酵第 8 d、8 d 和 20 d pH 值下降到最低点，为 6.59、6.58 和 6.66；随后 pH 缓慢升高至 7 左右；在发酵第 44 d 左右 pH 逐渐缓慢降低，直至试验结束。图 3 显示，35℃、25℃条件下 pH 值变化很相似，原因可能与鸡粪组相同。

3.2 产沼气性能

从图 4 可以看出，猪粪、鸡粪与牛粪在 35℃、25℃条件下日产气变化曲线趋势相同，都基本呈现出双峰变化趋势，而常温则出现多峰变化趋势，与常温条件下水温变化曲线相关联。

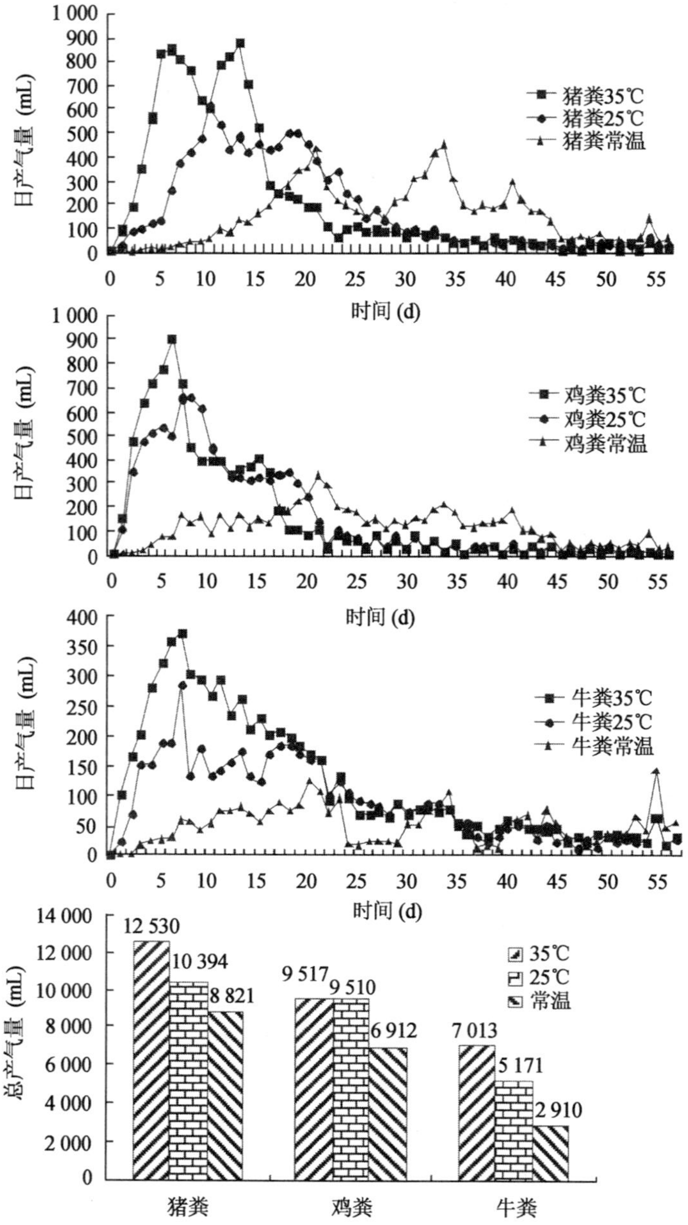

Figure 4　The effect of temperature on biogas production

图 4　温度对沼气产量的影响

3.2.1 猪粪发酵液

各温度条件下发酵前几天日产气量较少，随后开始逐渐升高，35℃、25℃和常温组分别在发酵第 6 d、10 d、21 d 出现第 1 次高峰，为 855 mL、612 mL、430 mL；随后逐渐降低，在发酵第 10 d、14 d、26 d 降低到 603 mL、423 mL、148 mL；随后又开始升高，在发酵第 13 d、18 d、34 d 出现第 2 次高峰，达到 878 mL、500 mL、448 mL；随后日产气量逐渐减小，但常温组在第 39 d 降至 180 mL，在第 41 d 又上升至 292 mL，第 53 d 降至 33 mL，第 55 d 上升至 138 mL，随后逐渐降低至试验结束。35℃、25℃和常温组的总产气分别为 12 530 mL、10 394 mL 和 8 821 mL，TS 产气率分别为 481 mL/g、399 mL/g、339 mL/g。综上，猪粪厌氧发酵在 35℃条件下具有明显优势。这与刘荣厚[7]以猪粪为发酵原料、以中温厌氧发酵瓶的底物为接种物，在自制的小型厌氧发酵装置上，研究了温度条件对猪粪厌氧发酵产气特性的影响的试验结果相似，在发酵初、中期，35℃组显现了明显的优势。

3.2.2 鸡粪发酵液

发酵前期各温度条件下日产气量较少，随后开始逐渐升高，35℃、25℃和常温组分别在发酵第 6 d、8 d、21 d 出现第 1 次高峰，为 902 mL、663 mL、333 mL；随后逐渐降低，在发酵第 12 d、14 d、30 d，产气量降低到 338 mL、313 mL、126 mL；随后又开始升高，在发酵第 15 d、18 d、34 d 出现第 2 次高峰，达到 410 mL、350 mL、218 mL；随后日产气量逐渐减小，但常温组在以后的发酵中陆续出现高峰，直至试验结束。35℃、25℃和常温组的总产气分别为 9 517 mL、9 510 mL 和 6 912 mL，TS 产气率分别为 578 mL/g、574 mL/g 和 417 mL/g。综上，鸡粪厌氧发酵在 35℃条件下具有明显优势。

3.2.3 牛粪发酵液

随着厌氧发酵进行，牛粪日产气量开始逐渐升高，并 35℃、25℃和常温组分别在第 7 d、7 d、20 d 出现第 1 次高峰，达到 373 mL、286 mL、125 mL；随后经过一段无规律的产气震荡期，这可能与牛粪中纤维含量高易形成浮渣层使底物与接种物接触不够充分有关，随后在发酵第 13 d、17 d、34 d 再出现小高峰，产沼气量达到 265 mL、185 mL、108 mL；随后日产气量逐渐减小，但常温组在以后的发酵中陆续出现高峰，直至试验结束。35℃、25℃和常温组的总产气量分别为 7 013 mL、5 171 mL 和 2 910 mL，TS 产气率分别为 263 mL/g、194 mL/g、109 mL/g。综上，牛粪厌氧发酵在 35℃条件下具有明显优势。

3.3 发酵料液上清液 COD 变化

猪粪、鸡粪、牛粪发酵液的初始 COD 分别为 33 553 mg/L、29 980 mg/L、30 136 mg/L。从图 5 可以看

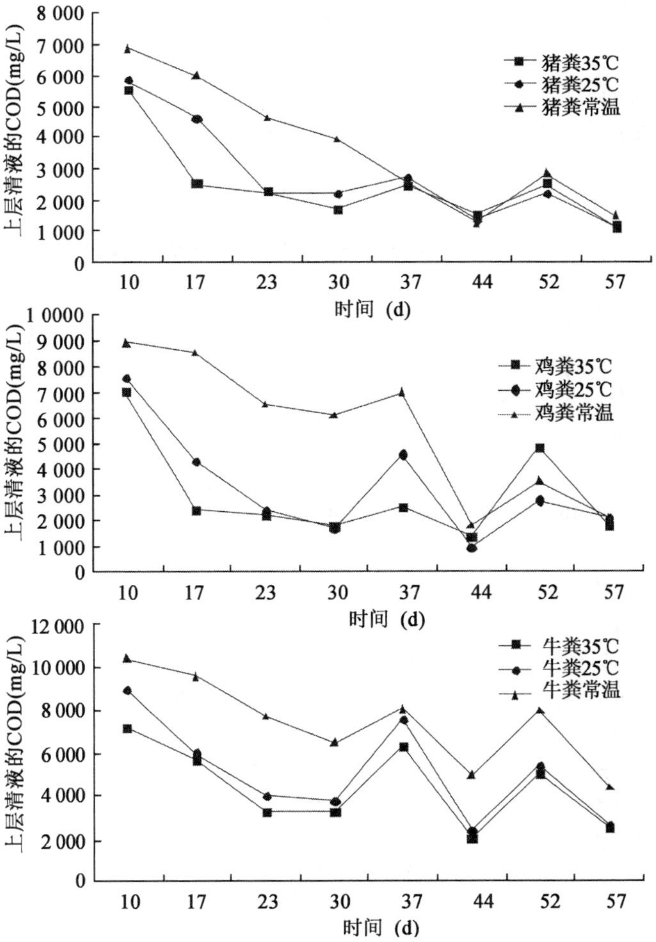

Figure 5 The effect of temperature on COD in supernatant

图 5 温度对上清液 COD 的影响

出，各温度组上清液 COD 总体趋势都是下降的，并且变化趋势基本相同。这是由于在厌氧发酵过程中，发酵体系内各种微生物类群利用营养物质的直接结果，发酵微生物利用发酵物料中的各种营养元素转化成部分气体的同时，自身的代谢也消耗了部分能量，使得整个发酵系统内及发酵上清液的 COD 降低。35℃组在发酵初期上清液 COD 下降最快，这是由于温度通过对厌氧微生物细胞内某些酶的活性的影响，进而影响微生物的生长速率和微生物对基质的代谢速率[8]，温度的下降会使 COD 去除率下降[9]。

3.4 发酵料液上清液 NH_4^+-N 变化

猪粪、鸡粪、牛粪发酵液的初始 NH_4^+-N 分别为 713 mg/L、1 504 mg/L、721 mg/L。从图 6 可以看出，猪粪组上清液 NH_4^+-N 总体变化趋势都是缓慢上升的，并且变化趋势基本相同，这是由于微生物氨化作用及有机氮的矿化分解，水溶性 NH_4^+-N 含量增加[10]，对于畜禽养殖粪便污水（发酵液）中的 NH_4^+-N 去除效果不明显[11]；其上清液 NH_4^+-N 在对应的时间点，35℃组最高，25℃组次之，常温组最低，这是由于温度较低时，微生物生长繁殖速度较慢，基质利用率不高，而在较高温度下，发酵罐中微生物不断分解基质，体系中生化反应较在室温时剧烈，其上清液 NH_3-N 变化也较明显[12~23]；35℃组上清液 NH_3-N 在发酵 52 d 后缓慢下降，是因为发酵罐中的微生物因外部环境（食物缺乏、pH 值上升等）不宜进入衰亡期，产生较少的氨态氮，并且温度条件较高时其氮元素的损失也会较高[14]，但是又因为厌氧发酵过程中没有暴露在空气中，而是在密闭的窗口中进行，既产生氨气又产生有机酸，形成有机酸的氨盐从而保护了氨，由此也减少了氮的损失，因此发酵

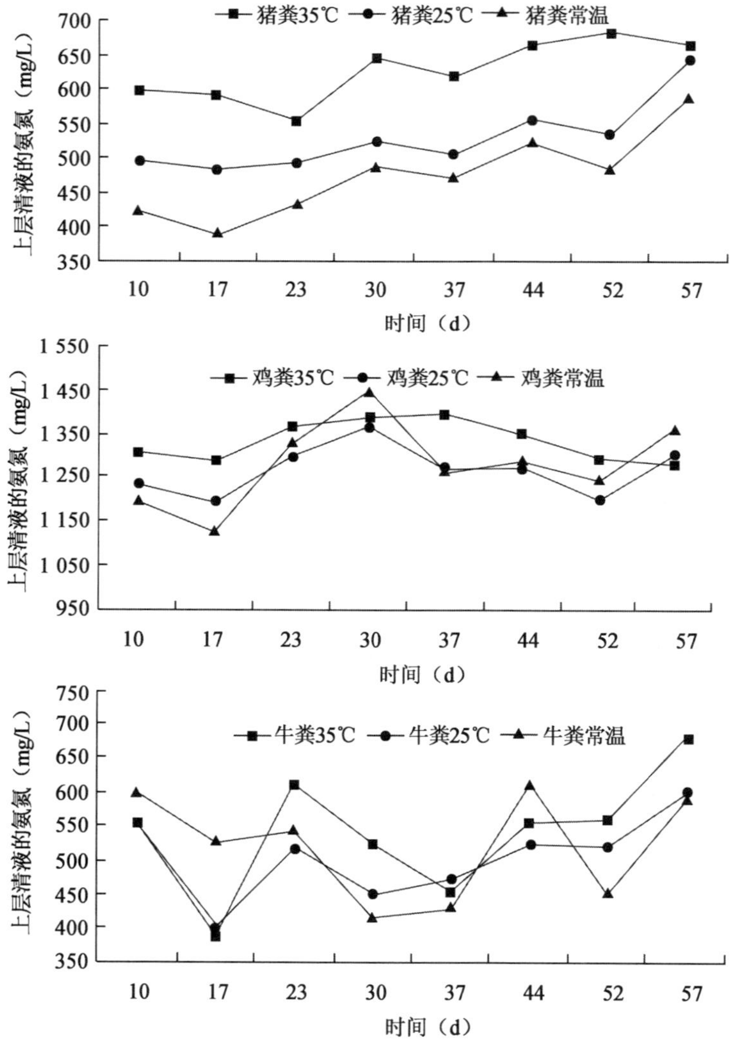

Figure 6 The effect of temperature on ammonium nitrogen in supernatant

图 6 温度对料液上清液氨氮的影响

处理猪粪减少了氮元素的损失[15]。鸡粪组上清液 NH_4^+-N 总体变化趋势都是上升的，并且上清液 NH_4^+-N 都比较高，这是因为新鲜鸡粪的粗蛋白高达 32.3%（以干物质计）[16]，还有较多氨、尿素、尿酸、肌酸和肌酸肝等非蛋白态氮素化合物[17]；35℃和 25℃组上清液 NH_4^+-N 变化趋势基本相同，并且在对应的时间点，35℃组上清液 NH_4^+-N 浓度较 25℃组高；常温组起伏较大，是因为氨气在水中的溶解度随温度升高而降低，在发酵第 30 d 上清液 NH_4^+-N 有一个高峰。牛粪在各温度条件下上清液 NH_4^+-N 的变化趋势基本相同，到试验结束时上清液的 NH_4^+-N 都升高。

4 结论

通过在 35℃、25℃和常温条件下厌氧发酵的对比，猪、鸡、牛粪便在各温度下发酵的 pH 值均呈先下降后升高趋势，并且变化速率随温度升高而加快，发酵温度降低，发酵液的 pH 值降到最低点的时间随之推迟，升高幅度随之减小，各粪便组厌氧发酵系统正常运行；各粪便组厌氧发酵在 35℃条件下具有明显优势，随发酵温度的降低，日产沼气达到高峰的时间随之延长，总沼气产量及 TS 产气率也随之减小，厌氧发酵性能也降低；料液上清液 COD 去除率 35℃条件下最高，随着发酵温度的降低，发酵上清液 COD 变化小，且较滞后；料液上清液 NH_4^+-N 总体变化趋势都是上升的，随着发酵温度的降低，发酵变化小，且较滞后；鸡粪较猪粪与牛粪容易酸化，TS 产气率也最高，刚好与 3 种不同消化吸收系统的特点符合。

项目及基金支持

"十一五"国家科技支撑计划"新型高效规模化沼气工程"重大项目课题(2008BADC4B10)

参考文献

[1] 于晓章，彭晓英，周朴华. 温度对厌氧嗜热菌群产甲烷能力的影响[J]. 湖南农业大学学报(自然科学版)，2005，31(4)：422-426.

[2] Kettunen R. H.，Rintala J. A. The effect of low temperature (5～20℃) and adaptation on the methanogenic activity of biomass [J]. Applied Microbiological Biotechnology，1997,48:570-576.

[3] 周德庆. 微生物学教程[M]. 北京:高等教育出版社，2002:275-276.

[4] 周岭. 有机废弃物厌氧发酵特性的研究[D]. 东北农业大学，2003.

[5] 徐曾符. 沼气工艺学[M]. 北京:农业出版社，1981.

[6] 周孟津. 沼气生产利用技术[M]. 北京:中国农业大学出版社，2005.

[7] 刘荣厚，郝元元，武丽娟. 温度条件对猪粪厌氧发酵沼气产气特性的影响[J]. 可再生能源，2006(5)：32-35.

[8] 周孟津，张榕林. 沼气实用技术[M]. 北京:化学工业出版社，2004.

[9] 夏雪芬，苏鸿洋. HABR 反应器常温下处理生活污水运行特性[J]. 安全与环境工程，2005，12(1):46-49.

[10] 李承强，魏源送，樊耀波，等. 堆肥腐熟度的研究进展[J]. 环境科学进展，1999，7(6):1-12.

[11] 顾夏生. 废水生物处理工程[M]. 北京:清华大学出版社，1985.

[12] 姚燕. 用畜禽粪便为原料生产优质厌氧发酵液工艺条件的研究[D]. 河南农业大学，2003.

[13] 雷廷宙，张全国，范振山，等. 沼液中全氮含量的影响因素试验研究[J]. 太阳能学报，2005，26(3)：396-399.

[14] 郝元元. 猪粪厌氧发酵工艺条件对产气特性及沼液营养成分的影响[D]. 沈阳农业大学，2006.

[15] 周孟津. 沼气生产利用技术[M]. 北京:中国农业大学出版社，2005.

[16] 钱永清，许大新，孙智锋，等. 鸡粪的营养成分分析[J]. 上海市农业学报，1994，10(增刊):37-40.

[17] 孙希琪. 鸡粪转化研究的进展[J]. 山东农业大学学报，1992，23(3):15-16.

猪粪和水果垃圾联合厌氧发酵:移动式中试运行结果

L. M. Ferreira, E. A. Duarte, D. Figueiredo

(葡萄牙里斯本科技大学农学院农业与环境化学系,Lisboa)

【摘要】 葡萄牙的生猪生产(特别是从产崽到出栏方式)的特点是产生大量稀释了的粪污,总固体(TS)含量为1.5%～2.5%,其中挥发性固体(VS)占67%。这种情况影响了厌氧发酵工艺的经济性。因此在一些大规模沼气工程中,采用其他底物与猪粪废水联合发酵以提高产气率有着重要的意义。而大部分发酵罐的使用时间较长,产气率也较低,其主要是由于使用者技术的缺乏,发酵罐设计上的不足,或是没有定期进行维护造成的。

为了更好地实现规模化农场中分散式沼气生产和生物垃圾管理,使用联合发酵技术,开发了移动式厌氧发酵中试装置。移动式发酵装置的特点是可以根据不同区域的生物垃圾组成,现场测试并运行不同底物的混合发酵工艺。

本文主要研究内容是利用水果仓库与水果分销系统丢弃的水果垃圾和猪粪废水联合发酵,开发规模化农场的分散式沼气生产工艺。最主要的目标是在实际运行中利用当地生产的水果垃圾(苹果和梨)来生产沼气,并与实验室结果(1),(2)进行比较,最终预测将来大规模应用时可能出现的问题。

移动式厌氧发酵装置可以通过一个标准集装箱式装载货车运输,它包括一个半自动装置,一个配有加热装置和机械搅拌器的不锈钢发酵罐($V=2\ m^3$),一个混合罐和一些用来储存粪污并用泵输送底物和发酵残余物的储罐,每个储罐都配有搅拌系统和输送泵。这套装置可以采用连续发酵工艺全自动运行,平均可运行10 d不需要重新添加底物。

在水力停留时间(HRT)为16 d的连续中温发酵中试条件下,研究了有机负荷为1.62～2.27 kg CQO/(m^3 · d)的猪粪废水(PS),有机负荷为4.8～5.1 kg CQO/(m^3 · d)的猪粪废水和水果垃圾(FW)3∶7的混合物时的产气情况。

主要监测的工艺运行参数有甲烷、二氧化碳、硫化氢和沼液沼渣中COD、氮、磷的含量。在处理完三个反应器容积底物之后,即系统被认为达到稳定运行状态时,得到运行结果。结果显示,当水果垃圾生物转化液和猪粪废水联合发酵时,容积产气率有明显增加。预计其产气率可以从0.56 m^3沼气/(m^3消化池容积·d)上升到1.84 m^3沼气/(m^3消化池容积·d),尽管其所产沼气中甲烷含量略有下降,但仍然可达到69%。沼气中的硫化氢含量大幅度降低,其主要原因是水果垃圾生物转化液中硫含量很低。并通过监测pH和总挥发性脂肪酸/碳酸氢盐碱度(T-VFA/BA)来观察工艺的稳定性,结果显示中试试验的稳定性强于之前的实验室实验。

产沼气能力方面,采用同样的混合物进料,中试试验的结果优于之前的实验室实验,其原因可能在实验室的操作、猪粪废水的保存时间、发酵罐的设计、粪污的采样等方面。中试结果显示采用水果垃圾联合发酵,发酵罐产沼气能力可提高300%。

【关键词】 厌氧消化;混合发酵;猪粪;水果垃圾

ANAEROBIC CO-DIGESTION OF PIG SLURRY WITH FRUIT WASTES-ON SITE OPERATION WITH A MOBILE PILOT PLANT*

L. M. Ferreira, E. A. Duarte, D. Figueiredo

Technical University of Lisbon

Contact: Luís Ferreira, Researcher, Dpt. Agricultural and Environmental Chemistry, Superior Institute of

---

* This paper was initially presented at the Orbit 2009 CHINA. 本文已发表于 Orbit 2009 CHINA。

Agronomy, Technical University of Lisbon, Tapada da Ajuda 1349-017 Lisboa, Portugal, Tel.: +351-2136-53424, E-Mail: lferreira@isa. utl. pt

EXECUTIVE SUMMARY

Pig production units (particularly farrow-to-finish operations) in Portugal are characterized by generating very dilute slurries, 1.5%-2.5% total solids (TS) with a total volatile solid (VS) content of 67%, and this represents a barrier to establish economically feasible Anaerobic Digestion (AD) processes. On the other hand there are several full scale farm biogas plants where could be interesting to introduce other substrates to co-digest with pig slurry, in order to raise biogas production. Most of these digesters are very old and operate in very inefficient conditions due to operators lack of technical skills, inadequate design of the plants or because reinvestments were not done in due time.

To demonstrate on site the capabilities and potential of farm scale co-digestion as a technological key tool for decentralized biogas production and biowaste management it was developed a mobile AD pilot plant. One of the advantages of the mobility characteristic of this plant is the possibility to test and demonstrate on site, the co-digestion process of different substrates according to the biowaste set from the different local economies.

This particular research field work, was focused on the utilization of fruit wastes (FW), rejected from centralised fruit storage and distribution facilities of producers as a co-substrate for the development of farm scale decentralised biogas production processes, with pig slurry (PS). Among other goals, the main objective of this investigation was to evaluate in real operation conditions the interest of the utilization of fruit wastes (FW) produced in the region (apples and pears), for biogas production, and to compare the results from lab scale trials of previous investigations (1), (2) and to anticipate problems for future full scale application.

The mobile AD plant, can be carried in a standard container loading truck, it consists of a semi-automatic device, equipped with a heated stainless steel digester ($V = 2\ m^3$) with a mechanical mixer, a mixing tank and tanks for slurry, pumpable feedstocks and digestate, all equipped with the respective pumping and mixing systems. This plant can operate continuous processes, fully automatic, during an average period of 10 days, before any refilling of substrates.

Dynamic mesophilic (35-37℃) continuous pilot trials with a HRT=16 days, were performed with pig slurry (PS) (OLR = 1.62-2.27 kg CQO/($m^3 \cdot d$)) and with a mixture of FW:PS with a volumetric composition of 30:70 (OLR = 4.8-5.1 kg CQO/($m^3 \cdot d$)).

The main process operational parameters (methane, carbon dioxide, H_2S and COD fractions of the digestate, nitrogen and phosphorous) has also been investigated. Results were obtained considering a steady-state achieved after three reactor volumes. Process performance results; show that specific biogas productivity when BFW is codigested with pig manure can be considerably raised. An increase from 0.56 to 1.84 m^3 biogas/ (m^3 digester · d) can be foreseen, nevertheless the biogas quality resulting from digestion of 30%BFW:70%PM mixture became slightly poor, although with a percentage in methane of 69%. The biogas quality regarding the H_2S content improved significantly, although the reduction achieved from the digestion of the mixture 30%BFW:70%PM, results mainly from the low S content of the BFW. Process stability was monitored measuring the pH and T-VFA/BA ratio. Both indicators, suggests that the co-digestion process developed in the pilot plant is more robust than the one developed in the lab scale.

With regards to biogas production performance, the pilot trial results were better than those obtained from AD lab trials using the same feedstock mixture of FW:PS. Lab manipulation, slurry conservation and storage period required to feed a continuous lab process, digester design and slurry sampling procedures, are some of the reasons identified that could justify the differences obtained. It was demonstrated the potential to increase the biogas production up to 300% through the use of FW as a co-substrate.
Key words: anaerobic digestion; co-digestion; pig slurry; fruit waste

1 Introduction

This research work was developed in the framework of a three years demonstration project, focused on the decentralised utilization of organic wastes from the food industry and agriculture, to produce biogas in farm scale plants. One of the main goals of the project was to involve the main stakeholders in one hand on the concept adoption and on the other to be aware about the technical and non technical barriers of farm scale systems, based in anaerobic codigestion for biogas production and organic waste management.

The management of different organic waste flows to process in biogas plants requires a proper logistic plan from suppliers and pre-storage, pre-processing and pos-storage facilities. To overcome the failure risks of these operations in an industrial farm plant, it was designed a mobile AD pilot plant capable to perform AD processes in a scale that was real enough to envisage and preview the needs of an industrial system operating in a farm plant.

The research work described in this paper is one part of the component of activities developed to upgrade biogas production.

1.1 Background

Pig production units (particularly farrow-to-finish systems) in Portugal are characterized by generating very dilute slurries, 1.5%-2.5% total solids (TS) with a total volatile solid (VS) content of 67%, and this represents a barrier to establish economically feasible Anaerobic Digestion (AD) processes. On the other hand there are several full scale farm biogas plants where could be interesting to introduce other substrates to co-digest with pig manure (PM), in order to raise biogas production. Most of these digesters are very old and operate in very inefficient conditions due to operators lack of technical skills, inadequate design of the plants or because reinvestments were not done in due time.

Fruit wastes (FW) are produced in large quantities in centralised fruit producers' storage and distribution facilities, during the selection and rejection processes before fruit enter into the market. The waste stream targeted by this project was originated from a group of apple and pear producers.

One of the most promising alternatives for managing these organic wet wastes is anaerobic digestion (AD). A major limitation of anaerobic digestion of FW is a rapid acidification of these wastes decreasing the pH in the reactor, and a larger volatile fatty acids production (VFA), which can stress or inhibit the methanogenic biomass activity (Bouallagui et al, 2005).

A previous research work concluded that a bioconversion process of fruit wastes before feeding a co-digestion process was considered advantageous, once the result of this bioconversion is a stable product (Bioconverted Fruit Wastes-BFW) in anoxic conditions, more convenient to be stored, handled and improves the stability of the AD process (Ferreira et al, 2007). The same authors studied the co-digestion of a mixture of BFW : PM, 30 : 70 (v : v) with an average organic loading rate (OLR) of 4 kg COD/(m^3 · d), operating at a hydraulic retention time (HRT) of 16 days.

A mobile AD pilot plant was developed to demonstrate on site the capabilities and potential of farm scale co-digestion as a technological key tool for decentralized biogas production and biowaste management. One of the advantages of the mobility characteristic of this plant is the possibility to test and demonstrate on site, the co-digestion process of different substrates according to the biowaste set from the different local economies.

1.2 Research objectives

Among other goals, the main objective of this investigation was to evaluate in real operation conditions the interest of the utilization of fruit wastes (FW) produced in the region (apples and pears), for biogas production. Furthermore, to compare the lab scale trials results from previous research work and to anticipate problems for future full scale application.

2 Methodology

2.1 Strategy

Before starting operating the mobile pilot plant, it was developed and studied at lab scale the anaerobic co-digestion process of fruit wastes with pig manure. It was concluded that a bioconversion process of fruit wastes before feeding a co-digestion process would be advantageous and a mixture (% v/v) of 30% of it with 70% pig manure would be advisable.

2.2 Origin of materials

The substrates used were screened pig manure (PM) from a farrow-to-finish pig farm and fruit wastes (FW) characterised by a mixture of refused flows of apples and pears. To perform the lab trials, samples of pig slurry were collected according to a procedure in order to get weekly composed samples. Fruit waste was pulped with a semi-industrial fruit mill with a 5 mm screen. Batchs of fruit pulp (aprox. 500 L) were stored in a 1 m^3 closed vessel with mechanical mixer at environmental temperature, in order to develop a spontaneous fermentation process. During this process, internal pressure was alleviated releasing the CO_2 produced, although due practical reasons it was not possible to measure it. After aprox. 10 days of fermentation, the pulp of bioconverted fruit wastes (BFW) was considered ready to be used as a co-substrate (*Figures 1a-1c*).

(a) (b) (c)

Figure 1 (a) Pig slurry rotating screener; (b) Refused apples and pears being pulped with a fruit mill/pulper with 5 mm screen; (c) Feeding fruit pulp to fermenter

Inoculum was obtained from a mesophilic (35℃) sewage sludge digester. Characteristics of BFW and PM are presented in *Table 1*.

Table 1 Initial characteristics of waste materials

	Lab trial		Pilot trial	
	PM	BFW	PM	BFW
pH	7.13-7.70	3.53-3.69	7.06-8.79	2.71-3.24
Conductivity (mS/cm)	10-15	3-4	16-20	3-4
TS (g/L)	15-21	56-59[a]	20-29	49-56[a]
VS (g/L)	10-14	49-57[b]	13-22	46-51[b]
COD (g/L)	17-36	168-206	25-36	161-181
CODsoluble (g/L)	9-23	132-164	13-15	119-141
T-P (mg/L)	342-668	106-109	525-552	107-120
TK-N (g/L)	1.8-1.9	0.64-0.71	2.07-2.34	0.62-0.69
NH_4^+-N (g/L)	1.04-1.34	0.12-0.14	1.52-1.71	0.11-0.13
T-VFA (g acetic acid/L)	n. d	5.5	n. d	6.0
Glucose (g/L)	n. d	0	n. d	0
Frutose (g/L)	n. d	0	n. d	0
Ethanol (g/L)	n. d	45.0	n. d	44.0

Note: n. d-not determined; [a] Non volatile fraction at 104℃;
[b] Volatile fraction at 550℃ of the non volatile fraction at 104℃

2.3 Lab and pilot set-up

Continuous lab trials, used a heated stainless steel digester (CSTR) with V= 11 L. Pilot scale trials were performed on site, in a farrow-to-finish pig farm. The mobile AD plant, was carried in a standard container loading truck, and it consists of a semi-automatic device, equipped with a heated stainless steel digester (V = 2 m^3) with a mechanical mixer, a mixing tank and tanks for slurry, pumpable feedstocks and digestate, all equipped with the respective pumping and mixing systems (*Figure 2*). This plant could operate the continuous process autonomously, fully automatic, during an average period of 10 days.

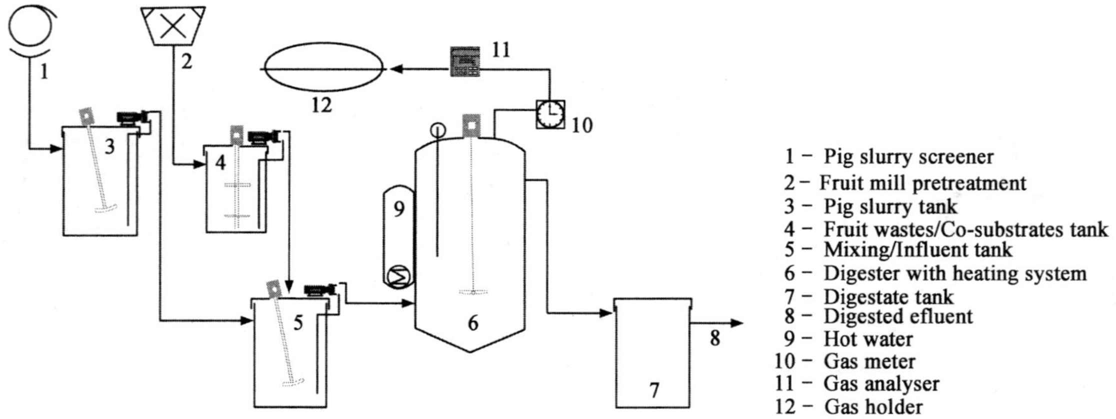

Figure 2 Diagram of the mobile pilot plant setup

2.4 Continuous trials

Considering the goals proposed, dynamic mesophilic (35-37℃) continuous lab and pilot-scale trials were performed. Pig manure AD lab trials operated with a OLR = 1.4-1.5 kg CQO/(m^3 • d) and a HRT=11 d, nevertheless it was possible to reach OLR = 1.6-2.3 kg CQO/(m^3 • d) with a HRT=16 d, on pilot plant trials. With regards to co-digestion trials with a mixture of FW : PM, a volumetric composition (% v : v) of 30 : 70 was used. Lab trials were operated at OLR = 3.8-4.2 kg CQO/(m^3 • d), although it was possible to reach na OLR = 4.8-5.0 kg CQO/(m^3 • d) with a HRT=16 d, on pilot plant trials.

The main process operational parameters (methane, carbon dioxide, H_2S and COD fractions of the digestate, nitrogen and phosphorous) has also been investigated. Results were obtained considering a steady-state achieved after three reactor volumes.

2.5 Analytical methods

The pH, chemical oxygen demand (COD), total solids (TS), volatile solids (VS), total kjeldahl nitrogen (TK-N), ammonical nitrogen (NH_4^+-N), total phosphorous (TP), total volatile fatty acids (T-VFA), Bicarbonate alkalinity were determined according to standards methods (APHA, 1995).

Glucose, fructose, ethanol and organic acids were determined by isocratic high performance liquid chromatography (HPLC) with IR or UV detection.

3 Results and discussion

3.1 Characterization of waste materials

Table 1, show the characteristics of the substrate flows used and differences can be observed on TS content of the pig manure used on the lab trials when compared with pig manure obtained on site during pilot scale trials. Composed sampling procedures and lab manipulation could be the reason for such differences. In comparison, fruit waste pulp had almost ten times more COD and 80% of it is soluble. Buffiere et al. (2005) reported very similar characteristics for apple wastes.

It can be foreseen that a better C/N ratio was achieved in the substrate mixture 30%BFW:70%PM when compared with 100% PM digestion. On the other hand *Figure 3*, shows the fruit wastes COD evolution during the bioconversion process and after a long storage period. A lost of 5% of the initial total COD was obtained which is a good indicator of the energy conservation during this process.

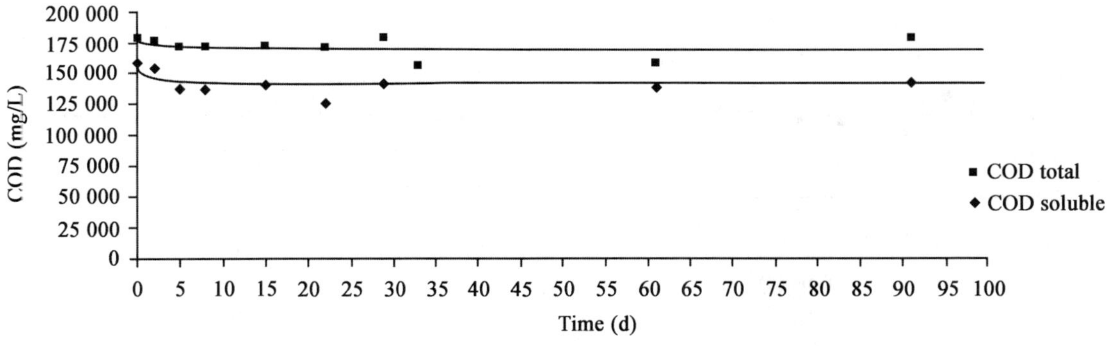

Figure 3 Chemical oxygen demand (COD) mg/L. Evolution during the bioconversion process and storage.

It can be discussed from *Table 2* in one hand the operating and process performances obtained between lab and pilot plant trials and on the other the results from the co-digestion of BFW with PM. The performance results obtained in the pilot plant were considerable better than those produced in lab scale. Some operational reasons could be argued, but the main reason is deeply related with the lower OLR achieved for the lab trials, conditioned by the PM quality. On site operation, enabled the use of pig manure on real conditions, avoiding sampling and lab manipulation interferences.

Process stability was monitored measuring the pH and T-VFA/BA ratio (*Figures 4a-b*, *5a-b*). A comparison between the results from lab trials and from pilot plant operation indicates a higher T-VFA/BA ratio was reached at lab scale, which would suggest a more stressed process. It was reported that without enough buffer capacity, values higher than 0.5 indicate stability problems in anaerobic digestion processes and this ratio should not be higher than 1 (Gerardi, 2003; Scharer, 2007). Both indicators, pH and T-VFA/BA ratio, suggests that the co-digestion process developed in the pilot plant is more robust than the one developed in the lab scale.

Table 2 Operating and process performance obtained from lab and pilot plant trials

Mixture BFW : PM (% v : v)	HRT (d)	OLR (kg COD/(m^3 · d)	m^3 biogas/(m^3 digester · d)	Biogas quality % CH_4	Biogas quality ppm H_2S	COD removal (%)	m^3 biogas/m^3 biomass
0 : 100[a]	11	1.4-1.5	0.47	73	>1 500	70	5.14
0 : 100[b]	16	1.6-2.3	0.56	75	>1 500	70	9.6
30 : 70[a]	16	3.8-4.2	1.57	67	605	75	24.31
30 : 70[b]	16	4.8-5.1	1.84	69	950	80	31.2

Note: [a] Lab trial; [b] Pilot plant trial

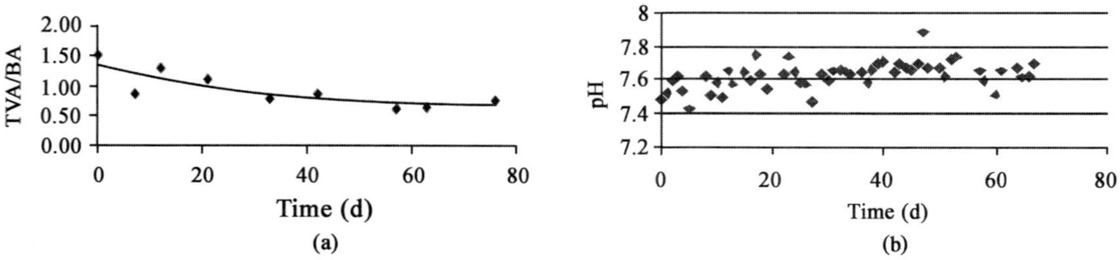

Figure 4 (a) TVA/BA evolution during the lab trial; (b) pH evolution during the lab trial

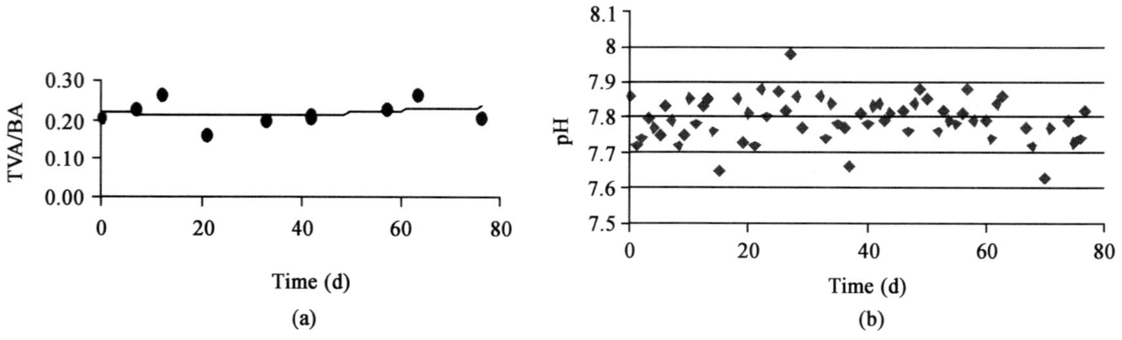

Figure 5 (a) TVA/BA evolution during the pilot-scale trial; (b) pH evolution during the pilot-scale trial

Process performance results; show that biogas productivity when BFW is codigested with pig manure can be considerably raised. An increase of 300% can be foreseen (*Figure 6*), nevertheless the biogas quality resulting from digestion of 30% BFW: 70% PM mixture became slightly poor in methane. The biogas quality regarding the H_2S content improved significantly, although the reduction achieved from the digestion of the mixture 30%BFW:70%PM, results mainly from the low S content of the BFW (results not shown).

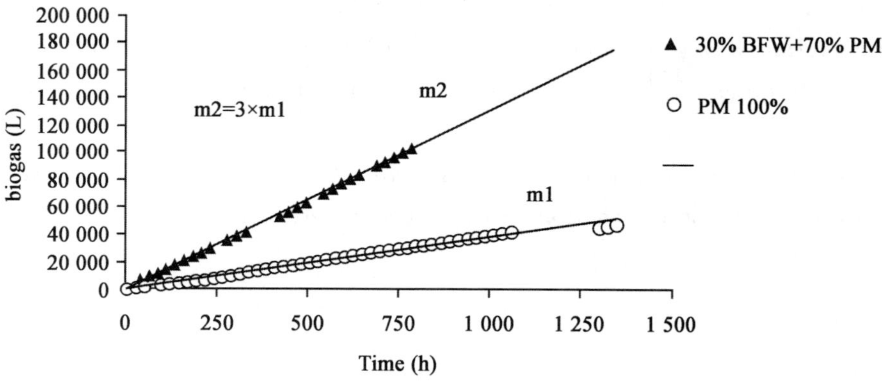

Figure 6 Biogas production rates performed during the pilot plant operation with the mixture of BFW:PM and PM

4 Conclusions

It was demonstrated the potential to increase the biogas productivity up to 300% through the use of bioconverted fruit wastes (BFW) as a co-substrate in anaerobic co-digestion processes with pig manure. Bioconversion of fruit wastes (apples and pears) was shown to be a feasible way of conserving methane potential and it is likely to offer a cost efficient solution for biogas production in farm scale biogas plants. The possibility to develop on site co-digestion processes, using versatile mobile pilot plant, enabled to obtain more realistic performance results than those obtained from lab scale trials. Particularly important were the results regarding process stability indicators measured, therefore the pilot plant operation was very useful to demonstrate the following:

(1) Conclusions concerning the process stability supported on the results from the lab scale co-digestion trials of a similar feedstock mixture of 30%BFW:70%PM were not definitive.

(2) The process was not unstable and it's robust enough to allow raising the OLR.

Lab manipulation, slurry conservation and storage period required to feed a continuous lab process, digester design and slurry sampling procedures, are some of the reasons identified that could justify the differences obtained.

Acknowledgements

We would like to thank the support given by the Program AGRO (Project 538-'CODIGANDES'), funded by EU and the Portuguese Institute for Biological Resources (INRB).

References

APHA, (1995): *Standard Methods for the Examination of Water and Wastewater*. American Public Health Association, Washington, DC, USA.

Buffiere P., Loisel D., Bernet N., Delgenes J. P., (2005): *Towards new indicators for the prediction of solid waste anaerobic digestion properties*. Proceedings of 4th International Symposium on Anaerobic Digestion of Solid Waste, September 2005, Copenhagen, Denmark.

Bouallagui H., Touhami, Y., Cheikh R. B., Hamdi M., (2005): *Bioreactor performance in anaerobic digestion of fruit and vegetables*. Process Biochemistry 40, pp. 989-995.

Gerardi M. H., (2003): *The microbiology of anaerobic digesters*. Hoboken, NJ: John Wiley & Sons, Inc.

Ferreira L., Duarte E. A., Silva C., Malfeito-Ferreira M., (2007): *Fruit wastes bioconversion for anaerobic co-digestion with pig manure. Process development for the recycling in decentralised farm scale plants.* Proceedings of the International Conference Progress in Biogas, Stuttgart, Germany, 2007, pp 135-140.

Scharer P., (2007): *Operating analytics of biogas plants to improve efficiency and to ensure process stability.* Proceedings of the International Conference Progress in Biogas, Stuttgart, Germany, 2007, pp. 135-140.

ANAEROBIC DIGESTION OF KITCHEN WASTE UNDER MESOPHILIC CONDITION

WANG Xinyi, DONG Renjie

China Agricultural University

Contact: Prof. Dr. DONG Renjie, China Agricultural University, No. 17 Qinghua Donglu, Haidian District, Beijing 100083, PR China, Tel.: +86-10-62737963, E-Mail: rjdong@cau.edu.cn

EXECUTIVE SUMMARY

In order to analyze the effects of different organic loading rates of 4 g VS/L, 8 g VS/L and 12 g VS/L, as well as to study the gas potential, daily biogas production and biogas yield, of kitchen waste, an anaerobic digestion experiment was carried out on kitchen waste of different feeding organic loading rates under 37℃. Results show that the effect was best when its organic loading rate was 4 g VS/L of kitchen waste concentrations on anaerobic digestion. At this organic loading rate, the gas production reached its peak at a digestion time (HRT) of 14 days and the final biogas yield amounted 528 mL/g VS.

Key words: Kitchen waste; different organic loading rate; anaerobic digestion

餐厨垃圾中温批式厌氧消化实验研究

王馨仪，董仁杰*

(中国农业大学，北京，100083)

【摘要】在 37℃温度下对不同 VS 浓度的餐厨垃圾进行批式厌氧发酵实验，分析不同 VS 浓度下(4 g VS/L、8 g VS/L 和 12 g VS/L)的厌氧产气规律(日产气量、沼气产率)。实验结果证明，当发酵液浓度是 4 g VS/L 时，餐厨垃圾厌氧处理的启动最快，在第 14 d 达到产气高峰；沼气产率最高，为 528 mL/g VS。随 VS 浓度的增加，发酵启动时间逐渐延长，持续的时间也延长。

【关键词】餐厨垃圾；不同的有机负载率；厌氧发酵

1 引言

餐厨垃圾是指居民日常生活以及除居民日常生活以外的食品加工、饮食服务、单位供餐等活动中产生的厨余垃圾和废弃食用油脂[1]。据统计，2009 年全国城市生活垃圾的年产量约为 1.57×10^{8} t[2]。餐厨垃圾是城市生活垃圾的重要组成部分，其所占比例为：北京 56%，天津 50%，上海 59%，深圳 57%，广州 57%[3]。如何有效安全地处理如此大量的餐厨垃圾，已成为现在面临的一大重要问题。

餐厨垃圾主要成分如表 1 所示。

Table 1 The main components of kitchen waste

表 1 餐厨垃圾的主要成分[4]

	水分	灰分	总碳	糖类	脂肪	蛋白质	盐分
各成分所占比例(%)	73.03	1.75	13.94	4.16	6.23	12.17	1.24

* 通信作者：董仁杰，男，博士，教授，中国农业大学，北京市海淀区清华东路 17 号中国农业大学 100083，rjdong@cau.edu.cn，62737963

城市生活垃圾可焚烧性有界限值，其数值为含水量≤50%、灰分≤60%、可燃成分≥25%[4]。以这个标准来衡量，虽然餐厨垃圾的灰分符合焚烧要求，但是其含水量和总碳（主要的可燃成分）均不在可焚烧标准范围内。另外，极高的含水量也降低了餐厨垃圾的热值。研究表明餐厨垃圾的单位热值为 2 100～3 100 kJ/kg，不能达到垃圾焚烧发电的热值要求（5 000 kJ/kg 以上）[5]，而且还能产生二噁英等毒害物质。因此餐厨垃圾不适合采用焚烧法处理。

如果进行填埋处理，餐厨垃圾的高水分含量又会带来垃圾渗滤水的问题。故对餐厨垃圾进行填埋处理不但填埋场本身会产生臭气，散播病菌，还会因为垃圾渗滤水的地表径流和渗透作用污染地表水和地下水[6]，并不是一个绿色环保的处理方法。由于垃圾填埋的成本低，填埋法依然是垃圾处理的主要方式。但长此以往会导致垃圾处理场占地不断增大，占用宝贵的土地资源，也增加了运输费用和处置负荷[7]。

用堆肥法处理垃圾虽然可以实现垃圾减量化并回收其中的营养成分，但是堆肥法占地和劳动强度都较大，而且生产出来的肥料肥效低，不能多次使用，否则其中的盐分会使土壤盐化[7]。

综上所述，通用的城市生活垃圾处理方式——填埋、焚烧和堆肥在应用到餐厨垃圾的处理上时都有一定局限性，因此寻求一种更好的处理厨余垃圾的方法势在必行。通过厌氧消化法处理餐厨垃圾是一种新兴的环境友好型处理方式。首先，厌氧技术可以使垃圾经厌氧消化后产生清洁能源——沼气；第二，消化最终物可作为高质量的有机肥料和土壤改良剂。第三，与好氧过程相比，厌氧消化无需氧气，降低动力消耗，因而使用成本低。第四厌氧消化减少了温室效应气体的排放量，因此做好厨余垃圾的厌氧发酵的研究可以真正实现垃圾处理的“无害化、减量化和资源化”[8]。

本试验采用批式厌氧发酵实验处理餐厨垃圾，并研究不同餐厨垃圾 VS 浓度下的厌氧产气规律。并确定餐厨垃圾的产气潜力。

2 材料与方法

2.1 实验装置

实验所用的反应器为 1 L 的厌氧瓶，有效容积为 0.6 L。瓶口以螺旋盖密封，上设出气口。气体由出气口上连接的硅胶管导出，用排水法收集。实验过程中反应器将置于恒温箱中，用温控仪控制温度为（37±1）℃。

2.2 实验材料

餐厨垃圾取自中国农业大学第一学生食堂。其主要成分包括米饭、肉类、蔬菜等。食堂的餐厨垃圾存放在餐具收集处的大铁桶中，为固液混合物的状态。为达到均匀取样的目的，首先用长柄勺对其进行搅拌，再取搅后较均匀的样品带回实验室中，剔除大的硬骨，木筷子和塑料袋，用小型榨汁机榨成浆状，放入－18℃的冰柜中冷冻。其主要性质见表 2。

Table 2 Sampling characteristics of kitchen waste and inoculum

表 2 原料与接种物性质

原料与接种物	TS(%)	VS(%)	pH
餐厨垃圾	21.46	19.96	4.69
接种物	5.51	3.74	7.57

接种物采用正常运行的猪粪厌氧消化反应器内取出的沼液，外观为灰黑色液体。其主要性质见表 2。

2.3 实验设计

实验在 12 个反应器中进行，分为 4 组，分别对应 0 g VS/L、4 g VS/L、8 g VS/L、12 g VS/L 的 VS 浓度，接种量均选取 50%。第一组为空白样品，不添加餐厨垃圾，300 mL 接种物作为底物进行中温厌氧批式消化实验，用来估算其他反应器的产气量中接种物的贡献率。第二、三、四组选取 3 个 VS 浓度 4 g VS/L、8 g VS/L、12 g VS/L 对餐厨垃圾进行批式中温厌氧消化实验，每个水平设置三个样本。300 mL 的接种物与相应浓度的原料混合作为发酵底物。各反应器进料量如表 3 所示。反应器置于恒温箱中，每隔 24 h 记录产气量，每天手动摇晃反应器 1 次，使餐厨垃圾与微生物充分接触。本批实验以 49 d 为观察周期。

Table 3 Design of experimental treatments

表 3 实验处理设计

	进料 VS 浓度(g VS/L)	进料量(g)	接种量(mL)	接种比例(%)	pH
R1	0	0	300	50	7.39
R2	4	12.57	300	50	7.38
R3	8	25.14	300	50	7.38
R4	12	37.71	300	50	7.34

2.4 测定指标和方法

总固体(TS)和挥发性固体物(VS)：烘干法[9]；pH 值采用 Orion 3-Star pH 计测定。

3 结果与分析

3.1 进料 VS 浓度对日产气量的影响

图 1 显示了餐厨垃圾在 3 个不同的 VS 浓度下的日沼气产量和接种物的日沼气产量在消化期内的变化情况。可以看出，不论是何种 VS 浓度，餐厨垃圾的厌氧消化都会在第 1 d 大量产气，接下来的 2～3 d 内迅速下降。这主要是因为餐厨垃圾内的米饭、肉类、蔬菜等经过蒸煮后，复杂的组织结构被破坏，水解产酸菌很快就可以将其降解生成大量的挥发性脂肪酸和 CO_2。空白样品在反应的前 12 d 内依然有一定量的产气。在前 9 d 产气量逐渐上升，到第 9 d 达到日产气高峰 100 mL，随后逐渐下降。因为接种物为运行中的猪粪厌氧发酵反应器中取出的新鲜沼液，本身还含有一定量的未降解有机物，所以在反应初期这部分有机物依然在产气。相比空白样品，添加了餐厨垃圾的其他样品组却在反应初期出现了不同程度的产气抑制。VS 浓度为 4 g VS/L 的样品组在经过前两天的产气高峰以后，第 3～7 d 产气量略低于空白样品，表现出了轻微的抑制现象；而 VS 浓度 8 g VS/L 的样品组在 3～7 d 产气基本停止，第 8 d 开始恢复；VS 浓度 12 g VS/L 从第 3 d 产气停止，一直到第 16 d 才开始恢复。这种现象可能是由于反应前两天原料迅速降解生成大量的 VFA，而产甲烷菌的世代时间低于产酸菌，还未大量增值，不能马上将这些 VFA 转变成甲烷，因此造成了 VFA 的积累，从而导致产甲烷菌的产气抑制。但是经过几天适应时间以后产甲烷细菌逐渐增殖，活性可以恢复。空白试样由于未添加餐厨垃圾原料，所以未出现此类抑制现象。4 g VS/L 的样品组在 14 d 出现第一个日产气高峰，其日沼气产量为 235 mL。该样品组在反应的全过程中仅出现此一个产气高峰，在三个 VS 浓度中它的产气高峰出现时间最早，到第 19 d 的时候反应基本结束。8 g VS/L 的样品组分别在 18 d 和 25 d 出现两个产气高峰，日沼气产量分别为 205 mL 和 162 mL，到第 27 d 的时候反应基本结束。12 g VS/L 的

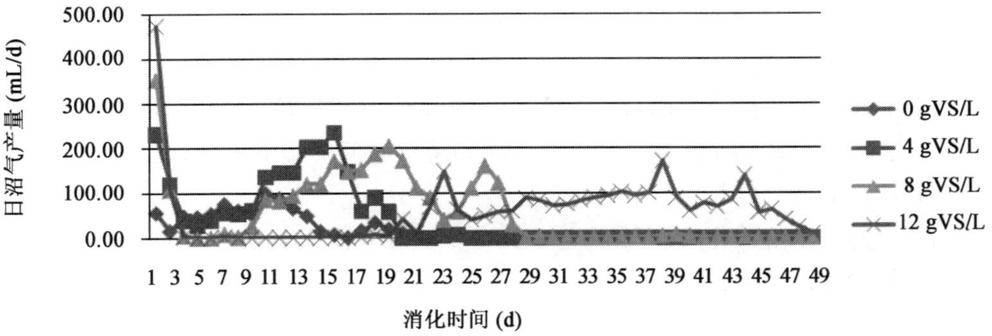

Figure 1 Daily biogas production vs digestion time during digestion of kitchen waste

图 1 中温厌氧消化餐厨垃圾的日产气量随时间变化图

样品组在 4 g VS/L 的样品组产气基本停止的时候才开始恢复，并分别在第 22 d，第 38 d 和第 44 d 出现三个产气高峰，其日产沼气量分别为 149 mL，172 mL，140 mL，一直到 49 d 反应才基本结束。对比三个 VS 浓度，从恢复产气到产气基本停止持续的时间分别为 11 d，19 d 和 33 d。由此可见在较低的 VS 浓度下餐厨垃圾批式厌氧消化启动快，反应周期短，日产气峰值高。随着 VS 浓度的提高，启动所需要的时间和反应的周期都会有一定程度的延长，日产气峰值也随之降低。在 12 g VS/L 的 VS 浓度下厌氧发酵需要相当长的恢复时间才可以启动，并且反应过程中出现的日产气峰值也要低于前两个 VS 浓度，反应持续时间较长，证明在此 VS 浓度下抑制情况较为严重，产甲烷菌活性不高。

3.2 进料 VS 浓度对沼气产率的影响

图 2 显示了餐厨垃圾在 3 个不同的 VS 浓度下的日沼气产率和接种物的日沼气产率在消化期内的变化情况。其中餐厨垃圾的沼气产率去除了接种物对产气的贡献。接种物的沼气产率为 73.48 mL/g VS。三个 VS 浓度 4 g VS/L、8 g VS/L、12 g VS/L 下的餐厨垃圾的沼气产率分别为 528 mL/g VS，454 mL/g VS 和 369 mL/g VS。可以看出在 4 g VS/L 的浓度下，沼气的产率最大；8 g VS/L 的浓度次之，12 g VS/L 的浓度最低。在消化完全的状态下，任何 VS 浓度下的餐厨垃圾沼气产率在理论上应该是相同的，这三组最终沼气产率出现一定的差异可能有以下几种原因：消化时间不够，特别是 12 g VS/L 的样品组反应结束时刚刚结束一个产气高峰，若继续观察，依然有可能产气，累积产气量还会再增加；排水集气法所使用的量筒精度不够，或者是量筒上有细微裂缝带来的气体计量误差；餐厨垃圾与猪粪沼液的 C：N 不同，本实验中不同 VS 浓度的餐厨垃圾接种同样量的猪粪沼液，造成各个样品组的反应器内总 C：N 不同，甲烷菌的营养状态不同，由此造成产气率的差异；厌氧发酵过程有机碳素转化为 CO_2 和 CH_4，生成沼气放出，有机氮如蛋白质和氨基酸等会被转化为氨氮留在沼液中，随着 VS 浓度的提高，厌氧发酵中会生成的氨氮量也会增加，因此在较高的 VS 浓度下甲烷菌可能受到氨氮抑制影响了产气。三个 VS 浓度的累积产气量达到总产气量 80%的时间分别为 14 d，23 d 和 41 d。可见过高的 VS 浓度会延长餐厨垃圾发酵完毕的时间。

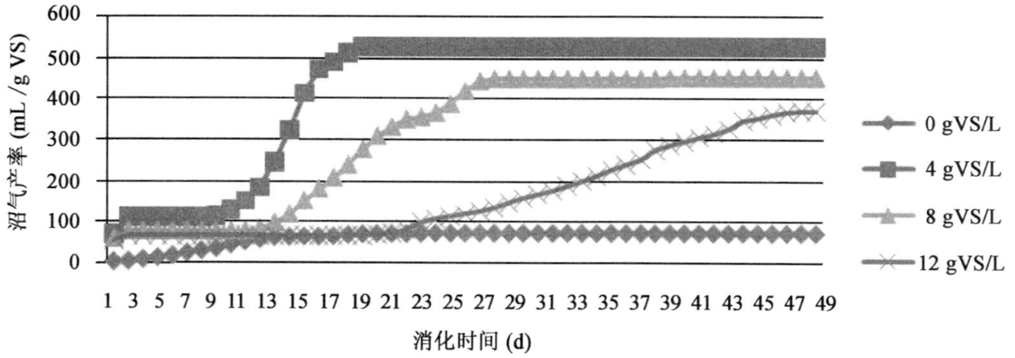

Figure 2 Biogas yield versus digestion time during digestion of kitchen waste

图 2 中温厌氧消化餐厨垃圾的沼气产率随时间变化图

4 结论

（1）餐厨垃圾具有良好的可生化性能，在 4 g VS/L 的 VS 浓度下沼气产率最高可达 528 mL/g VS。

（2）不同的 VS 浓度对餐厨垃圾厌氧发酵启动时间和持续时间的影响不同，随 VS 浓度的增加，发酵启动时间逐渐延长，持续的时间也延长。

（3）由于餐厨垃圾易于水解酸化，提高餐厨垃圾的 VS 浓度会使厌氧发酵的初期产生抑制，但随着发酵的进行，这种抑制作用会逐渐消除。

项目及基金支持

“十一五”国家科技支撑计划“规模化沼气工程沼液、沼渣高值利用技术研究与工程示范”课题

(2008BADC4B00)；北京市教育委员会科学研究与科研基地建设项目“规模化养殖场粪污双向清洁处理与利用研究”。

参考文献

[1] 方战强，吴坚. 餐厨垃圾处置方法探讨[J]. 华南师范大学学报(自然科学版)，2007，1：70-74.

[2] 中国统计年鉴，2010.

[3] 宋建利，石伟勇. 城市生活垃圾现状与资源化处理技术研究[J]. 河北农业科学，2009，13(7)：58-61.

[4] 潘丽爱，张贵林. 餐厨垃圾特性的试验研究[J]. 粮油加工，2009，9：154-156.

[5] 耿土锁. 食物性有机垃圾资源化方法[J]. 贵州环境保护，2002(12)：15-18.

[6] 汪群慧，马鸿志. 厨余垃圾的资源化技术[J]. 现代化工，2004，24(7)：56-59.

[7] 史琳娜，孔海民. 环境条件对 1 株降解厨余垃圾丝状真菌产酶活性的影响[J]. 浙江大学学报(农业与生命科学版)，2010，36(4)：393-398.

[8] 徐葳，冯磊. 厨余垃圾厌氧消化制取甲烷的影响因素研究[J]. 可再生能源，2008，26(6)：66-71.

[9] 贺延龄. 废水的厌氧生物处理[M]. 北京：中国轻工业出版社，1998.

沼气发酵罐内的微量元素浓度和生物活性——一种新测试方法的开发和应用

N. Engler[1], U. Merrettig-Bruns[2], A. Feher[3], T. Fritz[4], M. Nelles[1]

(1. 罗斯托克大学,罗斯托克 18051;2. Fraunhofer 研究院 UMSICHT,Oberhausen 46047;
3. HAWK 应用技术大学,哥廷根 31134;4. Schaumann Biotic 咨询公司,Pinneberg 25421)

【摘要】 本项目由德国西部奥伯豪森 Fraunhofer 研究院 UMSICHT 牵头与哥廷根应用科技大学以及德国罗斯托克大学共同合作,目的在于研发一种简单的测量方法来定量测定沼气发酵罐中生物活性并加以应用。在目前二期项目阶段,应用测量方法来探寻生物活性和微量营养素浓度之间的关系,筛选测试已经在德国近40个沼气工程开展。目前已进行了活性测试以及发酵样品的综合分析(微量营养素和微量元素),并得到生物活性和过程参数之间的相互关系,尤其是 NH_4^+-N 的参数影响。

本实验中测量的微量元素浓度变化范围较大,且对于单一微量元素而言,生物活性和微量元素浓度之间有着微弱的关系。

通过对一些样品的测定,发现极低浓度的单一微量元素就可能导致生物活性降低。该种添加剂对生物降解活性的影响可通过加入单一微量元素测试生物活性来确定,例如加入单一剂量的镍和锰到发酵样品中,在可以提高约6%的产气量。

【关键词】 微量元素;厌氧消化;分析方法;生物活性

TRACE ELEMENT CONCENTRATIONS AND BIOLOGICAL ACTIVITY IN BIOGAS DIGESTERS — DEVELOPMENT AND APPLICATION OF A NEW TEST METHOD

N. Engler[1], U. Merrettig-Bruns[2], A. Feher[3], T. Fritz[4], M. Nelles[1]

1. *University of Rostock (Germany), Faculty of Agricultural and Environmental Sciences, Department Waste Management*; 2. *Fraunhofer Institute for Environmental, Safety, and Energy Technology UMSICHT*; 3. *HAWK Fachhochschule Göttingen, Faculty of Resource Management*; 4. *Schaumann Biotic Consult GmbH*

Contact: Prof. Dr. Michael Nelles and Dipl.-Ing. Nils Engler, University of Rostock (Germany), Faculty of Agricultural and Environmental Sciences, Department Waste Management, Tannenweg 22, 18059 Rostock, Germany, Tel.: +49-381-403-4994, E-Mail: nils.engler@uni-rostock.de;

Dr. rer. nat. Ute Merrettig-Bruns, Fraunhofer Institute for Environmental, Safety, and Energy Technology UMSICHT, Oberhausen, Germany, Tel.: +49-208-8598-1229;

Dipl.-Ing. Adam Feher, HAWK Fachhochschule Göttingen, Faculty of Resource Management, Rudolf-Diesel Straße 12, 37075 Göttingen, Germany, Tel.: +49-551-30738-15;

Dr. Thomas Fritz, Schaumann Biotic Consult GmbH, An der Mühlenau 4, 25421 Pinneberg, Germany

EXECUTIVE SUMMARY

A research project under the leadership of Fraunhofer Institute UMSICHT at Oberhausen (Germany) in cooperation with University of Applied Science at Göttingen (Germany) and University of Rostock (Germany) aims to the development and implementation of an easy to use test method to quantify the biological activity in biogas digesters. In the currently running second project phase the test method is applied to find correlations between bioactivity and micronutrient concentrations. Screening tests at nearly 40 biogas plants in Germany were carried out. The developed activity test as well as comprehensive

analyses of the digester samples (macronutrients and trace elements) was performed. Correlations between measured bioactivity and process parameters, especially NH_4^+-N, were observed.
The measured concentrations of trace elements varied in a wide range and a weak correlation between measured bioactivity and trace element concentration was found for single trace elements.
For some samples, extremely low concentrations of single trace elements were suspected to cause low bioactivity. Bioactivity tests with adding single trace elements were performed to investigate the effect of such additives on biodegradation performance. Single dosing of e. g. Nickel and Manganese to the digester sample could improve gas yield during the methane-forming phase by approx 6%.
Key words: Trace elements; anaerobic digestion; analytical method; bioactivity

1 Introduction

The continuing growth of number and size of biogas plants in Germany and Europe will lead in medium-and long term to shortage of substrates. Thus, the optimized use of available biomasses will become more and more important. Furthermore it will be essential to make new biomasses, which can currently not be used in anaerobic digestion, accessible for biogas plants. For both tasks, optimization of biodegradation and accessibility of new biomasses, miscellaneous additives for biogas plants are available at the market. Especially trace element mixtures have reached certain importance at the Germen market. To evaluate the effect of these additives on the biogas process, a research project was initiated to develop an appropriate test method. The project is carried out under the leadership of Fraunhofer Institute UMSICHT at Oberhausen (Germany) in cooperation with University of Applied Science at Göttingen (Germany) and University of Rostock (Germany) and is still ongoing.

1.1 Background

The importance of trace elements for anaerobic digestion processes is commonly known, though defined requirements are still object of research. Methods to test the influence of trace elements to fermentation processes are mostly microbiological or molecular biological tests. These tests are comparatively expensive and time-consuming. Due to the high requirements concerning the materials, laboratory equipment and qualification and skills of the staff, they are not applicable to be used for process optimization on-site the biogas plant.
A cheap and short-term test method to detect the influence of additives on the biological activity of the biogas digester can contribute to improve the performance of biogas plants.

1.2 Research objectives

Aims of the project were the development of an easy to use test method to quantify the biological activity of biogas digesters and application of the test method to find correlations between bioactivity and micronutrient concentrations. The test had to be applicable on-site at biogas plants ant should deliver reliable results within a time period of approximately 5 days.
By means of the developed test method, screening tests at biogas plants in different regions of Germany were carried out. Samples from nearly 40 biogas plants in the countries of Lower Saxony, North Rhine Westphalia and Mecklenburg-West Pomerania were tested and comprehensive accompanying analysis of process parameters and trace element concentrations were performed. According to the locations of the three project partners, the method was named Oberhausen-Rostock-Göttingen Aktivity-Test (ORGA-Test).

2 Methodology

The test method is based on the assumption that the kinetics of biogas formation from a standardized easy degradable substrate indicates the performance of microbiological degradation.

The test is designed as an anaerobic digestion batch-test under standardized boundary conditions. A standardized synthetic substrate with a defined biogas yield is used. The metabolic kinetics of biogas production under standardized conditions is recorded at high temporal resolution. To enable a standardized presentation, the measured biogas production is referred to the synthetic substrate's theoretical maximum gas yield according to BUSWELL.

During the screening period, samples from 36 biogas plants in Germany were tested. Comprehensive analyses of the samples were carried out besides the bioactivity-test. Particular concentrations of trace elements (Co, Cu, Ni, Se, Mn, Mo, B) other metals (Fe, Zn) and macro-nutrients (N, P, S, K, Na, Ca) were analyzed.

2.1 Material and methods

The ANKOM *Gas Production System* is used to measure the kinetics of biogas formation. This system consists of digestion vessels with a volume of 500 mL and measuring modules at the top (*Figure 1*). Inside the measuring module is a pressure sensor, a purge valve and a pressure release valve. The sensor measures the pressure increase inside the digestion vessel (at constant temperature and constant volume). Data are transferred wireless via Bluetooth standard to the PC.

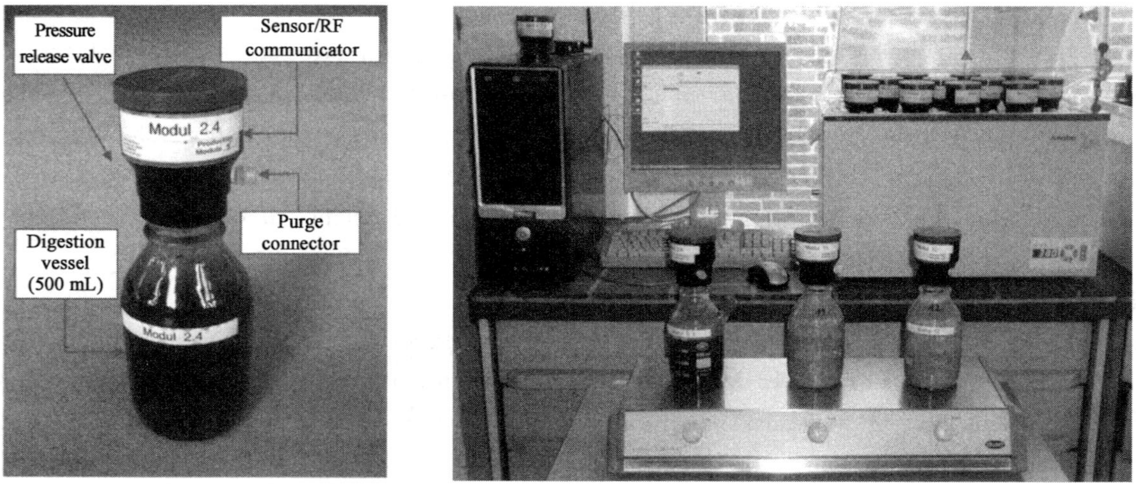

Figure 1 Single digestion vessel with measuring module (left) and digesting vessels with measuring modules in the water-bath resp. at the magnetic stirrer (right).

The purge connector allows to inert the headspace with Nitrogen. Maximum pressure inside the modules can be limited by defining a threshold, at which the pressure release valve opens. For the ORGA-Test, the maximum overpressure is predefined at 5 kPa (50 mbar). At this level, the pressure release valve opens for 250 ms. Therefore the gas can be assumed to behave as an ideal gas, and the error due to pressure dependent solubility of CO_2 and H_2S is negligible.

Additionally to the digestion modules, one so-called zero module measures the current ambient pressure. The appending software calculates the cumulative pressure increase for each digestion vessel, regarding changes in ambient pressure, and saves the data in a temporal resolution of 30 minutes.

Standardized substrates used for the test are:

• Acetate (as Acetic Acid) is directly available for methanogenous microorganisms. Kinetics of biogas formation from Acetate indicates the performance of the methanogenous phase.

• Pure maize starch has to be fully hydrolyzed. This allows a differentiated evaluation of the hydrolysis and methanogenous phase of the biogas formation process.

• A mixture of cellulose, starch, glucose, phosphate and Urea in a fixed formulation is used to simulate a complex substrate which is similar to maize silage concerning the ratio of C : N : P and the degradability. For the test, the samples are taken from the digester and put into the digestion vessels promptly. Then the substrate is added and the headspace of the digestion vessels is purged with Nitrogen. The digesting vessels are placed in a thermostatic bath at 38℃ to ensure constant temperature. After a time delay of 10 minutes to obtain uniform start conditions, the data recording is started. The test vessels are mixed once per day by means of a magnetic stirrer. The test period lasts from 5 to 10 days. Magnetic stirrer and the whole experimental setup are shown in *Figure 1*.

2.2 Data evaluation and interpretation

The amount of substance is directly proportional to pressure increase:

$$dn = dp \cdot \frac{V_0}{R \cdot T} \qquad (V_0 = \text{headspace of digestion vessels})$$

At pressures lower than 150 kPa (abs), the gas can be assumed to behave as an ideal gas with a molar volume of

$$V_0 = 22.4\,\frac{l(N)}{mol}$$

Thus, the volume of formed biogas can be calculated from the pressure increase:

$$dV = dp \cdot \frac{V_0 \cdot v_o}{R \cdot T}$$

Different criterions can be used to compare different samples and to evaluate the biological activity. The time span in which the biogas formation reached 50% of the maximum biogas yield according to BUSWELL can be defined as t_{50} and used as such criterion. Typical gas sum curves gained from the same digester sample using different substrates and the belonging criterions t_{50} are shown in *Figure 2*.

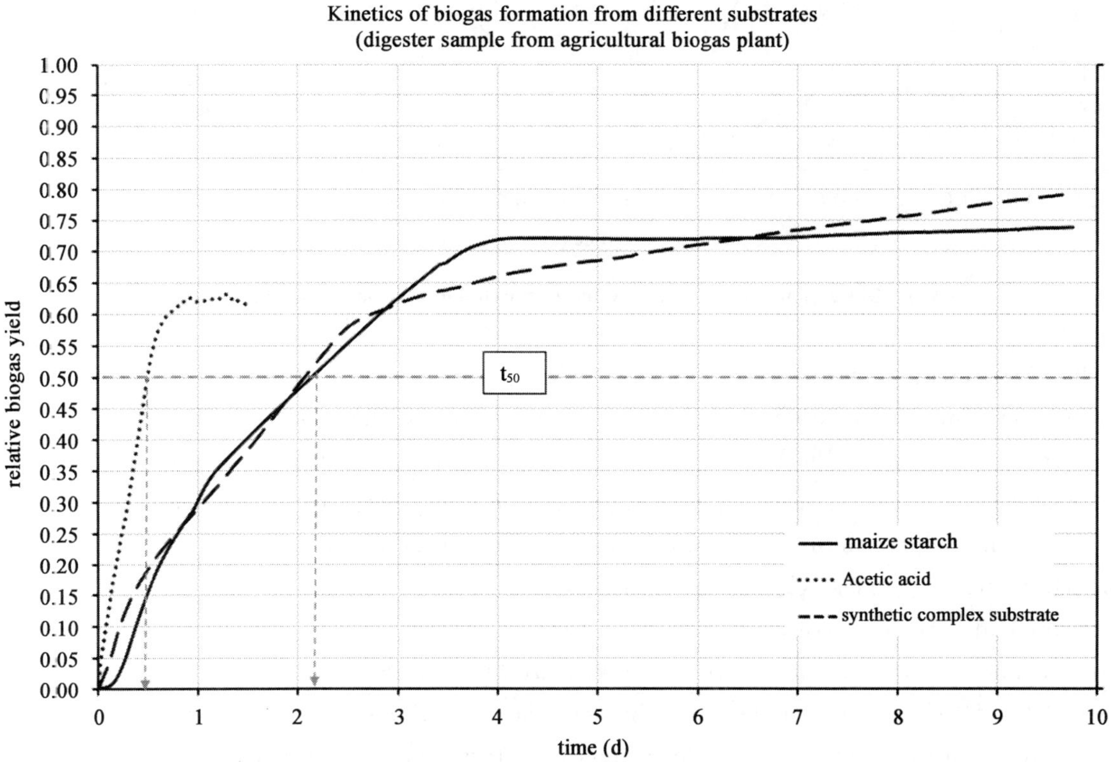

Figure 2 Typical biogas sum curves from different substrates and evaluation criterion t_{50}

3 First results

Gas sum curves from starch show two characteristically phases with nearly linear progression. *Figure 3* shows a typical gas sum curve from pure maize starch and the corresponding organic acids. During the first linear phase of the gas sum curve, the concentrations of acetic acid and butyric acid are rising whilst the pH is decreasing. In the second linear phase the acetic acid shows correspondingly an almost linear decrease. Propionic acid and butyric acid are degraded nearly to zero whilst the pH reaches its initial level. This indicates that the two linear phases represent the hydrolysis/acidogenesis (first phase) resp. the methanogenous (second) phase.

The slope of the biogas sum curve at these two phases curve can be used as an additional criterion to evaluate the biological activity.

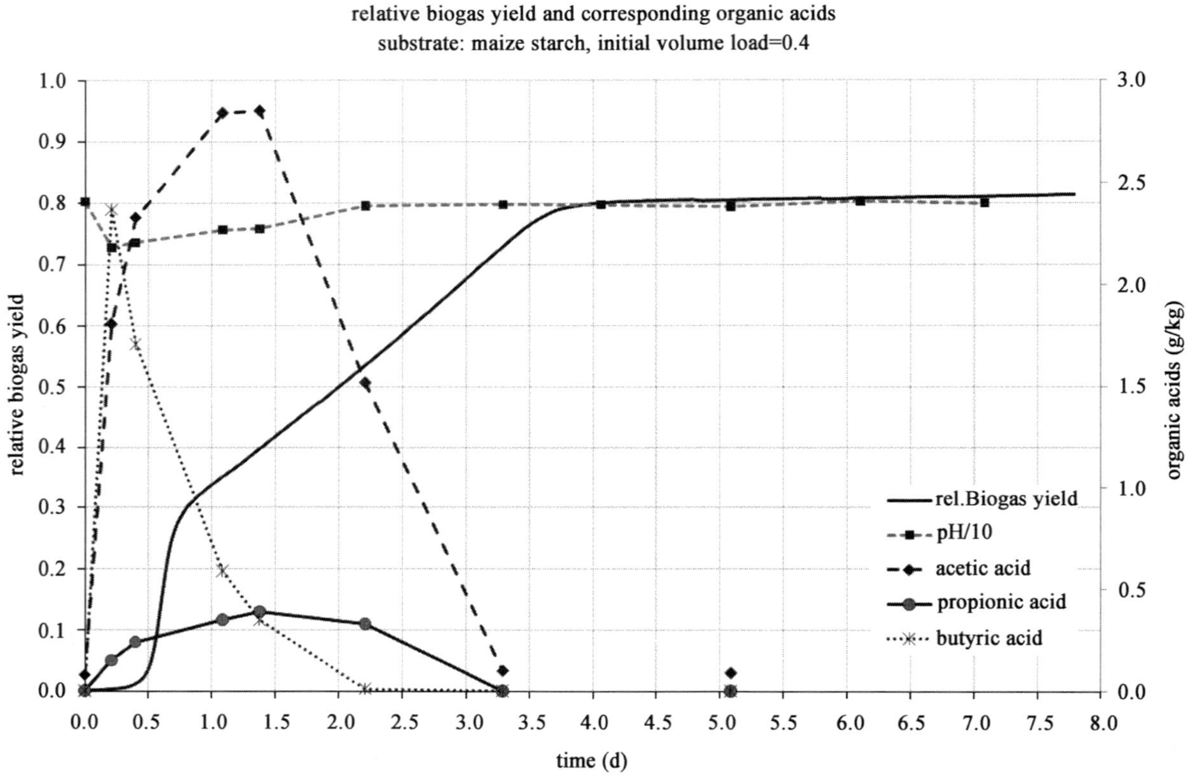

Figure 3 Biogas sum curve from pure maize starch and corresponding organic acids

3.1 Results of screening tests

During the research project mentioned above, nearly forty biogas plants situated in different regions of Germany were tested by the ORGA-Test. The tests were carried out with maize starch, acetic acid and dried and grinded maize silage for each digester sample. Additional, comprehensive analyses of the digester samples with special respect to macro-and micronutrients were performed:

- basic parameters: dry mass (DM), ash content, NH_4^+-N;
- macro-nutrients: N, P, S, K, Na, Ca, Mg, Fe;
- micro-nutrients or trace elements: B, Co, Cu, Mn, Mo, Ni, Se, V, W, Zn.

The results of the ORGA-Tests showed that the degradation kinetics varied in a wide range though test parameters as temperature, initial volume load, mixing interval etc. were identical. *Figure 4* shows the results from 36 samples measured with maize starch as substrate. The criterion t_{50} ranged between 0. 7 to 8 days. Two samples could not reach this criterion within the test period. None of the surveyed biogas plants showed obvious disturbances of the process or abnormal process behaviour at the time of sampling. The differences in the biodegradation performance of the microbial communities had to be caused by other factors. The concentration of NH_4^+-N was found to be most important for the test results. Samples with extraordinary high concentrations were excluded from the further considerations.

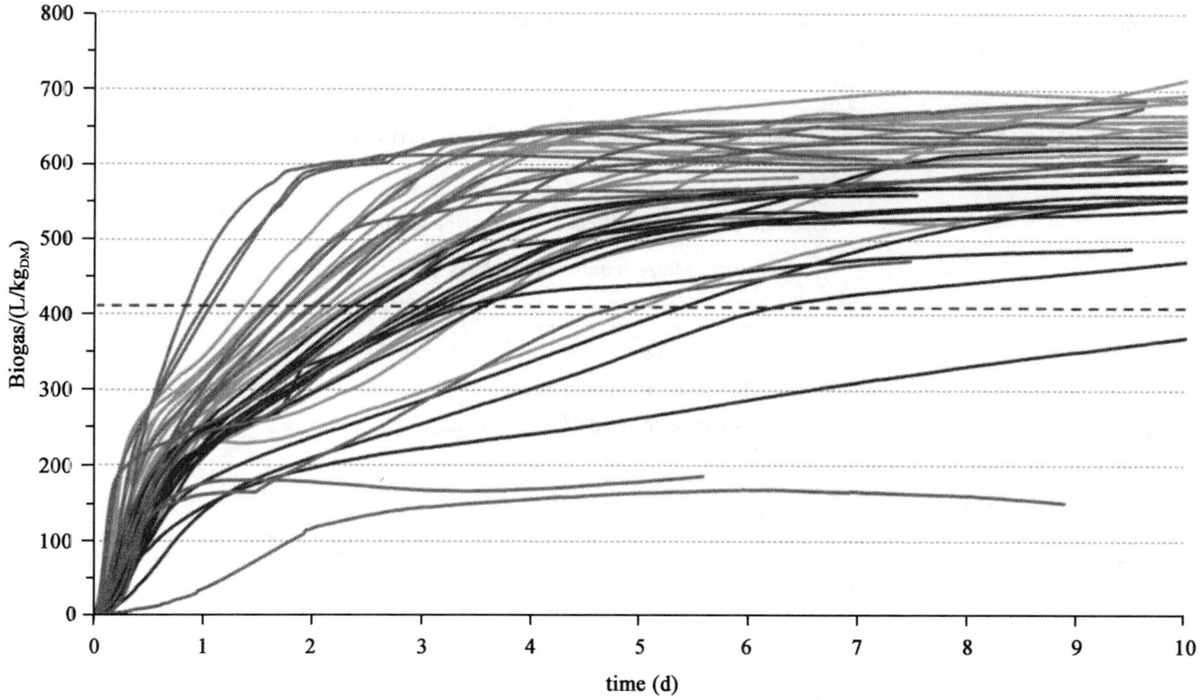

Figure 4 Kinetics of biogas formation from maize starch under identical test conditions, the broken line indicates the t_{50} criterion

The test results, expressed as t_{50} values, indicated a weak relation to several micronutrients. However a clear correlation between the concentration of a single micronutrient and the bioactivity was not found. A decrease of bioactivity is very often caused by multiple factors and simultaneously undersupply with several micronutrients.

As expected, the strongest relations were found for Ni and Mn using dried and grinded maize silage as substrate for the ORGA-Test. *Figure 5* shows the measured t_{50} and the found concentrations of Ni resp. Mn in the digester samples. Lower bioactivity is very often connected with lower concentration of Ni or Mn. However it has to be considered that low bioactivity despite sufficient supply with e. g. Ni can be caused by an undersupply with other essential trace elements.

For some digester samples with very low trace element concentrations an undersupply was suspected to be causal for a low bioactivity. For these digesters, tests with additionally dosing of single trace elements were performed. The digesters were sampled again and ORGA-Tests with and without additional dosing of trace elements were carried out simultaneously. The trace elements were added as EDTA-complex compounds. Starch was used as substrate to enable a differentiated evaluation of the different process phases.

One result is shown in *Figure 6*. The concentration of Ni was 2. 13 mg/kg DM in the untreated sample and was increased to 21. 05 mg/kg DM by the adding.

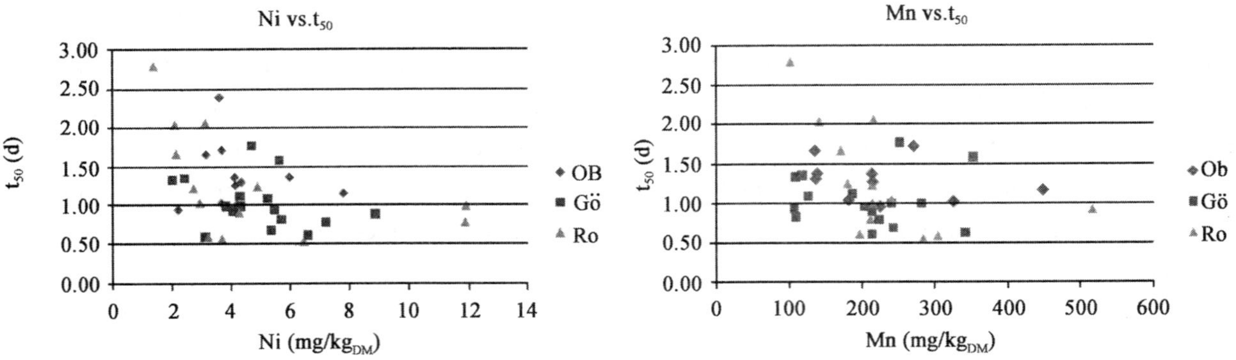

Figure 5 Trace element concentrations and measured bioactivity (ORGA-Test with maize silage)

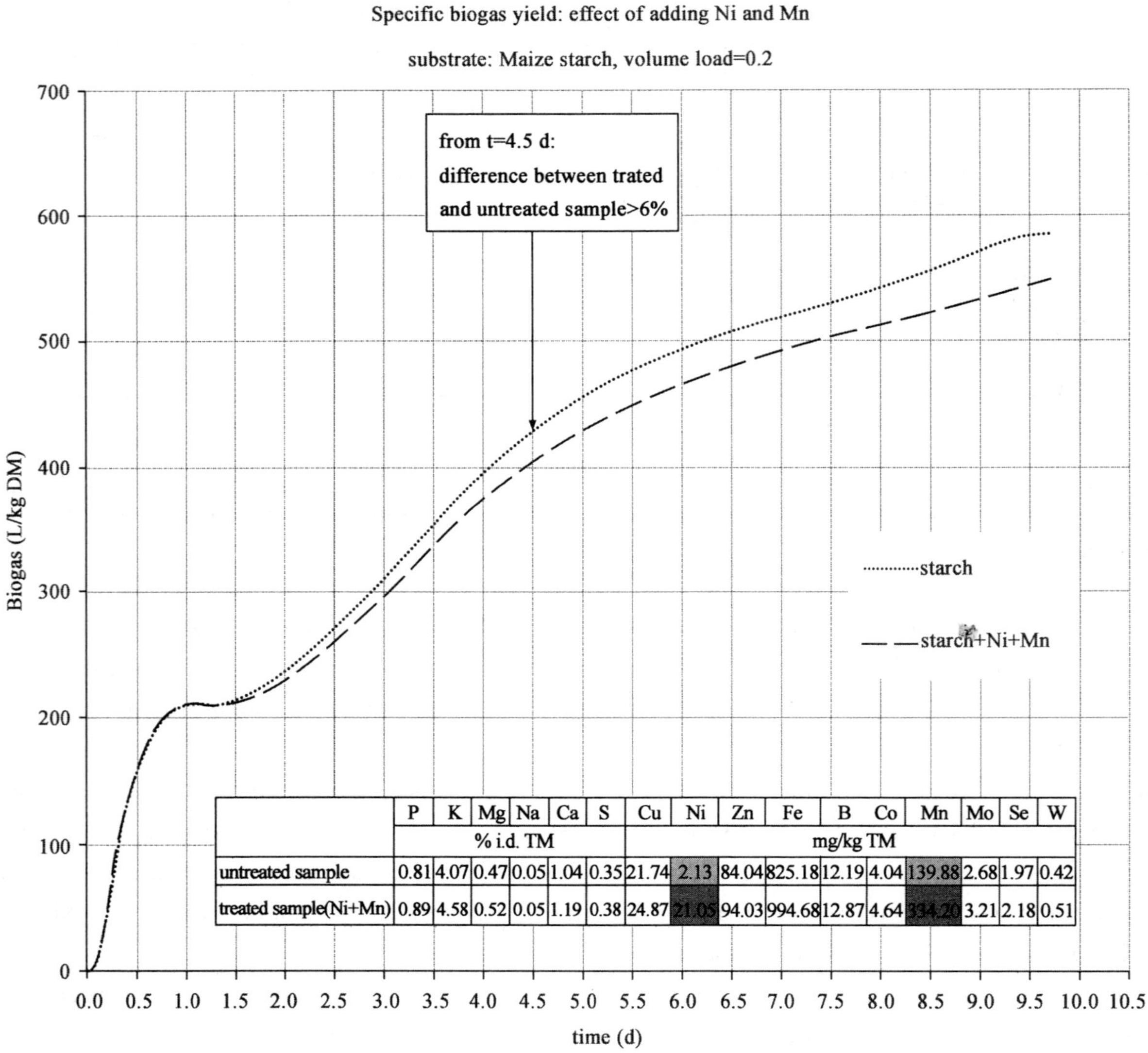

	P	K	Mg	Na	Ca	S	Cu	Ni	Zn	Fe	B	Co	Mn	Mo	Se	W
	% i.d. TM						mg/kg TM									
untreated sample	0.81	4.07	0.47	0.05	1.04	0.35	21.74	2.13	84.04	825.18	12.19	4.04	139.88	2.68	1.97	0.42
treated sample(Ni+Mn)	0.89	4.58	0.52	0.05	1.19	0.38	24.87	21.05	94.03	994.68	12.87	4.64	334.20	3.21	2.18	0.51

Figure 6 Influence of Ni and Mn-dosing on bioactivity

The hydrolysis (first slope) runs almost identically for the treated and the untreated sample. The influence of Ni and Mn is demonstrated in the second phase of biogas forming. After 4.5 days, the difference in specific biogas yield remained>6% between the treated and the untreated sample. This difference is far above the accuracy of the test method which was determined to 3.5%.

4 Conclusions

It could have shown that the developed test method is suitable for evaluating the biodegradation performance of the microbial community of biogas digesters.

A screening of nearly 40 biogas plants in Germany has revealed remarkable differences in bioactivity. The test field included manure-free driven biogas plants as well as plants with a high share of manure at the input. Although manure free plants showed trace element deficiency above average, also plants with a high share of manure were identified as undersupplied. A sufficient supply with essential trace elements seems to be essential for a good performance of microorganisms in the long term. Further screening of different biogas plants will lead to statistically firm target concentrations. It has however to be considered that the availability of trace elements for microorganisms is still matter of research.

Absence of Ni and Co showed to be important for a reduced biological activity. The results of the ORGA-Tests indicate that an undersupply with Ni causes reduced activity along the whole degradation chain whilst deficiency of Co only influences the activity of the methanogenous phase. The use of additives, especially trace element concentrates, in biogas plants can therefore help to avoid process disturbances.

Tests at large scale biogas plants are going to be carried out to study the effect of micronutrient dosing on process stability and biogas yield.

The project is still ongoing, further results are expected in the next months.

References

Buswell A. M., Mueller H. F., (1952): *Mechanism of Methane Fermentation.* Industrial and Engineering Chemistry, 44 (3), 550-552.

Hu et al., (2007): *Enhancement of methane fermentation in the presence of* Ni^{2+} *chelators.* Biochemicals Engineering Journal, 38 (1).

EFFECT OF TRACE ELEMENTS ON ANAEROBIC FERMENTATION: A REVIEW

ZHANG Wanqin, GUO Jianbin, CHENG Huicai, PANG Changle, DONG Renjie

China Agricultural University

Contact: Prof. Dr. DONG Renjie, China Agricultural University, College of Engineering, Qinghua East Road No. 17, Haidian District, Beijing 100083, PR China, Tel.: +86-10-6273-7885, E-Mail: rjdong@cau.edu.cn

EXECUTIVE SUMMARY

Trace elements as micro-nutrients are necessary for the anaerobic microbial metabolism and act as microbial agents responsible for the anaerobic digestion of organic material. They can promote the growth of the methanogenic bacteria and stimulate the activity of some enzymes, then further speed up the biosynthesis of methane. The stimulation and antagonism of trace elements play a very significant role on the performance and stability of biogas digesters. Trace elements deficiency often leads to anaerobic fermentation system operation instability and to a decline of the gas production. On one hand, the paper analyzes biological effects and requirements of trace elements on anaerobic methanogenic microorganisms, and the effect of trace elements on the metabolic pathways of methanogenesis. On the other hand, the previous and current literature is reviewed regarding the trace elements (Fe, Co, Ni) and how they influence anaerobic biogas digesters which are operated with solid organic substrates for production of methane. Trace elements such as Fe, Co, Ni can improve the efficiency of conversion of acetic acid, they eliminate the accumulation of volatile fatty acids and they improve the methane gas production. Trace elements can further antagonize the inhibitory effect induced by ammonia nitrogen and sodium ions.

Key words: Trace elements; anaerobic digestion; stimulation; antagonism; bioavailability

微量元素对厌氧发酵的影响研究进展

张万钦，郭建斌，程辉彩，庞昌乐，董仁杰*

（*中国农业大学，北京 100083*）

【摘要】 微量元素是厌氧微生物新陈代谢所必需的微量营养元素和厌氧发酵酶系统的重要组成成分，能够促进厌氧微生物的生长和激活酶的活性，加快甲烷的生物合成。本文一方面对微量金属元素在厌氧发酵系统中的生物效应，产甲烷菌对微量元素的需求，微量元素在甲烷合成途径中的作用进行了理论分析；另一方面对 Fe、Co、Ni 三种微量元素对提高乙酸的转化效率，消除挥发性脂肪酸的累积现象，提高甲烷产气量，拮抗氨氮、钠离子的抑制作用等方面的国内外已有研究进行了综述。

【关键词】 微量元素；厌氧发酵；刺激作用；拮抗作用；生物有效性

* 通信作者：董仁杰，男，博士，教授，中国农业大学工学工学院能源工程与低碳技术研究室，北京市海淀区清华东路 17 号 100083，rjdong@cau.edu.cn，010-62737885

1 前言

厌氧发酵技术处理有机废弃物的同时产生清洁能源——甲烷，实现沼气产业的工业化和商业化不仅可以解决环境污染的问题，而且能够缓解经济发展对化石能源的依赖。

要实现沼气产业的工业化和商业化，就必须考虑沼气工程的产甲烷效率和运行稳定性，微生物的营养需求是影响沼气工程稳定性的一个关键因素，营养不足可能会导致有机物降解不完全、系统运行不稳定等[1~4]。厌氧发酵系统和微生物生长除需要大量营养元素——C、N、P、S以外，还需要多种微量金属元素[5]。微量元素对沼气工程的高效运行具有重要意义[6]。

Frostell B. 等人认为，在正常操作条件下，若沼气工程运行不稳定，微量元素含量的缺乏及其生物有效性的降低是首要考虑的因素[7~9]。微量元素Fe、Ni、Co、Zn、Mo、Cu在废水厌氧处理过程中的刺激和抑制作用得到了广泛关注。如食品加工厂废水[10]、啤酒厂废水[11]、酿酒厂废水[12]、屠宰场废水[13]、糖蜜釜馏物、糖蜜加工厂废水[14]、制药厂废水[15]、高硫酸盐有机废水[16]等等。本文对国内外微量元素对厌氧发酵的影响进行了综述。

2 微量元素对厌氧发酵过程的调控作用

2.1 微量金属元素在厌氧发酵中的生物效应

微量金属元素在厌氧发酵过程中起重要作用，其促进或抑制效应一直是科研人员所关注的问题[17]。厌氧发酵中的微生物同其他高等生命体一样，都需要一定的营养进行生长繁殖。除了C、N、P、S等营养元素外，微量元素也是生命所必不可少的营养。微量元素对厌氧发酵的重要性主要体现在其常常作为辅酶、辅基、辅因子的成分出现在厌氧发酵微生物的酶系统中，并且对厌氧发酵的产甲烷阶段起着重要调控作用[5,19,20]。表1列出了几种重要的微量元素的功能及其影响的厌氧发酵微生物和金属酶。

Table1 The trace elements in enzymes of microbial conversions and its functions

表1 厌氧发酵微生物中的微量元素及其功能

元素	酶	微生物	参考文献	元素功能
Iron (Fe)	CO脱氢酶	产甲烷菌/产乙酸菌	[21]	Fe在合成产甲烷菌组织的金属元素中含量最高，还原性强在厌氧发酵过程中有重要作用，能够合成和激活多种酶的活性，能够形成硫化物沉淀，刺激胞外聚合物的分泌
	乙酰辅酶A合成酶	*Moorella thermoacetica*	[22]	
	氢化酶	*Desulfovibrio*, *Esherichia coli*	[23~25]	
	NO-还原酶	假单孢反硝化细菌	[26]	
	亚硝酸盐还原酶	*Pseudomonas stutzeri*	[26]	
	硝酸盐还原酶	*P. denitrificans*	[26]	
	固氮酶		[27]	
	甲烷单加氧酶		[28]	
	超氧化物歧化酶	产甲烷菌	[28]	
Cobalt (Co)	转甲基酶	产甲烷菌/产乙酸菌	[29]	Co存在于特定的酶中，同时是咕啉类物质的组成成分，参与构成一氧化碳脱氢酶(CODH)-该酶在乙酸的形成过程中起重要作用
	CO脱氢酶	产甲烷菌/产乙酸菌	[21]	
	类咕啉酶	*Methanosarcina barkeri* 巴氏甲烷八叠球菌	[30]	
Nickel (Ni)	CO脱氢酶	产甲烷菌/产乙酸菌	[21]	以CO_2和H_2为底物合成甲烷的产甲烷菌离不开Ni，Ni是F_{430}因子的重要组成元素，CODH是一种镍蛋白，Ni对硫酸盐还原菌具有重要作用
	乙酰辅酶A合成酶	*Moorella thermoacetica*	[22]	
	甲基辅酶合成酶	产甲烷菌	[31]	
	脲酶		[28]	
	氢化酶	产甲烷菌	[28]	

续表 1

元素	酶	微生物	参考文献	元素功能
Copper (Cu)	乙酰辅酶 A 合成酶	*Moorella thermoacetica*	[22]	Cu 是超氧化物歧化酶和氢化酶的组成成分，Cu 是构成产甲烷菌组织的重要元素之一
	亚硝酸盐还原酶	*Pseudomonas stutzeri*	[26]	
	氨单加氧酶	硝化细菌	[32]	
	超氧化物歧化酶	产甲烷菌	[33]	
	氢化酶	兼性厌氧菌	[33]	
Zinc (Zn)	甲酸盐脱氢酶	产甲烷菌	[33]	Zn 是多种产甲烷菌的组成成分，参与构成甲酸盐脱氢酶、超氧化物歧化酶、氢化酶
	超氧化物歧化酶	产甲烷菌	[33]	
	氢化酶	兼性厌氧菌	[28]	
Selenium (Se)	氢化酶	产甲烷菌	[33]	参与构成甲酸盐脱氢酶(FDH)
	甲酸盐脱氢酶	产甲烷菌	[28]	
Molybdenum (Mo)	甲酸盐脱氢酶	产甲烷菌	[33]	参与构成甲酸盐脱氢酶(FDH)，抑制硫酸盐还原菌的代谢，限制硫化物的生成
	硝酸盐还原酶	*P. denitrificans*	[26]	
	固氮酶		[27]	
	甲酸脱氢酶	甲基杆菌	[34]	
Tungsten (W)	甲酸盐脱氢酶	产甲烷菌	[33]	参与构成甲酸盐脱氢酶(FDH)，促进产甲烷菌利用 CO_2 和 H_2 合成甲烷
	甲基呋喃脱氢酶	产甲烷菌	[28]	

2.2 产甲烷菌的微量元素需求

产甲烷阶段是厌氧发酵过程中的最后环节，产甲烷菌的代谢强度决定着产甲烷效率，产甲烷菌的物质代谢和能量代谢都需要微量元素的参与，如微量元素 Fe 参与产甲烷菌体内细胞色素、细胞氧化酶的合成，还是胞内氧化还原反应的电子载体[35]。产甲烷菌生长所需要的微量元素的量非常少，但微量金属元素的缺乏却能够导致生物活力下降，进而影响整个厌氧反应器的运行效果和稳定性[36]。一些产甲烷菌的纯培养实验研究发现多种微量元素能够激活增强产甲烷菌的代谢强度(表 2)，并且这些产甲烷菌都需要 Fe、Co、Ni。美国著名厌氧生物学家 Speece R. E. 等人测定了 10 个产甲烷菌细胞中的微量元素含量，研究显示产甲烷菌细胞中重金属含量的顺序：Fe>Zn≥Ni>Co=Mo>Cu[37]同时 Speece R. E. 等人指出不同种类的产甲烷菌的元素组成有所不同，即使同种产甲烷菌在不同的环境条件下微量元素的含量和需求也有所不同。Speece 还特别指出，以 Fe、Co 、Ni 为序，只有当前一种营养元素足够时，后面一个才能对甲烷菌的生长起激活作用[38]。国内学者也测定了产甲烷菌细胞的元素组成(如表 3)[39]，结果与 Speece R. E. 等人测定的结果在 Ni 和 Zn 的含量浓度上存在差异，这可能与国内外学者所测定的产甲烷菌的生长环境不同有关。

Table 2 Metal stimulation of pure cultures of methanogens

表 2 纯培养的产甲烷菌的金属刺激浓度

微生物	底物	刺激元素及浓度(uM)	参考文献
Methanosarcina barkeri	甲醇	Fe^{2+} (35)	[40]
M. barkeri	甲醇	Co (1), Ni (1), Se (1), Mo (1)	[41]
Methanothrix soehngenii VNBF	乙酸	Fe (20-100), Co (2), Ni (2), Mo (2)	[42]
M. thermoautotrophicum	H_2/CO_2	Se (1), W (10)	[43]
M. thermoautotrophicum	H_2/CO_2	Fe (>5), Co (>0.01), Ni (>0.1), Mo (>0.01)	[44]
M. barkeri	甲醇	Mo (5)或 V (2)	[45]
Methanococcus ofnielli	甲酸盐	Se (1), W (100)	[46]
Methanospirillum hungatei GP1	H_2/CO_2	Mn (50)	[47]

Table 3 The element composition of methanogens (g/kg dry cell)

表 3 产甲烷菌细胞的元素组成/(g/kg 干细胞)[39]

元素	氮	碳	磷	硫	钾	钙	镁
含量	65	400	15	10	10	4	3
元素	铁	镍	钴	钼	锌	锰	铜
含量	1.8	0.10	0.075	0.060	0.060	0.020	0.010

2.3 微量元素对甲烷生物合成的影响

产甲烷菌通过甲烷的生物合成形成维持细胞生存所需的能量。甲烷生物合成过程中，甲烷的形成伴随着细胞膜内外化学梯度的形成，这种化学梯度驱动酶产生细胞内能量通货——ATP[48]。

目前发现的甲烷生物合成过程有 3 种途径，如图 1 所示。第 1 种途径为以乙酸为原料的甲烷生物合成，第 2 种为以氢气、二氧化碳为原料的甲烷生物合成，第 3 种是以甲基化合物为原料的甲烷生物合成过程，如：甲醇、甲基胺、甲基硫等。3 种途径最终都形成甲基辅酶，甲基辅酶在甲基辅酶还原酶催化下最终形成甲烷。其中 3 种途径中，以乙酸为底物的甲烷合成占自然界甲烷合成的约 70%，以氢气和二氧化碳为底物合成的甲烷约 30%，以甲基化合物合成的甲烷量很少。

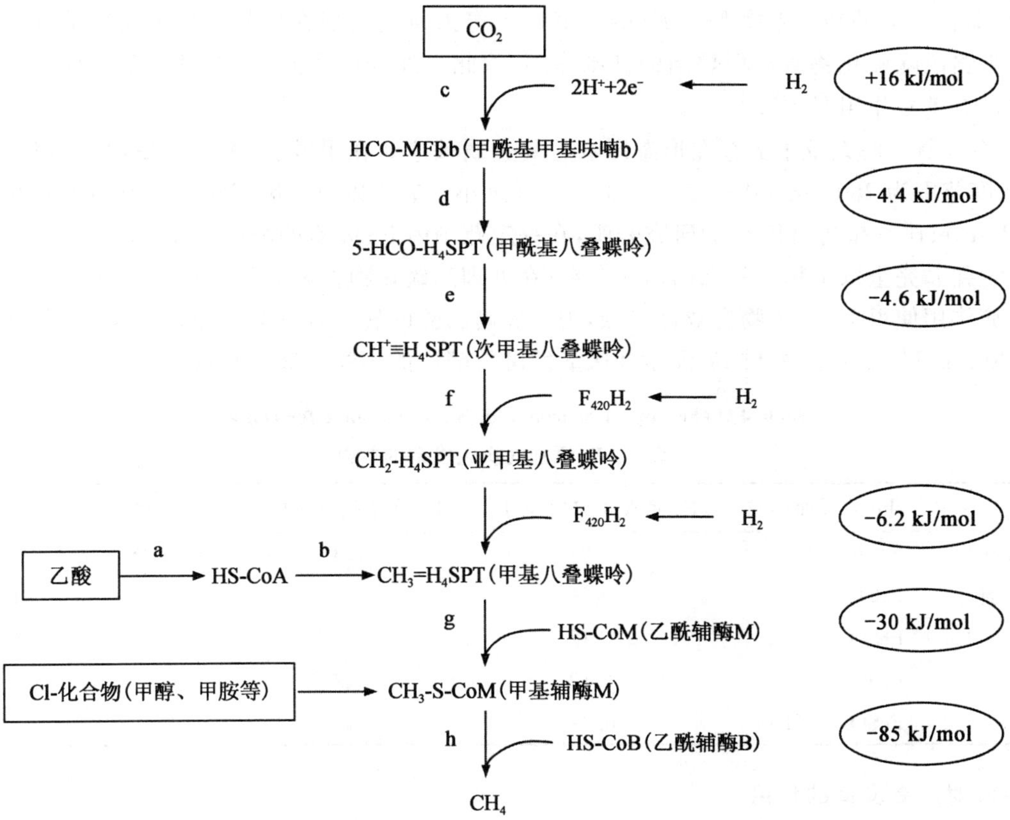

a. 乙酰辅酶 A 合成酶，b. 一氧化碳脱氢酶(CODH)，c. 甲酰基甲基呋喃脱氢酶，d. 甲基转移酶，e. 环化水解酶，f. $F_{420}H_2$ 氢化酶，g. CoM 甲基转移酶，h. 甲基辅酶 M 还原酶

Figure 1 The biosynthetic pathways of methane

图 1 甲烷的 3 中生物合成途径[29,48,56]

在甲烷合成的途径中，涉及多种酶、辅酶的催化作用，这些酶、辅酶大都含有 Fe、Co、Ni、Zn、Mo、W 等金属元素。催化甲基辅酶合成甲烷的甲基辅酶还原酶含有辅因子 F_{430}，而 Whit man 和 Wolfe[49]、Diekert[50] 等人发现，F_{430} 是第一种包含 Ni 的低分子量的生物物质。甲烷生物合成中的辅酶 M 甲基转移酶是一种含 Co

的金属酶[51];催化以乙酸为底物合成甲烷的关键酶为一氧化碳脱氢酶(CODH),一氧化碳脱氢酶分为两大类:存在于需氧微生物中的Mo-Fe-黄素酶类和存在于厌氧微生物中的Ni-Fe酶类,因此CODH在厌氧和好氧两种条件下都能起到一定的催化作用,在这两类金属酶中,位于活性位点起关键作用的金属分别是钼(Mo)和镍(Ni)[21],由此可见CODH是一个含有Fe、Co、Ni、Mo的金属酶,CODH除了催化乙酸合成甲烷外,也催化了同型产乙酸菌利用H_2和CO_2合成乙酸的反应[52];催化CO_2合成甲烷的甲酰基甲基呋喃脱氢酶、氢化酶分别是含Mo和Fe/Ni的金属酶[53~55],除甲烷的生物合成中的酶系统需要微量金属元素外,产甲烷菌中还含有许多由Fe^{2+}、Co^{2+}等金属离子组成的供无机物进入细胞的离子通道[48]。

3 Fe、Co、Ni对厌氧发酵过程的影响

由于Fe、Co、Ni在厌氧发酵过程所需微量元素中具有极其重要的作用,是国内外研究的热点课题,因而以下主要针对以上三种元素进行论述。

3.1 Fe对厌氧发酵的作用

Fe元素是产甲烷生长所必需的元素之一,其对厌氧发酵的影响得到了广泛的关注,表4列出了几组Fe的不同添加形式及添加量对甲烷发酵的影响。Fe元素的添加能显著提高沼气产量及甲烷浓度,主要是因为Fe的补充促进了挥发性脂肪酸的产生和提高了甲烷菌对乙酸的利用率,同时反应基质中有硫酸盐存在时,添加Fe能够激活乙酸的转化效率[61]。Hansen KH[62]等人研究发现在猪粪高温厌氧发酵过程中添加一定量的$FeCl_2$能够消除硫化物对产甲烷菌的抑制作用;除此之外,Fe元素对上流式厌氧污泥床反应器中颗粒污泥的形成也有重要作用[63~68]。

Fe元素在厌氧发酵系统中的存在形态影响着其金属效应。在甲醇、乙酸等为底物的甲烷发酵过程中,Fe的最佳浓度范围为280 mg/m^3~50.4 g/m^3[57],远远小于表4以农业废弃物为原料的厌氧发酵系统中Fe的添加量,P. Preeti[60]和刘莉莉[35]的研究表明,在一定范围内Fe元素的浓度越高,效果越好;这表明微量元素在发酵系统并非完全以生物有效态的形式存在;在处理高硫化物含量的甘蔗酒精釜馏物的反应器中,由于硫化物的沉淀作用使得Fe的生物有效性降低,为了提高乙酸的转化效率只有不断添加Fe的化合物使其浓度维持在600 mg/L[9]。因此探讨Fe元素在厌氧发酵发酵中的生物有效态有重要意义。

Table 4 The impact of iron additive on methane fermentation

表4 Fe添加对厌氧发酵的效果

原料	Fe的添加形式	Fe元素添加量(g/L)	沼气产量提高量(%)	甲烷浓度	参考文献
芒果加工废弃物	$FeCl_3$	1.38	123	提高26.5%	Raju NR[58]
水葫芦和牛粪	$FeCl_3$	适量	60	—	Patel VB[59]
牛粪	$FeSO_4$	1.12	40	—	P. Preeti[60]
鸡粪	$FeSO_4$	1.12	42	—	P. Preeti[60]
牛粪	$FeSO_4 \cdot 7H_2O$	0.41	18	—	刘莉莉[35]

3.2 Ni和Co对厌氧发酵的作用

微量金属元素Fe、Co、Ni对厌氧消化过程尤其甲烷发酵阶段有激活作用。关于产甲烷菌及厌氧反应器对Ni的需求,国内外学者作了大量研究。Williams[69]等人研究了Ni对鸡粪厌氧发酵的影响,当鸡粪沼液中Ni浓度为253 μM时,继续添加10 μM $NiCl_2$依然明显提高了沼气的产气量但对甲烷浓度无明显影响;这可能是因为鸡粪中的Ni以沉淀或其他非生物有效态的形式存在。Ni元素的添加能够提高牛粪厌氧发酵的总产气量和甲烷浓度也得到了证实[70,71]。van denBerg[72]等人证实在以乙酸为基质培养产甲烷菌的反应器中,添加Ni后反应器中乙酸利用率从2~4 g/(L·d)提高到10 g/(L·d)。Pobeheim H.等人研究了人工配置微量元素溶液对青贮玉米中温厌氧发酵的影响,结果表明在配置的微量元素溶液中去除Ni导致发酵系统运行不稳定和沼气产气量下降18%,在配置的微量元素溶液中去除Co会导致沼气产气量下降10%,在配

置的微量元素溶液中去除 Mo，但不去除 Ni 和 Co，发酵系统的稳定性和产气量变化不明显，即 Mo 的去除对发酵系统无明显影响，Pobeheim H. 等人的研究表明，在以青贮玉米为原料的发酵系统中，Ni 的促进效果优于 Co[73~75]，这与 Lebuhn M.[76]等人以青贮玉米为单一发酵原料的实验研究结果——Co 元素是最重要的金属元素不完全一致。国内农丽薇等人进行的“Ni 和 Co 对稻草厌氧消化影响”的研究结果表明 Co 在抑制酸积累上的激活作用不如同浓度的 Ni[77]，结论与 Pobeheim H. 等人的研究一致，即不同微量元素对某一发酵原料的作用效果可能不同。

3.3 Fe、Co、Ni 三种元素对甲烷发酵的协同效应

微量元素在促进微生物的新陈代谢中的作用并非孤立的，Speece[38]指出对于 Fe、Co、Ni 三种元素而言，缺乏某一种元素，甲烷发酵仍会进行，但速率会降低；按 Fe、Co、Ni 的顺序，只有前面一个元素充足时，后面一个才能对产甲烷菌的生长代谢起激活作用，即当 Fe 元素含量充足时，Co 才能表现出激活作用；同样，当 Co 的含量不足时，Ni 就不能表现激活作用，因此 Fe、Co、Ni 三种元素间的协同作用为维持高效的甲烷发酵有重要作用。国内外许多学者对 Fe、Co、Ni 三种元素对厌氧发酵的协同效应也进行了研究（表 5）。Fe、Co、Ni 三种元素对甲烷发酵的协同效应不仅表现在满足产甲烷菌的营养需求，Speece 在工业废水厌氧消化技术的研究中也指出，Fe、Co、Ni 的氯化物同时投加到反应器中可以促使反应器中甲烷菌向较为有利的优势菌种转移，即由甲烷丝菌占优势逐渐转变为甲烷八叠球菌占优势，后者的乙酸利用速率比前者高 3～5 倍，通过添加微量元素乙酸盐利用速率超过 30 kg/(m^3 · d) [37]，优势菌群的变化，提高了乙酸的利用率、甲烷量、整个厌氧发酵系统的运行效率。

Table 5 The synergistic effect of Fe, Co, Ni on the methane fermentation of several different substrates

表 5 Fe、Co、Ni 对几种不同底物甲烷发酵的协同效应

添加形式	添加量	原料	效果	参考文献
$FeCl_2 \cdot 4H_2O$ (Fe) $CoCl_2 \cdot 6H_2O$ (Co) $NiCl_2 \cdot 6H_2O$ (Ni)	1.0 mg/(L · d) (Fe) 0.1 mg/(L · d) (Co) 0.2 mg/(L · d) (Ni)	醋酸钙和乙醇	沼气产量分别提高了 24.6%和 25.4%。	[78]
		有机生活垃圾	沼气产量、甲烷浓度、COD 去除率分别提高了 43.4%、5.1%、10.2%。	[79]
		农贸市场废弃物	沼气产气量提高了 11.2%～25.4%，甲烷浓度从 52%提高到 57.5%，COD 的去除率从 33.8%提高到 44.1%。	[80]
	1.0 mg/(L · 7 d) (Fe) 0.15 mg/(L · 7 d) (Co) 0.4 mg/(L · 7 d) (Ni)	牛粪	沼气产气率由 0.85 mL/mg COD 增加到 1.14 mL/mg COD，总产气量提高了 35%，COD 去除率由 59.6%增加到了 80.5%；厌氧发酵系统中产甲烷菌的种类由杆菌占优势向球菌占优势转换。	[81,82]
	适量	玉米酒精废渣	沼气产气量提高了 97%	[83]
		青贮玉米	沼气产气量提高了 35%	[84]
$CoCl_2 \cdot 6H_2O$ (Co) $NiCl_2 \cdot 6H_2O$ (Ni)	0.2 mg/(L · d) (Co+Ni)	稻草	Co、Ni、Co+Ni 比未添加组沼气产气量分别提高 39.0%、74.5%、115.0%，Co 和 Ni 协同效应显著。	[77]
	10 nmol(Ni) 50 nmol(Co)	食品加工厂废水	甲烷产量提高 42%。	[85]

3.4 微量元素对毒性物质的拮抗作用

微量元素对维持沼气工程运行稳定性还体现在对毒性物质的拮抗作用上。McCarty[86]报道发酵液中 Na^+ 浓度为 3 500～5 000 mg/L 时，会对厌氧发酵系统产生中度抑制作用，当发酵液中 Na^+ 浓度为 8 000

mg/L 或者更高时，会产生强烈的抑制作用。李亚新[87]以人工配置的高钠盐(10 g/L)模拟废水为原料，研究了微量元素对 Na^+ 的拮抗作用，实验结果表明同时添加 Fe、Co、Ni 可以加快产甲烷菌对高 Na^+ 环境的适应，能缓解高高 Na^+ 对产甲烷菌的抑制和毒害作用，维持厌氧发酵系统的稳定运行。McCarty[86]报道发酵液中氨氮浓度为 50～200 mg/L 时，对厌氧发酵有促进作用，1 500～3 000 mg/L 时，对厌氧发酵系统开始产生抑制作用，当发酵液中氨氮浓度超过 3 000 mg/L 时，会对厌氧发酵产生毒害抑制作用。李秀芬[88]等人在高废水处理工艺中通过添加微量元素消除了高氨氮对厌氧发酵的抑制作用，其对氨氮的拮抗作用得到了证实。

4 讨论

微量元素对甲烷发酵的重要作用不容忽视，在温度、pH、HRT 等因素正常的条件下，若产甲烷过程不稳定或产气量下降，微量元素的缺乏是首要检查的因素。但微量元素在厌氧发酵系统中的存在形态及以何种形态发挥高效的生物效应依然是研究的重点和难点。

以能源作物(青贮玉米、能源草等)为单一发酵原料的沼气工程，常常因微量元素生物有效性的缺乏造成乙酸利用率下降，挥发性脂肪酸(VFA)积累，浓度到达 3～5 g/m^3，导致沼气工程运行稳定性和产气量下降。为缓解微量元素不足的情况，可以通过添加外源微量元素或根据不同的发酵工艺，使农业木质纤维素类有机废弃物与微量元素含量相对充足的城市固体垃圾、畜禽粪便等联合发酵，从而满足厌氧发酵系统对微量元素的需求。

一些研究表明在畜禽粪便的厌氧发酵系统中添加微量元素依然能够提高发酵系统的产气量，这说明畜禽粪便中的部分微量元素可能以非生物有效性的无机盐沉淀(磷酸盐、硫酸盐、硫化物、碳酸盐)形式存在或生物有效性很低。因为一些微生物可利用螯合剂来增强微量元素的生物有效性，因此进一步研究一种维持微量元素生物有效性的技术有重要价值。

虽然大量研究证实了微量元素对厌氧发酵的激活作用，但还需进一步探讨微量元素在厌氧发酵系统中的存在形态、不同形态下的生物有效性以及微量元素与微生物之间的相互作用机制。

项目及基金支持

1.“十一五”国家科技支撑计划“规模化沼气工程沼液、沼渣高值利用技术研究与工程示范”课题(2008BADC4B00)；

2.北京市教育委员会科学研究与科研基地建设项目“规模化养殖场粪污双向清洁处理与利用研究”。

参考文献

[1] 何荣玉，刘晓风，袁月祥，等. 沼气发酵外源添加物的研究进展[J]. 中国沼气，2007，25(5)：8-10.

[2] 杜倩. 厌氧发酵系统中消化液循环利用试验研究[D]. 东北农业大学，2010，6.

[3] 穆献中，刘炳义. 等. 新能源和可再生能源发展与产业化研究[M]. 北京：石油工业出版社. 5-25.

[4] M. Takashima，R. E. Speece. Mineral requirements for methane fermentation. Critical[J]. Reviews in Environmental Control，1990，19：465-479.

[5] Zandvoort M. H.，Hullebusch E. D.，Fermoso F. G.，et al. Trace metals in anaerobic granular sludge reactors：bioavailability and dosing strategies[J]. Eng Life Sci，2006，6(3)：293-301.

[6] Demirel P. Review：Trace element requirements of agricultural biogas digesters during biological conversion of renewable biomass to methane[J]. Biomass and bioenergy，2011，35，992-998.

[7] Frostell B. Process control in anaerobic wastewater treatment[J]. Wat Sci Technol，1985，17：173-189.

[8] Callander I. J.，Barford J. P. Precipitation，chelation，and the availability of metals as nutrients in

anaerobic digestion. I. Methodology[J]. Biotechnol Bioeng, 1983, 25: 1947-1957.

[9] Callander I. J., Barford J. P. Precipitation, chelation, and the availability of metals as nutrients in anaerobic digestion. II. Applications[J]. Biotechnol Bioeng, 1983, 25: 1959-1972.

[10] Oleszkiewicz J. A., Romanek A. Granulation in anaerobic sludge bed reactors treating food industry wastes[J]. Biol Waste, 1989, 27: 217-235.

[11] Kida K., Ikbal, Sonoda Y., Kawase M., Nomura T. Influence of mineral nutrients on high performance during anaerobic treatment of wastewater from a beer brewery[J]. Ferment Bioeng, 1991, 72: 54-57.

[12] Sharma J., Singh R. Effect of nutrients supplementation on anaerobic sludge development and activity for treating distillery effluent[J]. Bioresour Technol, 2001, 79: 203-206.

[13] Ivanov V. N., Stabnikova E. V., Stabnikov E. V., Kim I. S., Zubair A. Effects of iron compounds on the treatment of fat-containing wastewaters[J]. Appl Biochem Micro, 2002, 38: 255-258.

[14] Espinosa A., Rosas L., Ilangovan K., Noyola A. Effect of trace metals on the anaerobic degradation of volatile fatty acids in molasses stillage[J]. Environ. Technol., 1995, 32: 121-129.

[15] Seif H. A. A., Joshi S. G., Gupta S. K. Effect of temperature and cobalt addition on the performance of anaerobic thermophilic fixed film reactor in the treatment of pharmaceutical wastewater[J]. Environ. Technol., 1993, 14: 143-150.

[16] Patidar S. K., Tare V. Soluble microbial products formation and their effect on trace metal availability during anaerobic degradation of sulphate laden organics[J]. Wat Sci Technol, 2008, 58: 749-755.

[17] Bryant M P. Microbial methane production—theoretical aspects[J]. J Animal Science, 1979, 48(1): 193-201.

[18] Michael H. Gerardi. The Microbiology of Anaerobic Digestion[M]. 2003, 16.

[19] Scherer P., Lippert H., Wolff G. Composition of the major elements and trace elements of 10 methanogenic bacteria determined by inductively coupled plasma emission spectrometry[J]. Biol Trac Elem Res, 1983, 5: 149.

[20] William D. M., Lvanden B. Effects of Nickel, Cobalt, and Molybdenum on Performance of Methanogenic Fixed-Film Reactorst[J]. Appl Environ Microbio, 1981, 42(3): 502-505.

[21] Ferry J. G. Enzymology of one-arbon metabolism in methanogenic pathways[J]. FEMS Microbiol Rev, 1999, 23: 13-38.

[22] Seravalli J., Gu W., Tam A., et al. Functional copper at the acetyl-CoA synthase active site[J]. Chemistry, 2003, 100: 3689-3694.

[23] Fauque G., Peck H. D., Moura J. J. D., et al. The three classes of hydrogenases from sulfate-reducing bacteria of the genus Desulvovibrio[J]. FEMS Microbiol Rev, 1988, 54: 299-344.

[24] Albracht S. P. J. Nickel hydrogenases: in search of the active site[J]. Biochim BiophysActa, 1994, 11(88): 167-204.

[25] Sawers G. The hydrogenases and formate dehydrogenases of Escherichia coli[J]. Anthonie van Leeuwenhoek Int J Gen Mol Microbiol, 1994, 66: 57-88.

[26] Ferguson S. J. Denitrification and its control[J]. Antonie vail Leeuwenhoek Int J Gen Mol Microbiol, 1994, 66: 89-110.

[27] Schindelin H., Kisker Cl Schlessmanl et al. Structure of ADP-AIF_4 stabilized nitrogenase complex

and its implications for signal transduction. Nature, 1997, 387: 370-376.

[28] Fernando G. Fermoso, Jan Bartacek, Stefan Jansenc, Piet N. L. Lens. Metal supplementation to UASB bioreactors: from cell-metal interactions to full-scale application[J]. Science of the total environment, 2009(407): 3652-3667.

[29] Shima S., Warkentin E., Thauer R. K., et al. Structure and function of enzymes involved in the methanogenic pathway utilizing carbon dioxide and molecular hydrogen[J]. J Biosci Bioeng, 2002, 6: 519-530.

[30] Stefan J., Fraukje S., Wimf T., et al. Speciation of Co(Ⅱ) and Ni(Ⅱ) in anaerobic bioreactors measured by competitive ligand exchange-adsorptive stripping voltammetry[J]. Environ Sci Technol, 2005, 39: 9493-9499.

[31] Hausinger R. P. Nickel enzymes in microbes[J]. Sci Total Environ, 1994, 148:157-166.

[32] Ensign S. A., Hyman M. R., Arp D. J. In vitro activation of ammonia monooxygenase from Nitrosomonas europaea by copper[J]. J Bacteriol, 1993, 175: 1971-1980.

[33] Matainaho F. T. Comparisons of the anaerobic treatment of soluble wastewaters using various configurations under mesophilic and thermophilic conditions[D] Vanderbilt University, 2001.

[34] Girio F. M., Marcos J. C., Amaral C. M. T. Transition metal requirement to express high level NAD dependent formate dehydrogenase from aserine-type methylotrophic bacterium[J]. FEMS Microbiol Lett, 1992,97: 161-166.

[35] 刘莉莉，王敦球，关占良. 3种添加剂对牛粪厌氧发酵的影响[J]. 江西农业学报，2007，19(5)：119-120.

[36] 胡庆昊. 镍及其螯合物对甲烷发酵的影响及机理研究[D]. 江南大学，2008.

[37] Jarrell K. F., Kalmokoff M. I.. Nutritional requirements of the methanogenic archaebacteria[J]. Can. J. Microbiol., 1988, 4, 557-576.

[38] Takashima M., Speece R. E. Mineral nutrient requirements for high rate methane fermentation of acetate at low SRT[J]. J. water Pollut. Control Fed, 1989, 61(11-12): 1645-1650.

[39] Speece R. E. Nutrient requirements in anaerobic digestion of biogas[M]. London: USA Elsevier Applied Sciences Publication, 1987: 109-128.

[40] 贺延龄. 水的厌氧生物处理[M]. 北京：中国轻工业出版社，1998：105-107.

[41] Lin D., Kakinozo T., Nishio N., Nagai S.. Enhanced cytochrome formation and stimulated methanogenesis rate by the increased ferrous concentration in Methanosarcina barkeri culture[J]. FEMS Microbiol. Lett., 1990, 68, 89-92.

[42] Scherer P., Sahm H.. Effect of trace elements and vitamins on the growth of Methanosarcina barkeri [J]. Acta Biotechnol., 1981, 1: 57-65.

[43] Fathepure B. Z.. Factors affecting the methanogenic activity of Methanothrix soehngenni VNBF[J], Appl. Environ. Microbiol. 1987, 53: 2978-2982.

[44] Gerhard E., Butsch B. M., Marison I. W., von Stockar U.. Improved growth and methane production conditions for Methanobacterium thermoautotrophicum [J]. Appl. Microbiol. Biotechnol., 1993, 40, 432-437.

[45] Schönheit P., Moll J., Thauer R. K., Nickel, Cobalt, and molybdenum requirement for growth of Methanobacterium thermoautotrophicum[J]. Arch. Microbiol., 1979, 123:105-107.

[46] Scherer P.. Vanadium and molybdenum requirement for the fixation of molecular nitrogen by two

Methanosarcina strains[J]. Arch. Microbiol., 1989, 151: 44-48.

[47] Jones J. B., Stadtman T. C. Methanococcus vannielii: effects of selenium and tungsten on growth [J]. J. Bacteriol., 1977, 130:1404-1406.

[48] Pankhania I. P., Robinson J. P. Heavy Metal inhibition of methanogenesis by Methanospirillum hungatei GP1[J]. FEMS Microbiol. Lett., 1984, 22: 277-281.

[49] 单丽伟，冯贵颖，范三红．产甲烷菌研究进展[J]．微生物学杂志，2003，23(6)：42-46.

[50] Whit man. W. B., Wolfe R. S. Presence of nickel in factor F_{430} from Methanobacterium bryant[J]. Biochem Biophys Res Commum, 1982 (92): 1196-1201.

[51] Dieker G., Jaenchen R., Thauer R. K. Biosynthetic evidence for nickel therapy role structure of factor F_{430} from Methanobacterium thermo autitrophicum[J]. ArchMicrobio, 1980 (124): 103-106.

[52] Thauer R. K., Biochemistry of methanogenis: a tribute to Majory Stephenson[J]. Microbiology, 1998, 144: 2377-2406.

[53] Bainoitti A. E., Nishio N. Growth kinetics of Acetobacterium sp. on methanol-formate in continuous culture[J]. J. Appl. Microbiol., 2000, 88: 191-201.

[54] Bertram P. A., Thauer R. K. Thermodynamics of the formylmethanofuran dehydrogenase reaction in Methanobacterium thermoautotrophicum[J]. Eur. J. Biochem. 1994, 226: 811-818.

[55] Wasserfallen A. Formylmethanofuran synthesis by formylmethanofuran dehydrogenase from Methanobacterium thermoautothrophic Marburg[J]. Biochem. Biophys. Res. Commun., 1994, 199: 1256-1261.

[56] Michel R., Massanz C., Kostka S., Richer M., K. Fiebig. Biochemical characterization of the 8-hydroxy-5-deazaflavin hydrogenase from Methanosarcina barkeri[J]. Eur. J. Biochem. 1995, 233: 727-735.

[57] 祖波，祖建，周富春，汪林．产甲烷菌的生理生化特性[J]．环境科学与技术，2008，31(3)：5-8.

[58] Patel G. B., Khan A. W., Roth L. A. Optimum levels of sulfate and iron for the cultivation of pure cultures of methanogens in synthetic media[J]. Journal of Applied Bacteriology, 1978, 45: 347-356.

[59] RajuN. R, SumithraDevi S., Nand K. Influence of trace elements on biogas production from mango processing waste in 1.5 m^3 KVIC digesters[J]. Biotechnol Lett, 1991, 13: 461-464.

[60] Patel V. B., Patel A. R., Patel M. C. Madamwar DB. Effect of metals on anaerobic digestion of water hyacinth-cattle dung[J]. Appl Biochem Biotech, 1993, 43: 45-50.

[61] Preeti P. Rao, Seenayya G. Improvement of methanogenesis from cow dung and poultry litter waste digesters by addition of iron[J]. World Journal of Microbiology & Biotechnology, 1994, 10: 211-214.

[62] Van den Berg L., Lamb K. A., Murray W. D., Armstrong D. W. Effects of sulphate, iron and hydrogen on the microbiological conversion of acetic acid to methane[J]. J Appl Bacteriol, 1980, 48: 437-447.

[63] Hansen K. H., Angelidaki I., Ahring B. K. Improving thermophilic anaerobic digestion of swine manure[J]. Water Res, 1999, 33: 1805-1810.

[64] Lee H., Shoda M. Stimulation of anaerobic digestion of thickened sewage sludge by iron-rich sludge produced by the fenton method[J]. J Biosci Bioeng, 2008, 106: 107-110.

[65] Smith J. A., Carliel-Marquet C. M. A novel laboratorymethod to determine the biogas potential of iron-dosed activated sludge[J]. Bioresour Technol 2009, 100: 1767-1774.

[66]Smith J. A., Carliel-Marquet C. M. The digestibility of iron dosed activated sludge[J]. Bioresour Technol, 2009, 99: 8585-8592.

[67] Xu H., Aiyuk S., Zhang Y., Chen G., Pieters J., Verstraete W. Stimulation of methanogenesis in a laboratory-scale UASB reactor treating domestic sewage by Fe(0) application[J]. Environ Technol, 2004,25:613-619.

[68] Vlyssides A., Barampouti E. M., Mai S. Influence of ferrous ion on the granularity of a UASB reactor[J]. Chem Eng J, 2009, 146: 49-56.

[69] Yu H. Q., Fang H. H. P., Tay J. H. Effects of Fe^{2+} on sludge granulation in upflow anaerobic sludge blanket reactors[J]. Water Science and Technology, 2000, 41(12): 199-205.

[70] Williams C. M., Shih J. C. H., Spears J. W. Effect of nickel on biological methane generation from a laboratory poultry waste digester[J]. Biotechnol Bioeng 1986, 18: 1608-1610.

[71] Hassan D., Tandon S. M. Response of a cattle dung methane fermentation to nickel[J]. Biol Waste, 1987, 22: 261-268.

[72] 潘云锋，李文哲. 微量金属元素对牛粪厌氧发酵产气特性的影响[J]. 江苏环境科技，2006，19(4)：4-6.

[73] van den Berg L. Factors affecting rate of methane fermentation from Acetic acid by enriched methanogenic cultures[J]. Can SocMicrobiol, 1975, 22: 1313-1319.

[74] Pobeheim H., Munk B., Lindorfer H., Guebitz G. M. Impact of nickel and cobalt on biogas production and process stability during semi-continuous anaerobic fermentation of a model substrate for maize silage[J]. Water Res, 2011, 45: 778-787.

[75] Pobeheim H., Munk B., Muller H., Berg G., Guebitz G. M. Characterization of an anaerobic population digesting a model substrate for maize in the presence of trace metals[J]. Chemosphere, 2010, 80: 829-836.

[76] Pobeheim H., Munk B., Johansson J., Guebitz G. M. Influence of trace elements on methane formation from a synthetic model substrate for maize silage[J]. Bioresour Technol, 2010, 101: 836-839.

[77] Lebuhn M., Liu F., Heuwinkel H., Gronauer A. Biogas production from mono-digestion of maize silage-long term process stability and requirements[J]. Wat Sci Technol, 2008, 58:1645-1651.

[78] 农丽薇，徐龙君，谢金连，等. 微量 Co 和 Ni 对稻草厌氧消化的影响[J]. 环境科学研究，2008，21(2):163-167.

[79] 李亚新，董春娟. 激活甲烷菌的微量元素及其补充量的确定[J]. 环境污染与防治，2001，23(3)：116-118.

[80] 陈朝猛，曾光明，张碧波，等. 城市有机垃圾厌氧消化痕量激活剂的促进作用及产能研究[J]. 南华大学学报(理工版)，2004，18(1)：12-16.

[81] 谢金连，徐龙君，吴江，等. 痕量金属对农贸市场废弃物厌氧消化的影响[J]. 重庆大学学报（自然科学版），2007，30(5)：23-26.

[82] 张涵，李文哲. 微量金属元素添加频率对牛粪厌氧发酵特性的影响[J]. 东北农业大学学报，2006，37(4)：202-205.

[83] 张涵，李文哲. 微量金属元素添加频率对牛粪厌氧发酵细菌种类的影响[J]. 农机化研究，2006 (6)：173-175.

[84] 吕映辉. 提高产甲烷菌活性的应用研究[D]. 山东轻工业学院，2007.

[85] Hinken L., Urban I., Haun E., Urban I., Weichgrebe D., Rosenwinkel K. H. The valuation of malnutrition in the mono-digestion of maize silage by batch tests[J]. Wat Sci Technol, 2008, 58: 1453-1459.

[86] Murray W. D., vanden Berg L. Effect of nickel, cobalt and molybdenum on performance of methanogenic fixed-film reactors[J]. Appl Environ Microbiol, 1981, 42: 502-505.

[87] McCarty P. L. Anaerobic waste treatment fundamentals-part Ⅲ-toxic materials and their control[J]. Public Works, 1964, 95: 91-94.

[88] LI Yaxin, Speece. R. E. Stimulation effect of trace metals on anaerobic digestion of high sodium content substrate[J]. Water Treatment, 1995, (10): 145-154.

[89] 李秀芬，胡庆昊，陈坚. 微量金属元素及其螯合剂对厌氧消化产甲烷过程的影响[J]. 安全与环境学报，2009，9(2)：66-68.

厌氧发酵罐中的高效节能搅拌方式

E. Balssen

(ITT 公司,Langenhagen 30855)

【摘要】为了得到理想的沼气产量,沼气池中的高效搅拌是十分重要的。这种搅拌形式往往非常耗能。因此,ITT 公司在优化搅拌器的设计、以及投加原料的流变性数据方面进行了研究,已取得了十分可观的成绩。第一阶段的研究已成功节约能源 50%左右。我们期待通过进一步的研究,更好的优化工艺过程。

此外,新型搅拌器的更新换代将有助于进一步减少混合过程中的能耗。目前我们搅拌所需能耗占 CHP 产生的电能的比例从 2%降到了 1%。

【关键词】沼气发酵罐;搅拌系统;能源效率

EFFICIENT MIXING TO REDUCE ENERGY CONSUMPTION IN A DIGESTER

E. Balssen

ITT Water & Wastewater AB

Contact: Dr. Eilert Balssen, ITT Water & Wastewater AB. Bayernstraße 11, 30855 Langenhagen, Germany, Tel.: +49-511-7800-193, E-Mail: eilert.balssen@itt.com

EXECUTIVE SUMMARY

Effective mixing in a biogas digester is important to create a good gas production. This mixing often is a energy consuming technology. Therefore ITT is investing in researches to optimize mixer applications, mixer designs and rheological data of the input materials. The results today already make us very optimistic because we can show good first energy savings of about 50%. Through further upcoming researches we are expecting to optimize the process more and more.

Also the development of new mixer generations will help to reach a further reduction of energy consumption for mixing. We already have reduced the power input for mixing from 2% of the produced energy in a CHP to 1%.

Key words: Biogas digester; mixing systems; energy efficiency

1 Introduction

Well mixing in a digester means avoiding blind zones and crusting to create the best possible gas production. This well mixing is a basic which requires a respective part of the produced energy.

ITT targeted to optimize mixing efforts and is investing in their own researches as well as supporting scientific analyses. The results and the state of the art technology are presented in this article.

2 Energy-efficient mixing

The mainly used technology of mixing today is submersible mixers. Compact mixers with 300-500 rpm and propeller diameters up to 800 mm are producing the thrust for the movement and the turbulences to prevent crusting.

Both fewer revolutions per minute and a bigger diameter are giving more thrust per kW input power while the turbulences are declining.

In a first test in a biogas plant we replaced 2 compact mixers by 2 Banana mixers with a propeller diameter of 2.5 m. *Figure 1* and *Table 1* show details.

Figure 1 Installation of compact mixer together with 2.5 m Banana mixers

Table 1 Digester data for comparison test

Digester-Diameter 25 m/ filling height 5.5 m			
Installation	Number(kW)	Propellerø(mm)	Speed(rpm)
A	4×13	766	365
B	2×13	766	365
	2×5.7	2 500	42

Table 2 Result of test with different mixers

	A	B	Reduction (%)
Volume load (kg DM/(m^3 · d))	6.5	7.7	More gas
Energy consumption by volume (Wh/(m^3 · d))	129	73	43
Energy consumption by volume load (Wh/kg)	19.8	9.4	52

The results in *Table 2* show that the combination of the slow running thrust optimized Banana mixer for movement together with the compact mixer for crust avoiding is reducing the energy input by 50%. The effect of this combination is based on the different tasks. The banana mixers are creating an effective flow in relation to the power input and the compact mixers are preventing the build-up of a crust. The compact mixers also are the trouble shooters in case of problems.

The energy consumption for mixing in a biogas plant is varying a lot. In our research we measured power consumptions of up to 10% of the production (input energy into the grid through CHP). The average when using compact mixers is about 2%.

Until now we have reduced this power input to 1% only.

3 Viscosity of substrates

In the biogas process we use organic material like manure and organic wastes. These materials are structurally viscose media and belong to the Non-Newtonian fluids. For the mixing process this means the viscosity is changing during mixing. The higher the flow the lower is the viscosity. This is shown in *Figure 2*. The higher the shear rate (which in a simpler explanation can be compared with the input of thrust or the flow) the lower the viscosity is.

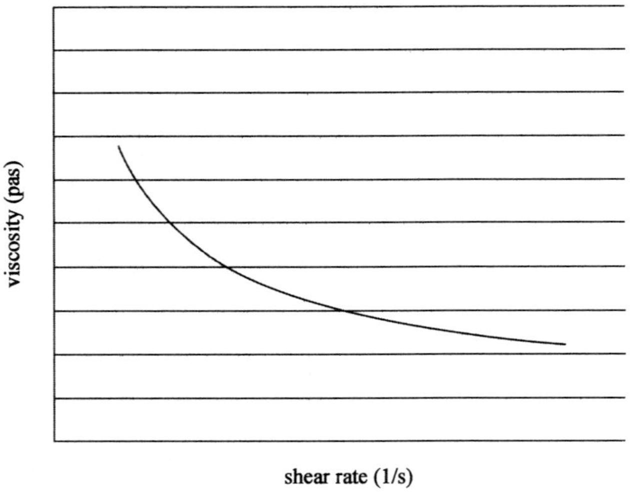

Figure 2 Reduction of viscosity by shear rate (Source: HSH - Dirk Müller)

If the input of thrust is too small the medium is flowing on a short way back to the back vanes of the propeller (*Figure 3*). This effect we call bulk flow. It can be solved by increasing the thrust through higher speed or by installing more mixers. In a biogas digester we usually install up to 4 mixers of different sizes or to say it more precisely with different thrusts depending on input material, quantity and digester size. Normally it is better to install 2 or more mixers to achieve a better performance if the place of thrust input is different. We also get more safety in case of problems.

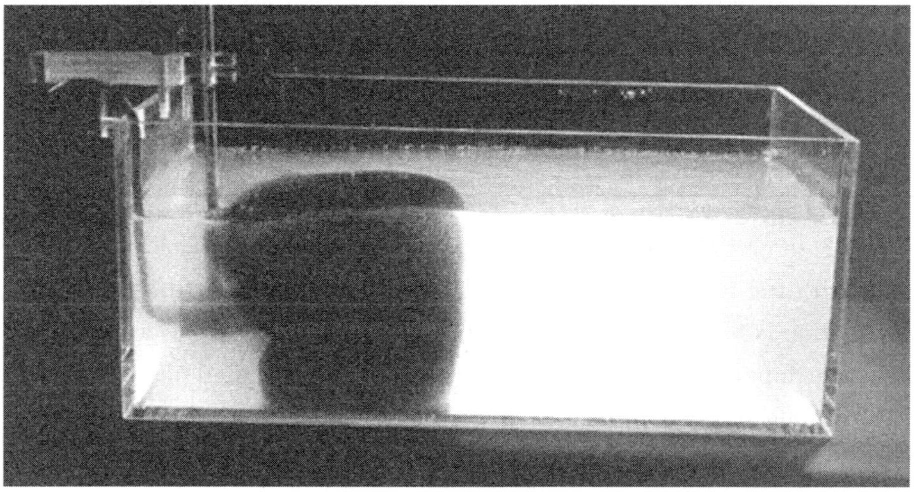

Figure 3 Effect of yield stress - a too small mixer (Source: Flygt laboratory)

4 Selection of mixers

We used these experiences to create a simple selection tool. In the relation of volume and thrust, shown as number and type of mixers, the lines are representing different substrates. This diagram also shows the big difference between different input materials in a biogas digester.

5 New mixer generation

Based on this experience ITT has developed a new Flygt mixer generation which is combining the benefits of compact mixers and the Banana mixer. It is a mixer with a 1.4 m propeller combined with a 7.5 kW or 11 kW motor. The speed is 67 rpm or 78 rpm. In a practical test with these mixers we could reduce the power consumption by 49% compared to compact mixers. This is the same as with the combination of Banana with compact mixers.

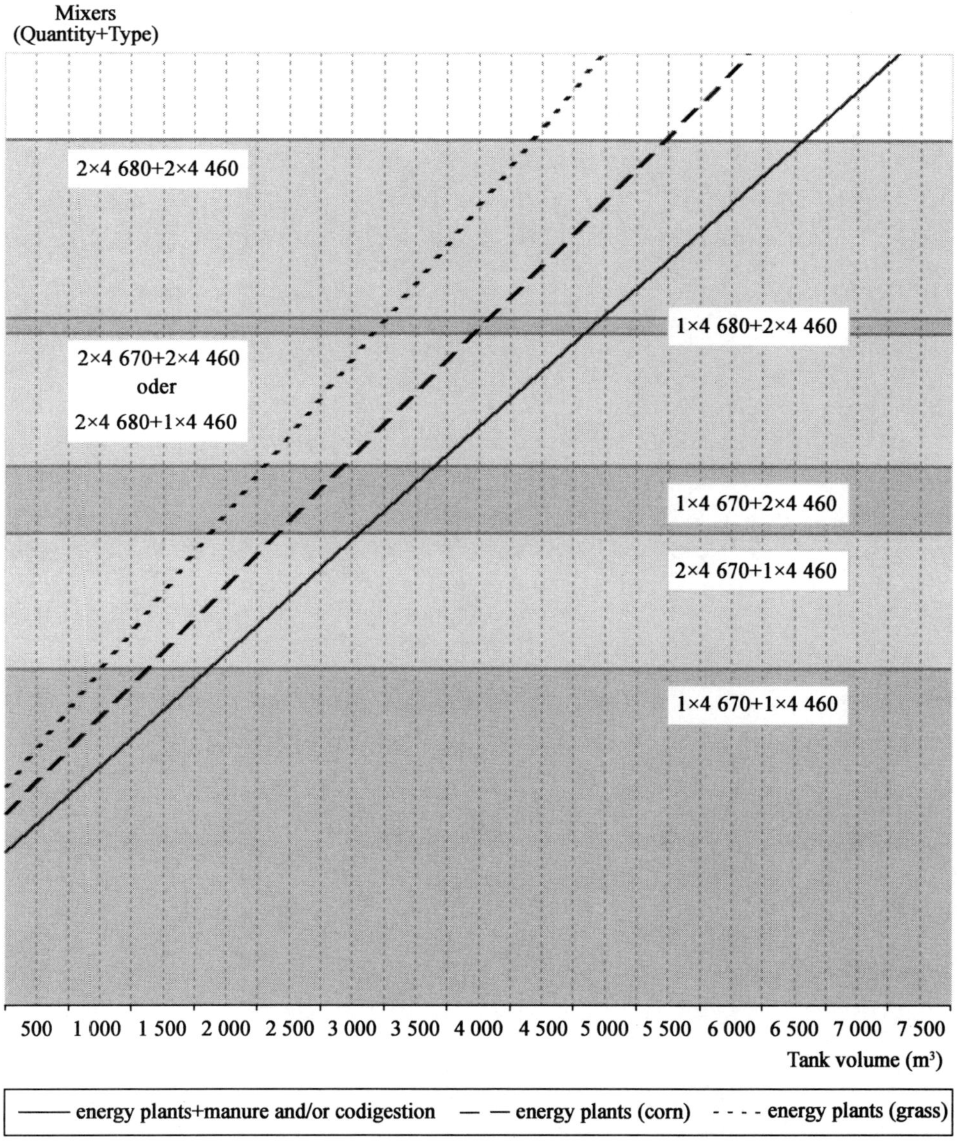

Figure 4 The flygt selection chart

6 Conclusions

Through our ongoing research in rheological basics about different input materials and the performance of different mixer types we already have reduced the power consumption in a biogas plant by 50%. We are positive to increase our knowledge about the rheological data of different input materials and to reach even more reduction of input power in the near future.

用于厌氧发酵的沼气发电机选择标准

R. Lattermann

（曼海姆能源系统技术有限公司，新加坡 62876）

【摘要】本文介绍了厌氧发酵中沼气组分预测的重要性，一般能提供精准预测的多是那些有多年生产经验的发酵罐供应商。优良的气体清洁系统对于厌氧发酵相关设备的抗腐蚀性有着重要的作用，而且为了预防沼气发电机的重要部件受到腐蚀，对沼气清洁系统运行状况的连续监测也变得尤为重要，通过对经沼气清洁系统处理的沼气组分分析，可以对沼气发电机的规格进行确定。

虽然连续的沼气气流可以供给沼气发电机，但在大多数情况下，沼气的生产受季节或其他循环的影响存在波动性，如日夜交替和垃圾填埋应用的长期循环。这就需要选择短期气体储存装置或是用多个发电机组的安装来缓解该问题。本文将提供一个判断最佳方案的方法。

本文阐述了不同技术下沼气发电机组的规格。根据不同应用条件下的沼气组分选择合适的设备规格尤为重要。

最后，文章介绍了一些以植物残渣和畜禽粪便为原料的城市污水处理系统、工农业污水处理系统及垃圾填埋处理应用沼气发电机组的实例。

【关键词】沼气发电机；应用；沼气

SELECTION CRITERIA OF BIOGAS ENGINES FOR ANAEROBIC DIGESTERS

R. Lattermann

MWM Asia Pacific Pte. Ltd.

Contact: Dr. Ruprecht Lattermann, MWM Asia Pacific Pte. Ltd., 11 Kian Teck Road, Singapore 62876, Tel.: +65-6660-1820, E-Mail: ruprecht.lattermann@mwm.net

EXECUTIVE SUMMARY

In this presentation, we aim to highlight the importance of a good prediction of the biogas composition that is expected from the anaerobic digester. Such accurate prediction is only possible by renowned and experienced digester suppliers. Furthermore, we will put emphasis on the importance of a good gas-conditioning system that eliminates corrosive components in the biogas generated. It is also important to have the performance of the gas-conditioning system monitored continuously, in order to prevent excessive corrosion of vital engine components. Based on the gas analysis of the gas compositions after the gas conditioning system, the exact sizing of the biogas generator set can start.

Some processes provide a constant stream of biogas for the combustion in biogas generator sets. In many cases, however, the biogas generation is subject to seasonal fluctuation, or other cyclic fluctuations, that may also be affected by the day and night cycle, or a typical long-term cycle of a landfill application. Based on the fluctuation of the biogas supply, the decision for short-term gas storage facilities to buffer the fluctuations, or a multiple-genset installation to cater for other cyclic fluctuation can be taken. The presentation will give a guideline what needs to be considered for the optimum solution for a given project.

Typical biogas-genset sizes, with their respective technology behind, will be presented. Of utmost importance is the correct sizing of the gas train, catering for the expected biogas composition for the respective application.

Finally, the presentation will contain some typical examples of biogas generator sets used in municipal waste-water systems, waste-water systems in the agro industry, both from plant residues and from animal manure, as well as for landfill applications.

Key words: Biogas engines; application; biogas

1 Typical biogas compositions

Different wastewaters originating from different sources, like wastewater from the starch industry (either from corn, *Figure 1*, or from tapioca, etc.), cattle farms, pig farms, chicken farms, palm-oil industry, etc., in combination with certain anaerobic digesters result in different biogas compositions after the digester.

Figure 1 Corn, feedstock for biogas

It is possible that two starch factories at the same location, separated by only a fence, being fed with identical feedstock, having in principle identical processes in their factories, generating almost identical wastewater, however, when operating different types of anaerobic digesters, the biogases generated from the different digesters can be hugely different, both in quantity as well as in quality.

What applies to the starch factory, in principle, also applies to any other anaerobic digester integrated in an agro-technical process. Other examples of anaerobic digesters are the fermenters in municipal wastewater systems and landfills.

For sizing of the biogas generator sets, mainly the Methane content of the gas volume flow, in terms of $m_N^3/h\ CH_4$, is the most relevant criterion, as it defines the energy content of the biogas. The higher the Methane content, the less overall biogas volume flow is required to generate a certain amount of electricity.

Typical compositions of dry biogas as they are found in the field are shown in *Table 1*.

Table 1 Typical biogas compositions

Compounds	Range	Typical
Methane (CH_4) in Vol%	45-70	55
Carbon Dioxide (CO_2) in Vol%	20-55	35
Nitrogen (N_2) in Vol%	0.01-10.0	4.5
Oxygen (O_2) in Vol%	0.01-2.0	0.3
Hydrogen Sulfide (H_2S) in $mg/m_N{}^3$	5-3 000	500
Ammonia (NH_3) in $mg/m_N{}^3$	5-1 500	250

Typical compositions of municipal sewage gas and typical compositions of landfill gas are shown in *Table 2* and *Table 3*.

Table 2 Typical sewage gas compositions

Compounds	Range	Typical
Methane (CH_4) in Vol%	57-72	65
Carbon Dioxide (CO_2) in Vol%	28-43	34
Nitrogen (N_2) in Vol%	0.0-20.0	1.0
Oxygen (O_2) in Vol%	0.01-2.0	0.3
Hydrogen Sulfide (H_2S) in $mg/m_N{}^3$	10-5 000	500
Ammonia (NH_3) in $mg/m_N{}^3$	0.1-2.5	0.15
Halogens (CKW) in $mg/m_N{}^3$	10-600	300
Siloxanes (Si) in $mg/m_N{}^3$	3-300	150

Table 3 Typical landfill gas compositions

Compounds	Range	Typical
Methane (CH_4) in Vol%	45-65	55
Carbon Dioxide (CO_2) in Vol%	20-45	27
Nitrogen (N_2) in Vol%	10-34	23
Oxygen (O_2) in Vol%	0-10	2
Hydrogen Sulfide (H_2S) in $mg/m_N{}^3$	10-5 000	500
Ammonia (NH_3) in $mg/m_N{}^3$	0.1-2.5	0.15
Halogens (CKW) in $mg/m_N{}^3$	10-600	300
Siloxanes (Si) in $mg/m_N{}^3$	3-300	150

From the above it is clear that the target must be to maximize the Methane (CH_4) content. It is also clear that a fire-extinguishing agent like Carbon Dioxide (CO_2) is not really desirable for the combustion, but it can be tolerated. Nitrogen (N_2) and Oxygen (O_2), especially at small quantities are harmless, but they may serve as an indicator whether the anaerobic digestion is in fact anaerobic or not. Larger quantities of Nitrogen and Oxygen may indicate a leak in the system, where ambient air is being ingested into the system, especially in landfill applications.

Hydrogen Sulfide (H_2S) and Ammonia (NH_3), as well as other potential contaminants, depending on the feedstock used in the process, have no sensible influence on the energy content of the gas or in the sizing of the biogas generator sets, but they are regarded harmful substances for the biogas generators and must be washed out of the gas prior to combustion in the engine.

Not indicated in the above table is the humidity of the biogas. Normally the humidity is quite high and the water fraction has to be extracted from the biogas prior to combustion.

2 Prediction of biogas quantity & quality

Both quantity and quality of the biogas has an influence on the selection of the size of the biogas generator set. Equally important like the optimization of the anaerobic digester is also the accurate prediction of the generated biogas quantity and quality, at least within a certain accuracy band, in order to be able to have the total capacity of the power plant designed correctly.

Normally, the anaerobic digesters and the biogas generator sets are being procured concurrently or at least pretty much at the same time. This means that at a time the anaerobic digester does not even exist, an accurate biogas quantity and quality is required from the designer/manufacturer of the digester. With a myriad of different factors having an influence on the performance of the anaerobic digester, it is clear that only suppliers with vast experience comprising different feedstock, different ambient conditions, etc. will be in the position to perform such prediction with sufficient accuracy.

The preliminary design of the entire system, including the total capacity of the power plant, as well as the design of the gas conditioning system must be based on the predicted values of gas quantity and quality. However, it is of utmost importance after the anaerobic digester has entered operation, to have the gas quantity and quality (detailed gas analysis) measured regularly (CH_4 and H_2S preferably measured continuously), to be able to optimize the gas conditioning system and maintain it in excellent operating condition.

3 Gas conditioning systems

Gas conditioning systems are not the topic of this particular presentation; however, we would like to highlight that for any satisfactory operation of biogas generator sets, primarily the gas conditioning system is key.

Figure 2 shows a typical biogas system where the main system components, the anaerobic digester (fermenter), the gas conditioning system (with compressor unit and emergency flare), the biogas generator set, open type or containerized, with or without cogeneration/trigeneration (besides electrical power, generation of heat or heat and cooling) with subsystems, switchboards, transformer and switchyard for feeding the (excess) electricity to the public grid.

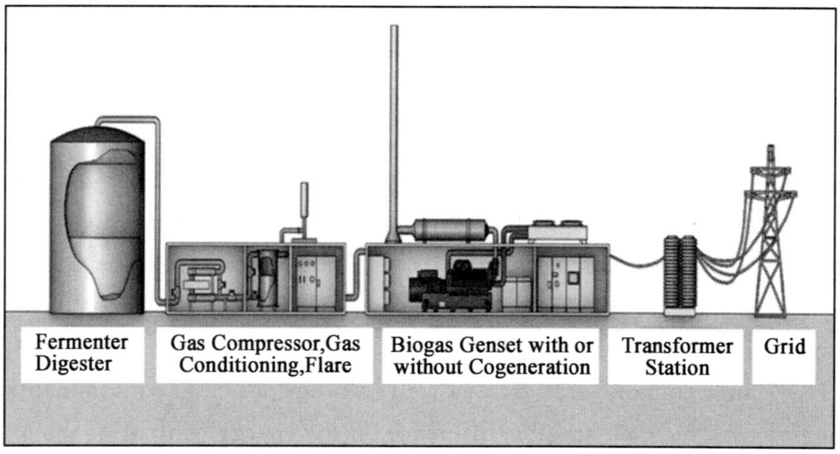

Figure 2 Typical biogas system components

It is not only of utmost importance to have the gas conditioning system designed correctly with filters, dryers and scrubbers, etc., in accordance with the contaminations expected and found in the gas, having in mind the contamination limits, but it is imperative that the performance of the gas conditioning system is being monitored continuously, so corrective action can be taken early, before corrosive attack can have a negative influence on vital engine components like turbocharger, mixture cooler, combustion chamber, etc.

In *Table 4*, typical permissible levels of residual contaminants in the combustion gas after the gas conditioning system are listed.

Table 4 Typical contamination Limit Values

Compounds	Permissible Limit Values
Total Sulfur (S)	<2 200 mg/10 kWh
Hydrogen Sulfide (H_2S)	<0.15 Vol%/10 kWh
	<1 500 ppm/10 kWh
Chlorine (Cl)	<100 mg/10 kWh
Fluoride (F)	<50 mg/10 kWh
Total Chlorine & Fluorine	<100 mg/10 kWh
Ammonia (NH_3)	<30 mg/10 kWh
Siloxanes/Silicon (Si)	<20 mg/10 kWh
Dust (2-10 micron)	<10 mg/10 kWh
Rel. Humidity (H_2O)	<80% (at the lowest temperature in the entire gas piping system)

Please note that the limit values are referenced to a standard gas with a lower calorific value LCV = 10 kWh/m_N^3. For gases with lower or higher LCV, the limit values have to be scaled accordingly. For example, at a lower LCV of the gas, a higher total gas volume flow is required to generate the same amount of electricity, so the permissible concentration of contaminants needs to be reduced, accordingly.

Once again, we would like to highlight the importance of an effective gas conditioning system that is continuously monitored and maintained.

4 Fluctuation of the biogas production

Depending on the feedstock and its availability, the production rate of the anaerobic digester is subject to a cycle. This cycle may be a daily cycle, determined by day and night, a seasonal cycle, determined by the availability of crops or the respective feedstock to the digester, or a long term life cycle of a landfill, where over a number of years, the landfill gas production increases, and for a subsequent number of years, the landfill gas production goes down until the Methane levels have reached so low values, that it is not longer economical to run a power plant on it, but to flare off the residual gas, as long as the Methane content can still sustain a stable flame in the flare.

During the practical operation, the actual biogas generation may be determined by an overlay of daily, weekly, seasonal and long-cyclic cycles that requires serious consideration with respect to sizing of the biogas generator sets.

In order to cope with fluctuations of day and night cycles or daily cycles, the biogas flow can be smoothened by introducing gas-storage facilities. There are many solutions available in the market; a possible solution is shown in *Figure 3*.

Figure 3 Biogas storage facility

The size of the storage tank must be designed to accommodate the daily fluctuations (peak minus average plus some safety margin). Depending on the operational requirements, the system can be designed for 24 hours continuous operation of the biogas generator sets, or only for 8, 10 or 12 hours of the generator sets. In this case, the formula for the required storage capacity needs to be amended, accordingly.

For seasonal fluctuations, storage of the biogas may not be the solution. For these cases, the correct split of the total power capacity of the power plant into multiple smaller units comes into play. *Figure 4* shows a multiple genset installation in a municipal wastewater plant.

Figure 4 Multiple genset installation

Anyway, in most cases and even without any fluctuation of the biogas supply, it is recommended to have minimum two biogas generator sets in operation; ideally always n+1 numbers of generators, with n being the number of generators to sustain the minimum electricity demand. In many cases, the biogas generator sets are running in cogeneration or trigeneration mode with hot-water or steam generation, potentially in combination with cooling. In this case, the individual sizes of the biogas generators and the requirement to run them may be determined by the hot-water or steam requirement rather than the electricity demand, or a commitment to deliver a committed number of kW's to the public grid.

The smallest unit of biogas generator sets of a multiple-genset installation must be designed in a way that at the minimum gas flow, the biogas generator set is minimum loaded 50% of its rated capacity. Prolonged operation of biogas generator sets below the 50% threshold will lead to undesirable deposits in the combustion chamber that may lead to increased wear and tear and knocking, when more gas is available and the power output is being ramped up.

5 Flexibility of containerized gensets

In case of landfill application, a preferred solution is containerized smaller-capacity landfill generator sets that can be installed as and when the landfill-gas production allows the installation of the next unit and so on. Once the gas production decreases after a number of years, the genset containers can easily be removed and installed at another landfill site, where the gas production is still on the increase.

Figure 5 shows a typical containerized genset solution with all genset-related auxiliary systems like gas train, cooling-water pumps, radiator system, genset ventilation system, cogeneration/trigeneration system, monitoring and control system, switchboards, and all related piping and cabling work prefabricated for instant installation and commissioning.

Figure 5 Typical containerized biogas genset

Containerized generator sets require a simple slab foundation, only, and no additional powerhouse is required.

6 Biogas generator sets from MWM

MWM, Germany, is a manufacturer of gas generator sets and complete power and cogeneration/trigeneration systems, thereof, for natural gas, biogas and a variety of special gases, like coal mine Methane, coking gas, etc., for a multitude of different applications.

MWM was founded by Mr. Carl Benz as 'Mechanical Workshop Mannheim' in 1871, and Carl Benz built the first engine of the world running on city gas. Today, MWM is one of the leading manufacturers of gas engines and gas generator sets, worldwide with subsidiaries and an extensive network of sales and service partners, worldwide.

Figure 6 shows the range of gas engines/gensets for 50-cycle applications. All genset models, TCG 2016, TCG 2020 and TCG 2032, in all cylinder variations are available for natural gas and biogas applications. Depending on the gas composition, the power output for biogas may require adjustment.

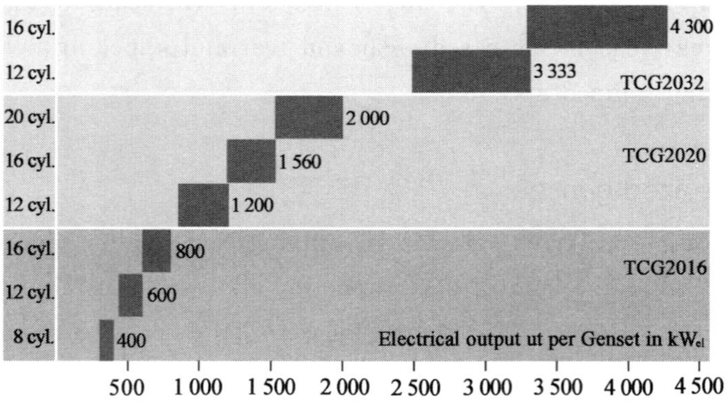

Figure 6 MWM Gas genset range (50 Hz)

With this wide range of available biogas generator sets, MWM is able to fulfill most of today's market requirements. Depending on the result of the biogas genset selection process, starting from singe 400 kW_{el} installation, through multiple 400 kW_{el} installations up to single and multiple, 4 000 kW_{el} installations, total electrical power-plant capacities of 400 kW_{el} to 20 MWel and bigger can be built.

The gas engines from MWM, disregarding the various engine families, are all based on similar leading-edge technological principles:

- State-of-the-art, multiple-cylinder V-engines;
- Turbocharging and two-stage intercooling;
- Single, individual cylinder heads with four-valve technology;
- Non-wearing high-voltage ignition system;
- Microprocessor-controlled high-voltage ignition system;
- One ignition coil per every cylinder;
- Electronic control and monitoring of genset operation through TEM;
- Exhaust emissions controlled according to combustion chamber temperature.

In *Figure 7* engine model TCG 2016 is shown. In *Figure 8* engine model TCG 2020 is shown.

Figure 7 TCG 2016 (400-800 kW_{el})

Figure 8 TCG 2020 (1 125-2 070 kW_{el})

In *Figure 9* engine model TCG 2032 is shown.

Figure 9 TCG 2032 (3 000-4 300 kW_{el})

7 Total electronic management system

Special emphasis is given on the MWM electronic monitoring and control system (TEM) that allows for maximum system flexibility while maintaining a simple, easy to understand, interface for the operator with touch-screen technology.

The following features allow for the high flexibility of the system, especially for multiple-genset installations

- Wide range of control and monitoring options;
- Easily adjustable to local conditions;
- Simple and secure operational concept;
- Convenient remote diagnosis and remote control;
- Functions allowing optimized economical efficiency;
- A modular system to govern all auxiliary functions;
- Multiple language documentation and handling;
- Available for the engine series TCG 2016 to TCG 2032.

In *Figure 10*, the touch screen of the TEM system is shown.

Figure 10 TEM Touch-screen system

Based on the MWM biogas generator sets in combination with the versatile TEM electronic management system, a high degree of flexibility in the power-plant design is given.

8 Gas train sizing

Depending on the CH_4 content of the combustion gas, different mixture ratios between combustion gas and air are required to arrive at a combustible gas-air mixture.

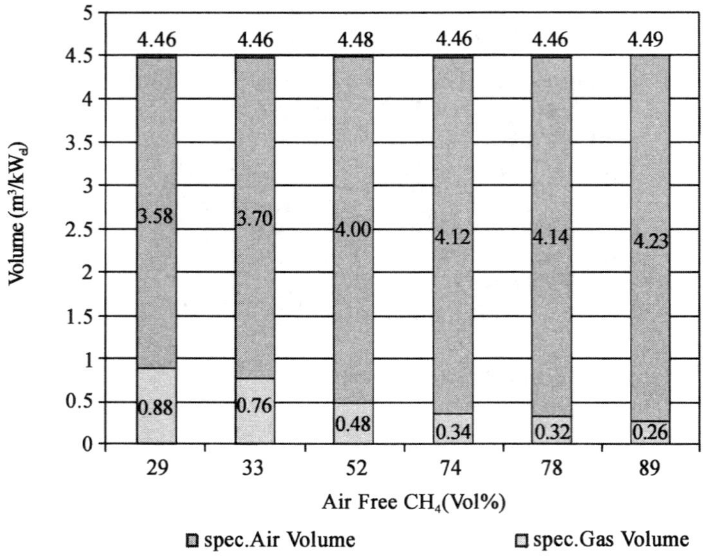

Figure 11 Gas/Air ratios for different gases

Figure 11 shows examples for the specific volumes of air and gas required per produced kWh of electricity as a function of the air-free CH_4 content in the gas.

Variations in the gas properties can be balanced by selection of the correct gas train and the mixture-formation process of the MWM gas-mixture system.

Figure 12 shows the principle design of the MWM mixture forming process.

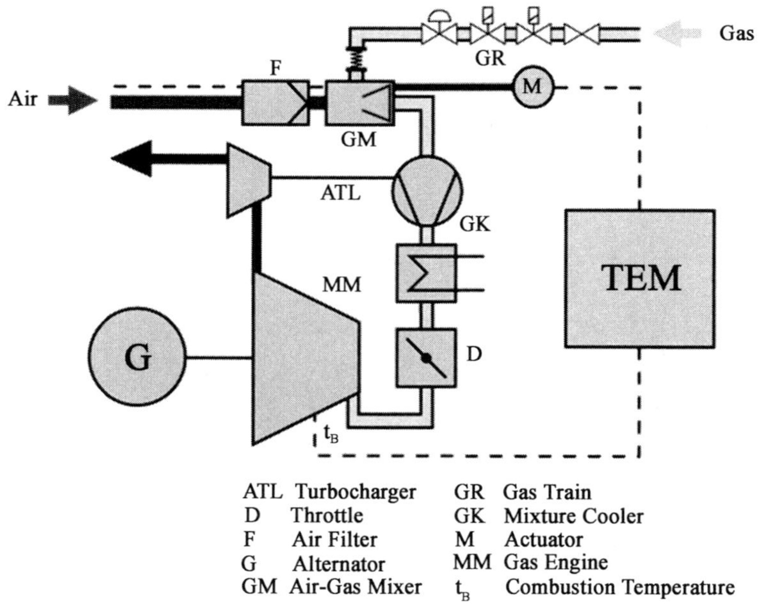

Figure 12 MWM Mixture-forming process

In the air-gas mixer (GM) the air fuel ratio is established: constant flow, constant calorific value independent of the CH_4 fluctuations in the gas. With the TEM Total Electronic Management, e. g. the regulation of the gas flow is automatically adjusted via an actuator (M) taking into account the average combustion temperature (t_B) of the engine. The mixture is compressed in the exhaust driven turbo charger (ATL), cooled down, throttled and burnt in the cylinders of the gas engine.
Figure 13 shows a closer view of the MWM gas mixer.

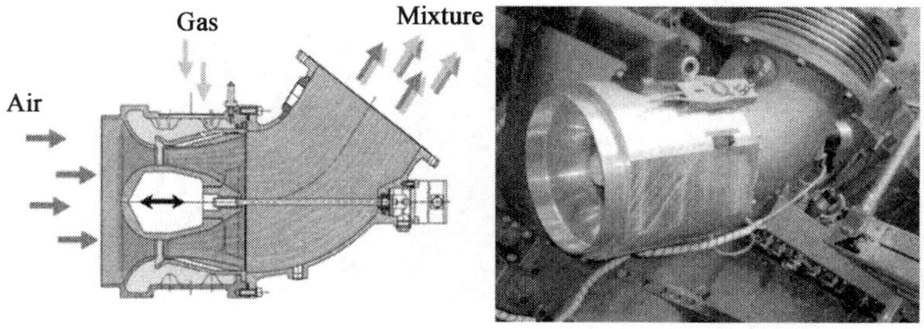

Figure 13 Operating principle of the gas mixer

The MWM air gas mixer works using the Venturi principle. The inner 'Zeppelin' is driven by the engine control via an actuator depending on setting limits and the gas quality. For easy start, the air-gas mixer has an automatic special starting position.
Typically, the gas composition of biogas is not absolutely stable but is subjected to variations within a certain tolerance band. CH_4 concentration fluctuations flow along the gas pipeline without significant inner mixing. When a new gas source (for example a gas storage tank) is being connected to the gas pipeline, it is only a matter of time when the engine has to handle gas with a sudden change in calorific value. Such a change would normally lead into heavy knocking, instant shut down, or even engine damage. The MWM TEM system, however, detects changes in combustion temperature and reacts immediately to such changes, adjusting the mixture and avoiding engine knocking, shut down, or damage.

9 Typical references for biogas gensets

MWM has extensive experience with gas generating sets for biogases originating from anaerobic digestion, including sewage gas and landfill gas, both in non-tropical countries as well as in tropical countries.
A variety of different references are being presented for each MWM engines series.

9.1 Anderlingen-Ohrel, Germany

Genset/Engine:
1 × TCG 2016 V08C, containerized
1 × TCG 2016 V16B
Segment/fuel type:
Biogas
Customer:
Anderlingen-Ohrel, Germany
Total output:
400 kW_{el} — 716 kW_{th}
Installation/Commissioning:
2009 + 2007

A TCG 2016 V08 C, packed in a container provides 3 200 MWh electrical and 2 552 MWh thermal energy per year that is used for the fermentation process. A biogas-fired TCG 2016 V16B, installed in earlier days, that uses the same gas, supplies heat to a district heating network. The integrated biogas conditioning from MWM saves the technology-bonus out of the renewable energy law.

9.2 Nong Bua Farm, Thailand

Genset/Engine:

2 × TCG 2016 V16

Segment/fuel type:

Biogas

Customer/Operator:

Nong Bua Farm, Thailand

Total output:

1.4 MW_{el}

Installation/Commissioning:

2006

Powered with this fuel, each of the units provides 700 kW of electricity for the farm. The thermal output of 707 kW per unit is used to heat water to 90℃. Thus, the plant's total efficiency is higher than 80 percent.

9.3 Schwallungen, Germany

Genset/Engine:

2 × TCG 2020 V12

Segment/fuel type:

Biogas

Customer/Operator:

Schwallungen, Germany

Total output:

2.0 MW_{el}

Installation/Commissioning:

2006

The Power plant Schwallungen is with approx. six Million Euros capital expenditure the largest biogas plant in Thüringen. The plant utilises leftovers from approx. 60 000 collection points and produces biogas that is converted in electrical and thermal power.

9.4 Sewage Plant Mannheim, Germany

Genset/Engine:

1×TCG 2020 V20, containerized

1×TBG 616 V16

3×TBG 620 V20 K

Segment/fuel type:

Natural gas/sewage gas

Customer:

Sewage plant, Mannheim

Total output:

4.5 MW_{el}

Installation/Commissioning:

2009

The sewage plant in Mannheim, Germany, expands its asset with a TCG 2020 V20 packed in a container. The generated waste heat is used for heating the digestion towers. The produced electrical power of 1 MW_{el} is used for the electrical consumption of the plant. The sewage plant with 5 gensets at total has now an electrical output of 4.5MW_{el}.

9.5 Eastern Creek & Grange, Australia

Genset/Engine:

5 × TBG 620 V16 K

Segment/fuel type:

Landfill gas

Customer/Operator:

Eastern Creek & Grange, Australia

Total output:

6.3 MW_{el}

Installation/Commissioning:

2002

With its five gensets MWM could reach more than 130 000 running hours within 5 years time, providing high reliability and availability while utilising a low quality landfill gas. Results to date have been sound with availability as high as 99% in some months.

9.6 Rahim Steel, Bangladesh

Genset/Engine:

8 × TCG 2032 V16

Segment/fuel type:

Natural gas

Customer:

Rahim Steel, Bangladesh

Total output:

31.2 MW_{el}

Installation/Commissioning:

2008

These MWM gensets supply a steel factory in Bangladesh with electrical power. Excess electrical energy is exported to the grid. The crucial factors are the requirements for fast step-load responses. Accordingly, the engines are equipped with special features to ensure customized load responses. This is one of the first MWM power plants using these engineering options at a large scale.

9.7 Precision Energy, Bangladesh

Genset/Engine:

15 × TCG 2032 V16

Segment/fuel type:

Cogeneration/natural gas

Customer:

Precision Energy

Total output:

60.0 MW_{el}

Installation/Commissioning:

2009/2010

In 2009 MWM won a contract for 15 TCG 2032 engines to be delivered within 3 months to Precision Energy in Bangladesh. These engines provide 60 MW_{el} of reliable power for the national grid. At present the engines are installed at a powerhouse built especially for this purpose.

9.8 Ämmässuo Filandll, Finland

Genset/Engine:

4 × TCG 2032 V16

Segment/fuel type:

Landfill gas

Customer:

Ämmässuo, Finland

Total output:

15.0 MW_{el}

Installation/Commissioning:

2010

The landfill in Ämmässuo, Finland, about 30 ha in size uses the landfill gas collected from the garbage fermentation in 4 units TCG 2032 V16. Because of the low calorific value of the landfill gas, two gas control trains per engine had to be installed. Each gas control train supplies one cylinder bank with gas. For the 4 gensets the existing powerhouse was rebuilt.

10 Conclusions

MWM has a long history in gas-genset development and manufacturing reaching back to the year 1871. MWM offers a product range of gas gensets for biogas applications with unit sizes between 400 kW_{el} to 4 300 kW_{el}, highly efficient for straight electric power generation, cogeneration or tri-generation systems. The selection of the right system solution depends on the demand of electricity and/or the demand for heating or cooling. Furthermore, the selection is determined by potential fluctuation of the gas quantity/quality over time.

MWM gas-mixing and combustion systems are well suited to cope with the requirements of biogas applications, especially with the requirements of fluctuating gas qualities.

MWM has a vast experience in biogas generator systems applications worldwide. With its worldwide presence, MWM is prepared to work in any part of the world and to provide local service.

MWM is looking forward to team up with new customers for the design and implementation of new, innovative biogas solutions.

沼气发电机的正确操作运行准则

T. Elsenbruch

（*GE Jenbacher 公司，Jenbach，6200* ）

【摘要】本文从燃气发电机制造商的角度阐述了如何正确操作运行沼气发电机。本文以颜巴赫沼气发电机（250～4 000 kW）为代表进行论述，其结论仍适用于其他燃气发电机。同时，本文针对气体组分、微量气体成分、气体湿度对发电机运行的影响以及相应的服务计划进行了说明。

德国和奥地利在生物质发酵方面的发展迅速，无论是数量上还是质量上，尤其是对关键因素的研究，都极大地推动了沼气运营方面技术的发展。本文对奥地利运行多年的沼气工程进行了分析，提出了如何提升沼气工程建设运营的重要理念。同时，还讨论了沼气组分的数量和质量及其对沼气工程的影响，特别是受到硫化氢污染的沼气会影响沼气发电机的运行和寿命。且沼气清洁系统的规格也会对沼气工程的经济可行性产生巨大影响。

【关键词】沼气发电机；应用；燃烧特点；热电联产；硫化氢

MAIN CRITERIA FOR A TROUBLE FREE OPERATION OF GAS ENGINES WITH BIOGAS

T. Elsenbruch

GE Jenbacher GmbH & Co. OG

Contact：Dipl. Ing. （FH） Thomas Elsenbruch，GE Jenbacher GmbH & CO OG，Achenseestraße 1-3，6200 Jenbach，Austria，Tel.：＋43-5244-600-2519，E-Mail：thomas. elsenbruch@ge. com

EXECUTIVE SUMMARY

In this paper the requirements of the various interfaces for a trouble free operation of gas engines with biogas from the point of view of a gas engine manufacturer will be presented. The presentation is in regard to Jenbacher products（power range from 0. 25 to 4 MW） with an emphasis on biogas engines，but generally for gas engines. Specifically，the subjects such as gas composition，gas trace components，moisture and their influence on the operation of the engine and the service plan will be presented.

The boom in the biomass fermentation segment in Germany and Austria has greatly increased the knowledge of biogas operation both qualitatively and quantitatively – especially with regard to the critical factors. Specifically，several sites in Austria that have been operational for a number of years have been analysed. Essential concepts to improve site installation and operation will be presented. Furthermore the qualitative and quantitative composition of the biogas and its effects will be discussed. Above all，the hydrogen sulphide contamination of the gas influences the availability of the engine and the adequate dimensioning of the gas cleaning system thus has a large influence on the economic feasibility of the project.

Key words：Biogas engines；application；combustion characteristics；CHP；hydrogen sulphide

1 Introduction

Increased ecological consciousness and the knowledge of limited reserves of primary energy in the form of fossil fuels make it necessary to utilize all available energy sources economically. Cogeneration plants with gas engines produce electricity and heat at decentralized locations, where they are required. They offer optimal efficiency in the utilization of energy with minimum environmental burden.

Gas engines are presently powered primarily with natural gas, biogas or propane. Especially biogas from anaerobic digestion processes are a very suitable kind of fuel for gas engines. A very wide spread and well known application is the utilisation of gases coming from the sludge tanks of waste water treatment plants. The same process of anaerobic digestion can be utilised for biomass products from agriculture, organic waste from households, food industry or slaughter houses.

2 Different anaerobic digestion processes

As a metabolic product of the participating methane bacteria, the prerequisites for its production are absence of oxygen, a pHvalue from 6. 5 to 7. 5 and a constant process temperature of 15℃ to 25℃ (psychrophile), 30℃ to 40℃ (mesophile) or 50℃ to 55℃ (thermophile). The fermentation period is approximately ten days for thermophile, 25 to 30 days for mesophile and 90 to 120 days for psychrophile bacteria. The fermentation systems of today operate largely within the mesophile temperature range.

The process of biogas generation is divided into three steps (*Figure 1*):

- Preparation of the organic input;
- Fermentation;
- Post-treatment of the residual material.

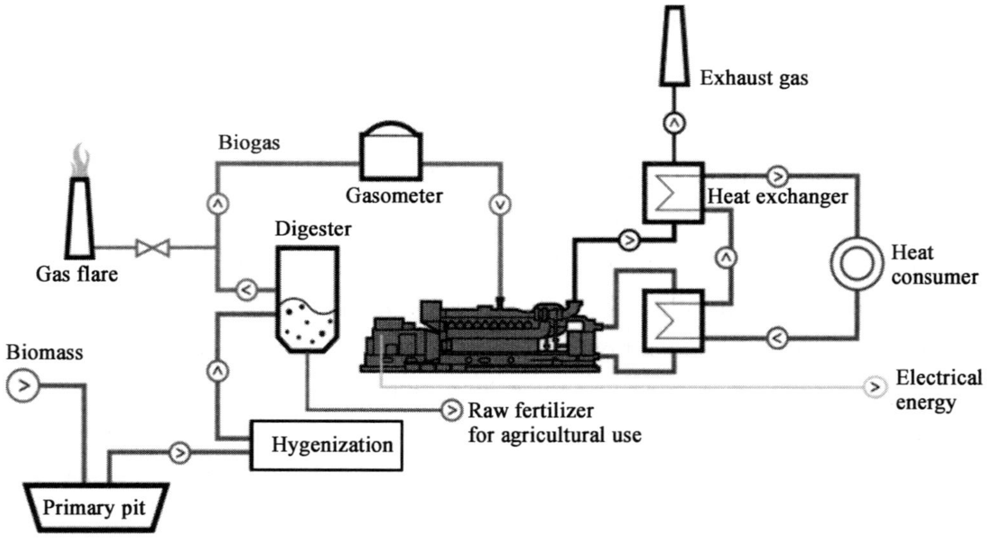

Figure 1 Anaerobic digestion process for biogas in CHP

At the start, the organic material is collected in a primary pit, sterilized to remove harmful germs in case of food waste and moved to the digester. The biogas produced in the digester is collected in a gas storage tank to ensure a continuous supply of gas, independent of fluctuations in the gas production. Finally, the biogas is fed into a gas engine. For safety reasons, the installation of a gas flare is recommended so that excess gas can be burned off in the event of excessive gas production. The end product from the fermentation of the biomass can be utilized as fertilizer. The gas mixture produced in the digester consists of 50% to 70% methane (CH_4) and 30% to 50% carbon dioxide (CO_2). This composition makes biogas well suited for combustion in gas engines.

The generated electrical energy can be utilized for the treatment plant as well as to supply the public power grid. The thermal energy can be used for heating the digester or to offset the heat requirements of the treatment plant.

3 Critical substances in biogases

Gases, respectively their constituents, have different properties which can be assessed through their characteristic values, such as methane number, heat value and laminar flame speed, to mention only a few (*Table 1*). To be able to achieve an ideal degree of energy conversion, these values have to be considered during design of the engine.

Table 1 Characteristics of different gases

Gas type	Composition	Density (kg/Nm^3)	LHV (kWh/Nm^3)	Methane Number	laminar flame speed (cm/s)
Hydrogen	H_2	0.089 9	2.996	0	302
Methane	CH_4	0.717	9.971	100	41
Propane	C_3H_8	2.003	26.00	33	45
Natural gas(example)	CH_4= 88.5% C_2H_6= 4.7% C_3H_8= 1.6% C_4H_{10}= 0.2% N_2= 5%	0.798	10.14	80	41
Landfill gas(example)	CH_4= 45% CO_2= 50% N_2= 5%	1.188	4.5	150	21
Biogas(example)	CH_4= 65% CO_2= 35%	1.158	6.5	135	27

The biogas from the anaerobic digestion plant consists mainly of a mixture of methane (about 50 to max. 65%) and CO_2. From the perspective of combustion inside the engine this fuel is very suitable for the engine. Nevertheless, one must pay attention to a number of singularities of the biogas, especially the trace substances—in particular H_2S. In the event of the simultaneous processing of industrial biomass waste in the co-fermentation process further substances (phosphorus compounds, siloxanes, etc.) can get into the biogas and perhaps cause problems in the engines or in the exhaust gas system.

3.1 Formation of hydrogen sulphide and its impact on gas engines

The anaerobic fermentation of biomass generally produces not only methane and carbon dioxide but also a certain amount of hydrogen sulphide (H_2S) from the protein component of the biomass. This gas is toxic, attacks ferrous materials and has an unpleasant smell of rotten eggs. Typical H_2S contents in biogases range from 0.01% to 0.5% by volume (corresponds to 150 to 7 500 mg/Nm^3 gas).

The relatively high content of hydrogen sulphide (H_2S) in biogas and sewage gas causes the development of SO_2 during the combustion. When biogas is used in gas engines, an excessive H_2S content has harmful effects:

- The lubricating oil acidifies and has to be changed very frequently;

- If the oil is not changed in good time, moving parts of the engine become excessively worn;
- When the H_2S content is very high (2 000-4 000ppm), some engine components suffer direct damage. Where condensates arise in the gas flow (due to insufficient drying of the biogas in the gas control system or during starting procedures in the exhaust gas heat exchanger), the result is an acid attack on metal parts.

By further oxidation especially in the catalytic converter SO_2 reacts to SO_3. Together with water it forms sulphuric acid (H_2SO_4) which is among the most aggressive acids. These acids can cause massive corrosion problems in exhaust gas heat exchangers. Therefore it is very important to avoid their condensation.

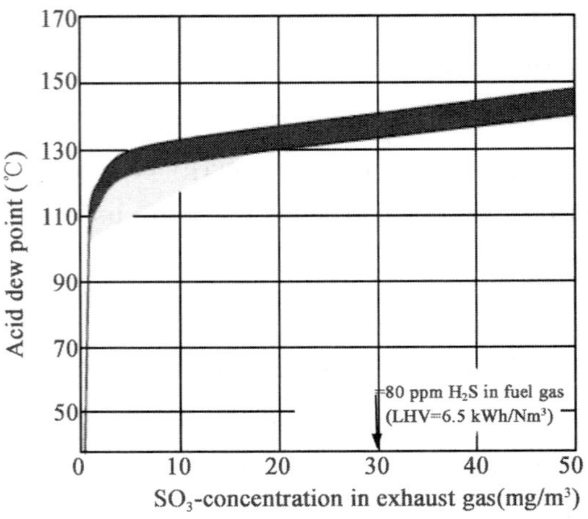

Figure 2 Acid dew point in dependence on SO_3-concentration

3.2 Different solutions to remove hydrogen sulphide from biogas

3.2.1 Binding the sulphur to iron salts in the digestion process

A method frequently employed to bind sulphur in biomass digesters is the use of iron salts (iron-Ⅱ-sulphate, iron-Ⅱ-chloride), which are mixed into the biomass in solid or liquid form. Here, the iron performs the function of binding the sulphur, thereby preventing the H_2S from escaping into the biogas. The following chemical reaction takes place:

$$Fe^{2+} + H_2S \longrightarrow FeS + 2H^+$$

The resulting iron sulphide salt and the elemental sulphur are present in insoluble form in the fermentation substrate and are removed along with the putrefied material. In biogas plants the iron salt can be mixed into the organic substrate or liquid manure or even measured into the digester, in which case it is vitally important to mix it evenly through in order to prevent the release of H_2S as far as possible.

For small plants, the iron salt can be purchased in bags and a sufficient level of desulphurisation can be maintained by adding a few bags a day. For large plants an iron-Ⅱ-chloride solution should be added from a container by means of an automatic metering pump. This method can also be combined with the air-injection method (see 3. 2. 2) or with biological desulphurisation (see 3. 2. 3), thereby making it possible to further reduce the H_2S content and take the strain off the other methods.

3.2.2 Biological desulphurisation in the digestion process

The exhalation of H_2S is often prevented in biomass digesters by introducing (injecting) a controlled amount of air directly into the digester. This causes the autonomous growth of a certain number of micro-organisms which degrade H_2S. In terms of plant technology, this process is very straightforward but has to be followed precisely.

It reduces the methane yield as methane only forms well in anaerobic conditions (without oxygen). Individual cases have also been reported in which sulphuric-acid-forming micro-organisms occur and cause acid corrosion in the digester.

3.2.3 Biological desulphurisation in the downstream reactor

The biological degradation of H_2S described at 3. 2. 2 can also be produced in a separate reactor. The biogas is conducted from the digester through a solid bed reactor with tower packing. The tower packing is sprinkled with a nutrient solution. As described at 3, the controlled injection of air results in the autonomous growth of a population of H_2S-degrading micro-organisms. The pH value must be controlled. The advantage is that methane production is not impeded by oxygen. The disadvantage is that a typical reactor has to be built to a considerable size. The H_2S content should be below 3 000 (5 000) ppm.

3.2.4 Chemisorption in solid bed of activated carbon

The standard activated carbon used to remove organic silicon compounds (siloxane and others) can only bind a limited amount of sulphur and is therefore not suitable as a method for desulfurising biogas. However, doped special carbons, which bind up to 60% of their own weight to H_2S are supplied for selective desulphurisation. The process must be conducted in moist conditions and with a low O_2 content. The advantage is the small amount of required container capacity and the disadvantage is the high price of the special carbon (5~6 €/kg). This is not feasible for biogas if it is the only measure to remove H_2S. But it is a very good solution as a 'police filter' to increase operational safety.

4 Utilization of biogas in engines

Combustion characteristics of gases

Besides the heating value, the methane number and the ignition limits there are still two important criteria for assessment of gases for use in engines, namely the laminar flame speed and the mixture heating value (*Figure 3*). The great differences in the heating value have nearly no effect on a correctly controlled turbocharged gas engine, since the mixture heating value is the important variable for the power output. A so-called naturally aspirating engine with a fully open throttle and the weak biogas would have a power output loss of about 8%. However, the difference in quality has a definite effect on the concept of the gas mixer and the control concept. In the case of GE's Jenbacher gas engines the patented LEANOX control concept is used; here a change in the heating value during engine operation is detected directly and the gas mixer can be used immediately for correction of the air/fuel ratio (*Figure 4*). This method ensures that even under varying conditions of the biogas (pressure, CH_4 concentration) the reliable combustion can be ensured.

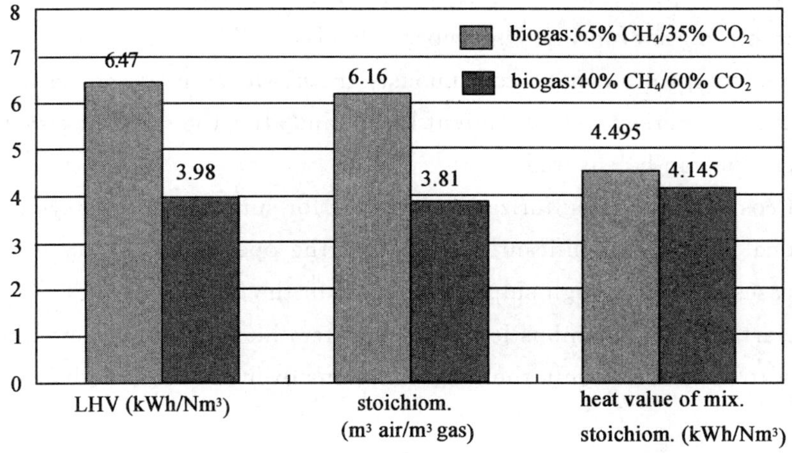

Figure 3 Parameters of biogas

As the air fuel ratio in a lean burn engines is increased to a level where a safe ignition of the mixture can still be maintained but on the other hand the combustion temperatures are minimized to obtain as low as possible NO_x-emissions, it is very important to have continuous and reliable control on the air/fuel ratio. By the means of the above described LEANOX lean-burn combustion control system continuous reliable

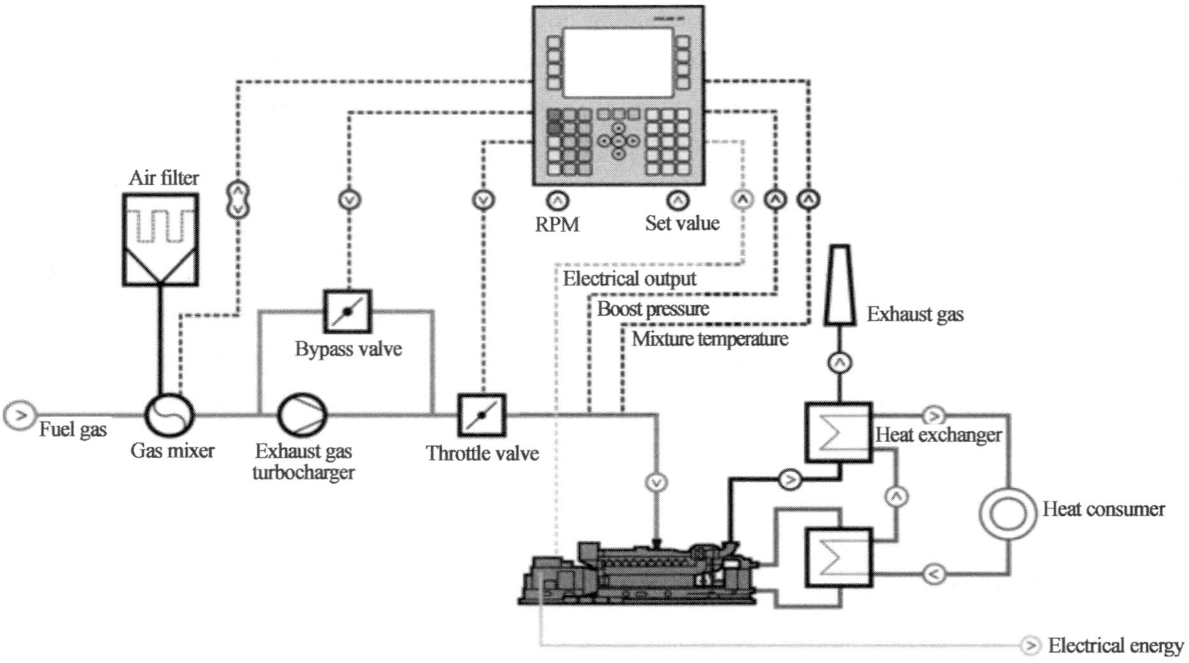

Figure 4 LEANOX control concept

control of the air/fuel ratio is ensured and thus NO_x-emissions can be kept under control under all varying frame conditions. Further this control system provides predictability of the life-time of all parts related to the combustion (pistons, cylinder-heads, valves, spark plugs) and thus the maintenance costs can easily be guaranteed.

5 Conclusions

The desulfurization of biogas is one of the most important steps in the biogas utilisation process. In most of the cases the method described under 3. 2. 3, the biological desulfurization in a separate reactor downstream the anaerobic digester brings the most reliable, stable and enduring results. Professional systems from specialized companies control continuously the pH level in the process, the O_2-concentration in the system and further of course the H_2S content before and after the scrubbing system. By these means long term reliable operation can be ensured.

The initial investment costs of such desulfurization costs are for sure higher than system described under 3. 2. 2, but the operational cost are significantly lower and the operational safety is extraordinarily much higher, especially in cases with very high sulphur contents in the biogas.

A reliable state of the art lean-burn combustion control system like the above mentioned LEANOX-system provides the capability to control not only the NO_x-emissions over the life-time but also helps to minimize operational expenses.

热电联产技术及其安全考虑

T. Fischer, K. Backes

(*Krieg & Fischer 工程有限公司*, *Göttingen 37085*)

【**摘要**】热电联产(CHP)机组通常利用沼气作为燃料并将其转化为电能和热能。目前市场上有多种热电联产技术,以内燃机耦合发电机组为主,其他还有微型燃气轮机,斯特林机和燃料电池等技术,但应用很少,本文不作论述。由于内燃机采用气体燃料,这就需要避免形成可燃气体。引擎同时生产电能和热能,产生的能量中大约 50%是热能,热能可以用来取暖、烘干、生产蒸汽或用作空调,合理使用这部分热能增加沼气工程的能源效率和经济效益。

沼气在热电联产前需经过预处理,特别是硫和水蒸气必须去除以避免腐蚀引擎,为减少引擎燃烧产生的尾气污染,尾气排放前需经过后处理。

本文介绍了典型的热电联产设计,对集装箱和建筑物内安装的热电联产技术进行了比较,并对一些特殊部件,例如鼓风机、进气口/出气口、余热利用和控制单元进行了介绍,控制系统不仅是操作人员的控制工具,同时也能够增加安全性。并对热电联产的维护管理和备品备件提供了建议并给出范例。本文还对两种常用的热电联产机组进行了比较:双燃料引擎用在 350 kW 以下的沼气工程,而燃气奥托机则适用于更大的沼气工程。同时,本文对热电联产的各个模块也进行了简单的介绍,并特别关注了其安全性。

【**关键词**】沼气内燃机;安全;热电联产;热利用

TECHNOLOGY AND SAFETY ASPECTS OF CHP MODULES

T. Fischer, K. Backes

Krieg & Fischer Ingenieure GmbH

Contact: Dr Katharina Backes, Krieg & Fischer Ingenieure GmbH, Bertha-von-Suttner-Straße 9, 37085 Göttingen, Germany, Tel.: +49-551-900-363-0, E-Mail: Backes@KriegFischer.de

EXECUTIVE SUMMARY

Biogas is most commonly used to fuel combined heat and power generator engines (CHP) which convert the gas into power and heat. Various types of CHP-technologies are available on the market. The most prevalent type is the combustion engine coupled with a generator. Furthermore micro gas turbines, Stirling motors and fuel cells exist, but they play a minor role and will not be considered in this paper. As the combustion engine operates with gas a key aspect is safety, therefore the development of an explosive atmosphere has to be avoided. The engine produces electricity and heat in combination. As about 50% of the energy produced is thermal energy, the use of this heat is an important factor in order to increase the efficiency of the plant and its economy. Heat can be used not only for heating but also for example for drying, steam production or air conditioning.

The biogas has to be pre-treated. Especially, sulphur and water vapour have to be removed to protect the engine from corrosion. An after-treatment of exhaust gases of the engine is needed to reduce emissions.

The design of a typical CHP used for biogas follows. The paper compares the installation of a CHP in a container and in a building. Information about special parts, such as, blower, air-inlet/outlet or heat utilisation and control will be given. The control system is not only a tool for the operator but also increase safety. Remarks are given regarding maintenance control and spare parts, and examples listed. The two common kinds of CHP are compared: the dual fuel engines are used for small plant up to 350 kW whilst bigger biogas plants usually use gas-otto engines. This article gives a short introduction to CHP modules with a special focus on increasing safety.

Key words: Biogas combustion engines; safety; CHP; heat utilization

1 Introduction

The use of combined heat and power generators (CHP) is the combined production of electricity and heat. It is the most common usage of biogas in biogas plants. Depending on the power generator technology in a CHP a high percentage of heat is generated in addition to the electricity.

The total efficiency is within a range of 85%-90%. In this case the loss of energy produced with biogas is about 10%-15%. The electrical efficiency of modern CHP's is about 40%. Therefore, from 1 m^3 biogas about 2.0-2.4 kW_{el} can be produced, depending on the content of methane.

Combustion engines coupled with a generator are used almost exclusively for CHPs. The motors run at a constant speed (1 500 rpm) for the directly coupled generator to provide electrical energy that is compatible with mains power. For the operation of the generator, i.e. for electricity generation, in addition to the conventional dual fuel engines and spark ignition gas engines, micro gas turbines, Stirling motors or fuel cells can be used. Since this technology is however still mostly in the developmental or prototype stages, we will concentrate mainly on CHPs with combustion engines.

2 Safety

Biogas is a gas mixture and consists mainly of methane, carbon dioxide, hydrogen sulfide and other trace gases. The density of the mixture amounts to 1.18 kg/m^3. The density in relation to air is only 0.9 (Eder and Schulz, 2006). That does however not mean that biogas will concentrate in a layer above the air in an enclosed space, as the heavier components, like carbon dioxide, may separate. Temperature also plays an important role. That is why gas detectors should not only be installed under ceilings, but also at floor level. The general integration of a combustion gas engine into a biogas plant can be seen in the previous paper (T. Elsenbruch, Jenbacher, 2011).

Table 1 Typicalcompositionof Biogas (Studie Einspeisung von Biogas in das Erdgasnetz, 2006)

	Symbol	Range
Methane	CH_4	45%-70%
Carbon dioxide	CO_2	25%-55%
Nitrogen	N_2	0.01%-5%
Oxygen	O_2	0.01%-2%
Ammonia	NH_3	0.01-2.5 mg/m^3
Hydrogen sulfide	H_2S	10-30 000 mg/m^3

In combination with atmospheric oxygen, the biogas gas mixture is to some extent explosive, which is why specific safety regulations must be observed during the construction and operation of a biogas plant. An explosion hazard exists in the range of 6 to 12 percent of volume of biogas. The required ignition temperature lies at 700℃. The max. follow-up flame speed in air is 0.25 m/s. The amount of air

theoretically required is 5.7 m^3 per cubic meter of biogas.
Three conditions must be met simultaneously before a fire or explosion can occur. A fuel (ie. combustible gas) and oxygen (air) must exist in certain proportions mentioned above, along with an ignition source, such as a spark or flame. The ratio of fuel and oxygen that is required varies with each combustible gas.
Ignition sources in biogas plants can be:

- Hot surfaces, e.g. on the turbocharger (> 500℃);
- Open flames (fire, flames, embers);
- Mechanically generated sparks (friction, impact, grinding);
- Electrically generated sparks (switching processes, loose contacts, balancing currents);
- Exothermic reactions (spontaneous combustion of dust);
- Lightning strike;
- Electrostatic discharge.

The minimum concentration of a particular combustible gas necessary to support its combustion in air is defined as the Lower Explosive Limit (LEL) for that gas. Below this level, the mixture is too 'lean' to burn. The maximum concentration of a gas that will burn in air is defined as the Upper Explosive Limit (UEL). Above this level, the mixture is too 'rich' to burn. The range between the LEL and UEL is known as the flammable range for that gas.

- LEL-Lower explosive limit 4.4-5 Vol% CH_4;
- UEL-Upper explosive limit 15-17 Vol% CH_4.

Any possible leakage of biogas in the machine room is monitored for safety and blocked by a room air monitoring system with the following safety—oriented functions, e.g. 20% of the LEL (lower explosive limit) (0.9 percent of volume CH_4) in the room air, followed by the safety measures:

- Optical and acoustic warning signal;
- Ventilation at 100% capacity.

e.g. 40% of the LEL (lower explosive limit) (1.8 percent of volume CH_4) in the room air, followed by the safety measures:

- Optical and acoustic warning signal;
- Ventilation at 100% capacity;
- Automatic deactivation of gas utilisation and shut-off of the gas supply outside the machine room, or deactivation of all non-ex-protected equipment in the engine room (e.g. disconnection of power supply).

Furthermore, additional dangers, e.g. suffocation or poisoning, and mechanical hazards exist.
Detailes are given in the Safety Rules for Biogas Systems (2008).

3 Container/Building

Engines can be installed in a container or in a building. Many engine manufacturers offer to install their motors directly in a container. The benefit here is that most of the installation work can be done in the factory. After arrival on site, the container is quickly connected and ready for operation. The container is modified for sound insulation. It comes equipped with a sump tray or an oil separator for the motor oil. Installations like the emergency cooler and the chimney can be positioned directly on the roof of the container. The container must be installed in such a way that its front side is accessible, in order to be able to extract the motor if necessary. Depending on motor size, a 6 m or 12 m container can be used. The width can vary between 2.5 m and 3 m. We recommend a 3 m container to have more space beside the engine in case of maintenance and repair works. The required control cabinets are usually installed in a second container.

In the installation inside a building, the CHP must be accessible from three sides. The installation room must be sound—insulated. The doors must open in flight direction. Just like in a container, a collection of any leaking oil must be ensured. The benefit here is that there is generally more space around the motor for repairs and maintenance inside a building than in a container.

4 Pretreatment of biogas

Biogas has to be treated before it is used in a gas engine. Biogas is saturated with water vapour, and contains in addition to methane and carbon dioxide also traces of hydrogen sulfide. The amount depends on the kind of input substrate. A mixture with water vapour results in the formation of sulphuric acid. The acid corrodes gas line, fittings, motors, exhaust pipes, etc. A desulphurisation of the biogas is therefore the most important measure to reduce corrosion. The sulphur comes from the degradation of substrates containing protein. The following desulphurisation processes can be employed:

- Chemical-physical process;
  - —Addition of iron salts to the fermentation substrate;
  - —Absorption by ferrous substances (dry filter);
  - —Activated carbon filter.
- Biological process (targeted injection of small volumes of external air into the gas space);
- Combination process.

Furthermore it is important to dewater the biogas saturated with water vapour in order to prevent condensation in the motor. An initial dewatering of the biogas occurs in the gas storage tank and in the pipes, while the biogas is cooled to ambient temperature. The pipes must therefore be installed frost-free and at a gradient. Water and hazardous gases can be extracted from the biogas in a cooling process, aided by a cooling unit. This will help protect the motor from premature wear.

5 Exhaust gas emission

Unfiltered exhaust gases usually contain air—polluting toxins, which are harmful to the environment and a health hazard. Each country has specific environmental protection laws to limit pollutants. In Germany, the emission of pollutants is governed by the Federal Immission Control Act (BImSchG) with the relevant Federal Air Pollution Control Ordinances (BImSchV), and/or the German Technical Guidelines on Air Quality Control (TA Air). Quite often that implies the use of emission control equipment to comply with the specified maximum threshold values. The use of a high chimney can also help control the propagation of emissions and aid their dilution.

In the exhaust emissions of motors powered with biogas, the following main components occur in addition to carbon dioxide:

- Dust
- Nitrogen oxides;
- Carbon monoxide;
- Formaldehyde;
- Sulphur dioxide.

An after-treatment of exhaust gases is needed to reduce formaldehyde emissions. Possible processes are e. g. thermal oxidation or the use of an oxidation catalyst (oxicat). Prerequisite is that the biogas has been filtered through an activated carbon filter. The sulphur contained in the biogas is extremely harmful for the catalyst.

6 Special parts

6.1 Special parts: Blower

The biogas must be pre-compressed before it can be supplied to the motor. Pre-compression is achieved with the help of a pressure booster blower.

The blower should be positioned in such a way that it does not affect the roof of the gas storage tank. The roof of the gas storage tank is very sensitive to underpressure, and therefore is equipped with a vacuum switch. The activation of the vacuum switch in an automatic control system effects the immediate shut-down of the motor. It can frequently lead to problems if the blower is positioned too close to the sensor. There may even occurunderpressure which is fatal for a gas holder roof.

Where several motors are used it makes sense for reasons of redundancy to equip each motor with its own individual blower instead of using one large blower. It must be ensured that one of the blowers has sufficient capacity to additionally power the emergency flare stack.

The gas blower must be ex-protected, as it is possible that external air can be introduced via a leak in the biogas line, which would result in an explosive air/gas mix in the CHP container.

The biogas is supplied to the motor by way of a safety-controlled pipe section. This section is equipped with a quick action stop valve, deflagration protection, a pressure regulator, two solenoid valves and a gas flow meter. The gas is then mixed with the combustion air shortly before it enters the motor.

6.2 Special parts: Air input output

The air input/output on the one hand supplies the motor with combustion air, and on the other hand cools the radiation of the engine. There are several systems available for cooling the engine room. Some companies equip the building with a fan that pushes air into the CHP room. Others suction the air out of the CHP room. Sufficient ventilation capacity must be ensured to avoid the generation of an explosive mixture in case of biogas leakage.

CHP ventilation is of particular importance in hot climates or on hot summer days, as high motor temperatures can lead to performance losses and finally to an engine shut-down due to overheating.

6.3 Special parts: Lubrication oil tanks

The motor must be lubricated continuously. Most CHP providers therefore place a container for used oil and for fresh oil by the motor.

It is beneficial if it is possible to pump out used oil directly to a tanker and provide fresh oil via an oil line. These tanks must be double-walled, as the oil is a water-polluting substance. The CHP building must be additionally equipped with an oil separator or a sump tray, which can hold the entire oil fill of the motor in case of a leak. In Germany, this measure is dictated by the Facility Ordinance on Handling Water Pollutants (VAwS).

6.4 Special parts: Heat utilisation

The waste heat of the motor cooling water and the exhaust gas heat are extracted via a heat exchanger. From there the heat is distributed within the biogas plant to the heat exchangers and to external heat consumers. Should the heat extraction be insufficient, then surplus heat can be released into the atmosphere via an emergency cooler, which can be installed next to or on the roof of the CHP container, and which is connected to the motor cooling circuit just like a consumer.

A motor generates heat in two places:

- Intercooler
  —Heat has to be used or discharged in an emergency cooler
- Exhaust gas heat exchanger
  —Optional, if heat can be used

The heat produced in cogeneration can be used for several heat utilization possibilities:

- Domestic or industrial heating purposes, distribution in a heat distribution grid;
- Water heating;
- Steam production;
- Air conditioning, cold production with heat; absorption cooling machines operate with various refrigerant systems, e. g. waterlithiumbromide, water-ammonia etc. and when integrated in a CHP plant can be driven either by heat from the water cooling or from the exhaust heat exchanger;
- Drying (wood, wood-chips, grain...);
- Thermal energy storage in a heat storage tank or thermo-chemical storage.

6.5 Special parts: Control

The control cabinet is installed in direct vicinity to the unit. Data are transmitted to the unit control via a highly fault free bus connection. The workstation can be places anywhere on the plant, at the unit or in the switch room, at the customer's choice. The basic functions of the control system are for example:

- The automatic program sequence for start and stop of the gas engine;
- Reliable control of the exhaust gas emission to low values;
- Integrated digital speed and power control;
- Monitoring of all sensors of the unit;
- Monitoring of the engine cooling water;
- Control and monitoring of the lubrication oil circuit inclusive oil pre- and after-lubrication as well as oil change;
- Electronic operation diary for recording any warning, malfunction;
- Electronic operation hours counter;
- Connection to a workstation;
- ...

The entire biogas plant is monitored at the most important measuring points. Measured values are captured and analysed in a central PLC. All important aggregates are connected as well. The individual units are displayed on a screen. Normal operation is automated as much as possible. Operating personnel can however directly access individual aggregates via the PLC in order to check or control the processes.

Specific signals, like safety signals (e. g. pressure switch-off) can be fed into the central controls of a networked system.

6.6 Noise emission

The requirements regarding noise insulation are described in the Federal Immission Control Act, specifically in TA Noise. Residential and industrial areas are distinguished in five categories, each of which has a prescribed immission value. A further distinction is made between day and night.

Sufficient noise protection is generally safeguarded as long as the overall noise pollution of a plant does not surpass this prescribed value at one of the immission points.

The CHP container or the building with the CHP can be additionally noise-insulated with mineral wool and perforated plate covering as needed to further reduce noise emissions. The ventilation system can be equipped with splitter attenuators. The exhaust gases are fed via a pre-muffler and a rear silencer and then exits via a vertical tail pipe into the atmosphere. The exhaust tail pipe should end approx. 10 m above the foundation level.

Noise emitting parts of the biogas plant are:

- Mixing devices with outside mounted engines;
- Engine of solid input device or others;
- Stack;
- Emergency cooler;
- Air input output;
- Blower of the engine;
- TA Noise (technical instruction on noise control).

The point of noise measurement is at the window of the nearest building.

The operator must prove that the noise levels produced by the operation of the biogas plant or the CHP do not surpass the relevant emission threshold values for the respective area.

7 Start-up

Operating instructions must be available at initial start-up. The start-up of a biogas plant is a special operation, which requires specific measures. A dangerous explosive atmosphere can occur in the gas room of the fermenter and the post-fermenter during start-up. Ignition sources (see section 2. 'Safety') must be eliminated. The biogas quality must be verified before the CHP start-up. The quality is sufficient and not an explosion hazard if the methane content of the biogas is higher than 30% and the oxygen content is <3%. The motor can be started, once a methane content of 45% has been reached.

Safety check: The gas supply must be checked before start-up. All connections and fittings must be checked for proper seal, and the gas pressure must be checked and set where needed. The hot water system must be installed properly and operationally. The entire external cabling must be checked and verified. The control system must also be checked for proper connection of all cables and lines.

An initial test run is carried out for 48 hours at full load. The thermal and electrical capacity is verified with relevant measurements. These values are used as final acceptance criteria for the CHP supplier.

Commissioning:

- Proof full electric power with 50% CH_4;
- Proof full thermal power with 50% CH_4;
- Proof noise emission;
- Proof exhaust gas emission;
- Commissioning regarding safety.

8 Maintenance, spare parts and maintenance control

Biogas plants, and in particular their motors require periodic check-ups. Daily tasks to be carried out include a tour of the plant and check for any faults; maintenance of the CHP includes:

- A gas meter reading and the recording of the operating hours of the motor;
- Checking the motor oil fill level;
- Checking for any fault lamps lighting on the control cabinet in the power room.
- Weekly checks include:
- Visual check of the motor and lines;

- Checking the solenoid gas valve for proper function and for contamination;
- Checking the gaps of the self-closing gas shut-off valves for proper seal.

Example for a simple maintenance service, normally done by the operator is oil change. Periodically analysis of the oil (pHvalue, sulfur) have do be done in a lab. The interval of oil change depends on:

- The concentration of sulphur in the gas/oil;
- The buffer capacity of the inserted oil;
- Others.

Every operator has to determine the necessary interval of oil change of his engine by himself!

If maintenance tasks or repairs are carried out with the motor at standstill, personnel must ensure for climate and environmental reasons that as little biogas as possible is released into the atmosphere. Methane is an important greenhouse gas. The global warming potential of 1 kg of methane is, calculated over a period of 100 years, 25 times higher than that of 1 kg of carbon dioxide. According to latest findings, this factor can be as high as 33, when considering the interaction with atmospheric aerosols.

Large gas storage tanks that can store generated biogas for several hours are an effective option for reducing gas emissions. A further option would be to increase the number of motors. If the gas storage tank has for example a storage capacity for 4 hours, and the only motor is switched off for repairs for 13 hours, then the gas storage tank will no longer be able to hold the generated gas after 4 hours. The gas is fed to the emergency flare stack, or— in the worst case scenario-vented into the atmosphere. If the plant however consists of two small CHPs, then at least half of the generated gas can be used by the motor. The gas storage tank will therefore last for 8 hours instead of only 4. After 8 hours the gas storage tank will however be so full that the motor will not be able to use up the generated gas volume. The gas must be fed to the emergency flare stack.

The downtime of the CHP in case of repairs will also be largely dependent on the reaction time of the operator. An operator, who is fast to react and well organised, will be able to increase the availability of his plant. Future-oriented planning for repair and maintenance tasks is a key factor. Frequently needed spare parts should be stored at the plant in advance in order to minimise waiting times for spare part deliveries.

CHP-supplier offer different sorts and timelines of service contracts. In general the customer can select between three different types of contracts:

- Per-call maintenance contracts;
- Scheduled maintenance contract;
- Full service maintenance contract;

(1)Per-call contract:

The first type is concluded for 'on demand' maintenance. These contracts are helpful for managing the stock of spare parts but every conduction of service/maintenance necessary for solving problems induces additional costs and will be accounted individually. In case of a well-qualified plant manager it can be a good choice, otherwise the risk assessment regarding financial aspects is very difficult for the client and the whole responsibility for the warranty claims is on his site.

(2)Scheduled maintenance contract:

This type of contract includes the service works and materials necessary for maintaining mainly the engine-generator assembly unit. For this purpose almost every engine manufacturer issues a maintenance plan on which to refer for maintaining the operational state. Only the costs of these scheduled maintenance stages are covered by the indicated price. In case of an emergency, this sort of contract causes additional costs for the customer concerning travel costs, labour costs and materials, like e. g. spare parts. The costs for this sort of contract are about 0. 7 Euro cent per installed kW and operation hour but you have to add costs for oil, spark plugs or other things which are not included in the contract.

(3) Full service contract:

The last type includes all service works and materials necessary to maintain the operational condition of the CHP unit and its peripheries, e. g. the gas flare, air systems, oil changes and so on. This happens to the responsibility of the CHP Company and to their own expenses on a fixed price base for the customer. Because there is a risk assessment on a highly degree involved in calculating this type of service contracts it is slightly more expensive than the others. On the other hand the customer gets the big advantage of having an all inclusive service package with no other responsibilities than to conduct a daily visual inspection and, if it occurs, to report a malfunction. Most suppliers recommend concluding this type of contract as a long term contract, due to the fact that every engine of a CHP has a limited lifespan of about 60 000 operating hours (8 years). After this period a major overhaul, which sets the engine into a 'like new' condition, is included. That means after 8 years, when this cycle ends, the customer is in possession on an engine with practically zero operating hours and everybody involved have a base for a clean economically calculation. Because of such a long contractual relationship and the potential risks it is better to reduce the contract to two years to keep flexible. The costs for this sort of contract are between 1. 0 and 1. 5 cent per installed kW and operation hour.

9 Mode of operation

Stand-alone operation: The gas engine is not connected to the electric grid. It can take over energy power supply for sensitive items.

Parallel to grid operation: The electricity produced in cogeneration is fed to the electric grid.

The voltage and frequency of the supplied power is defined by the grid. Within the European grid, the frequency is 50 Hz and the voltage for the low-voltage grid is 230 V/400 V.

10 Different types of CHP

Two types of gas engines are common. There is the gas-otto engine on the one hand which is mostly used for big engines and the dual fuel engine which is used for small plants. 80% of all engines used in biogas plants are gas engines. Only 20% are duel fuel engines. Most of them have less than 350 kW_{el}.

10.1 Gas-Otto engine

Gas-Otto engines are developed specifically for gas operation that works according to the Otto principle. The engines are operated as lean engines with a high over-supply of air to minimize carbon monoxide emissions. In lean operation, less fuel is used in the engine which leads to a reduction in engine performance. This is compensated with the supercharging of the engine using an exhaust turbo charger. Gas-Otto engines rely on a minimum content of methane in the biogas of 45%. At lower methane contents they can not operate. Otto engines between 100 and 1 000 kW_{el} are designed upon needs. For higher electrical performance, adapted diesel aggregates are used that are fitted out with spark plugs. If no biogas is available, Gas-Otto engines can also be operated with natural gas. This can be useful during the start up phase of the biogas plant in order to use the engine heat for the necessary heating process.

10.2 Dual fuel engine

Pilot injection or dual fuel engines work according to the diesel principle (i. e. no spark plug). These engines are not always built specifically for gas operation, being modified for it. Biogas is mixed in a gas mixer together with the combustion air and then passes through an injection system in the combustion chamber where it is ignited with the injected ignition oil. The proportion of ignition oil is 10% of the injected fuel performance. As a result of the relatively low quantity of injected ignition oil, there is a danger that injectors carbonize and thereby wear out faster. Pilot injection gas motors are also operated with a high excess of air.

References

Eder B., Schulz H., (2006): *Biogas-Praxis—Grundlagen Planung Analgenbau Beispiele Wirtschaftlichkeit*. Staufen bei Freiburg, pp. 238.

Leitfaden Biogas, (2010): *Leitfaden Biogas—von der Gewinnung zur Nutzung*. Fachagentur für nachwachsende Rohstoffe e. V. (FNR), Gülzow, 2010, pp. 272.

Bundesverband der landwirtschaftlichen Berufsgenossenschaften, (2008): *Sicherheitsregeln für Biogasanlagen (Safety Rules for Biogas Systems)*, Kassel, 2008, pp. 75.

Fachagentur für nachwachsende Rohstoffe e. V. FNR, (2006): Studie *Einspeisung von Biogas in das Erdgasnetz*. Leipzig, 2006, pp. 199.

TA Luft (TA Air), (2002): *Technische Anleitung zur Reinhaltung der Luft (Technical Instruction on Air Quality control) TA Luft: In GMBBI*. Gemeinsames Ministerialblatt —Joint Ministerial Gazette, pp. 511.

TA Lärm (TA Noise), (1998): *Technische Anleitung zum Schutz gegen Lärm (Technical Instruction on noise)*. Sechste Allgemeine Verwaltungsvorschrift zum Bundes-Immissionsschutzgesetz, GMBl. 1998, pp. 503.

从设计失败和沼气工程事故中得到的教训

T. Fischer, K. Backes

(*Krieg & Fischer 工程有限公司,Göttingen 37085*)

【摘要】本文描述了三个有代表性的沼气工程破坏性事件,并从中总结出如何避免类似事故发生的经验:因为沼气不但可以使人窒息和中毒,还可以引发爆炸,故一些特殊的破坏事故和教训都与沼气有关;缺少或者错误使用导致安全装置的破坏,如发酵罐的破坏及火灾破坏;大多数事故是由于建造的错误,或者缺少安全设备而导致的。因此,通过良好的规划和对运行人员正确的指导,不仅可以降低损失,还可以避免事故的发生。

【关键词】沼气工程事故;设计失败;安全

LESSONS LEARNED FROM DESIGN FAILURES AND BIOGAS PLANTS DISASTERS

T. Fischer, K. Backes

Krieg & Fischer Ingenieure GmbH

Contact: Dr Katharina Backes, Krieg & Fischer Ingenieure GmbH, Bertha-von-Suttner-Straße 9, 37085 Göttingen, Germany, Tel.: +49-551-900-363-0, E-Mail: Backes@KriegFischer.de

EXECUTIVE SUMMARY

Three interesting and informative events of real biogas plant damages are described and it will be concluded from the failures and accidents how to avoid them. Examples of special damages and the lessons learned will be presented: All cases deal with biogas. Biogas causes danger of suffocation and toxification on the one hand and is potentially explosive. Other topics are damages attributed to missing or wrong used safety devices, damages of tanks, and fire damages. Most accidents happen because of faults in construction or missing safety features and equipment. A lot of accidents can be avoided or the losses can be cut by a good planning and most important a good instruction of the operator.

Key words: Biogas plant disasters; design failures; safety

1 Introduction

Since 1993 Torsten Fischer is not only active as a biogas specialist in planning and construction of all kinds of biogas plants but his experience is also wanted to do expert opinions and due diligence analysis. His clients are mainly insurance companies, plant operators and more and more energy providers. In 2002 Torsten Fischer became active as a member of the working group 'Safety' of the German Biogas Association and contributes to better safety regulations. Despite increasing safety regulations, the number of failures and accidents, that are decided in court, increase with the number of biogas plants. Since 2009 Torsten Fischer is accredited as expert witness in the field of biogas, authorized by the Ingenieurkammer Niedersachsen, the professional association of engineers, and has dealed already with a lot of court cases.

2 Experience of insurance companies

In parallel to the increasing number of biogas plants, a growing number of insurance providers are now offering specialist insurances to cover biogas plant operations. In addition to insurance cover for personnel, damage and accidents or environmental damage, insurances now also offer downtime insurance. Insurance providers have learned which system parts are most often affected by damage, and what kind of pay outs they will be facing in case of an insurance claim. In contrast with regulatory requirements as part of the authorisation procedures, the insurer puts particular emphasis on the protection of fixed assets and damage due to down-times in addition to the general protection of personnel and environment.

Damages at different parts of the plant and payments related to different plant parts: The most damage sensitive part is the engine, followed by the solid input device and the mixing system (Marsh GmbH). In the last years damages causes by biological breakdowns became more important than material damages. The insurance of the digester biology becomes more important.

Comparing the frequency of damages more than 50% of all damages cause payments of less than 10 000 €. Payments of damages above 50 000 € are paid at 5% of all damages (Gothaer Allgemeine Versicherung). But about 80% of the insurance contracts are damage-free (lots of new plants still have warranty). The number of damages increases disproportionate to the age of plant.

3 Case studies

3.1 Case study 1: Fire damage CHP-Container

3.1.1 Plant description case 1

A typical biogas plant for the processing of renewable raw materials in connection with agricultural operations. The plant with a motor capacity of 191 kW_{el} has been in operation since summer 2008. The machine and control container is a 20 foot steel container with wooden flooring. The container rests with its reinforced corners on slightly raised point foundations.

The pipe that supplies biogas to the gas motor is a PEHD (yellow) DN 150 version, and leads from the fermenter/post-fermenter to the combined heat and power (CHP) unit container in the last section underneath the CHP container. The pipework is carried out flush with the terrain, without or at most with insignificant cover. Following the section of pipe running underneath the container, it then rises along the wall of the container to be led into the container at the top. The main gas tap is located here. Inside the container, the biogas line runs underneath the lid to the ex-protected condenser and the gas regulation train.

At about the half-way mark of the container, the gas motor unit is mounted in a frame on the floor of the container. The control cabinet of the CHP is mounted directly in front of the motor block frame. The smoke sensor and gas detector are installed above the control cabinet.

The exhaust pipe runs underneath the motor block, approx. 18 cm above the wooden floor. Insulation is installed only along the vertical section and not along the horizontal section.

All electric cabling is mounted on a cable support, which is affixed to the base structure of the gas motor.

3.1.2 Damage description case 1

In spring of 2009 the biogas plant suffered fire damage to the machine and control container. The incident resulted in damage to the generator/motor block of the CHP. As the damage was assessed as being considerable, Krieg & Fischer Ingenieure GmbH was commissioned to prepare an expert report.

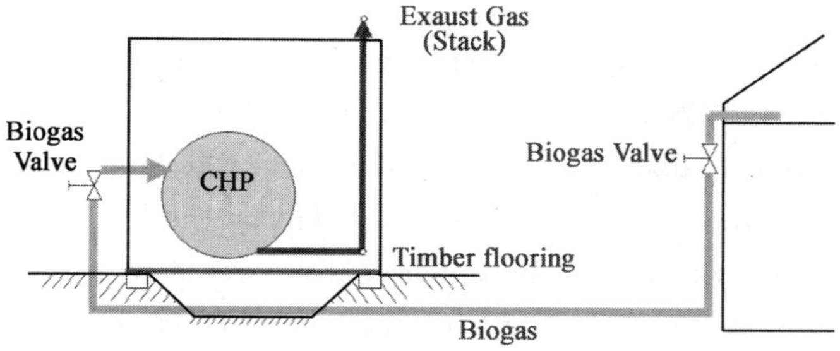

Figure 1 Drawing of the biogas plant (case 1)

The wooden floor underneath the base frame of the motor block was burned completely. All electric cabling on the cable support is charred and melted. The sump tray is located at a height of approx. 1 m above ground, directly underneath the motor block, and shows severe encrustation. Only a few of the cables above the sump tray are charred and melted. There is some blue plastic residue on the exhaust pipe. The PEHD biogas line was damaged underneath the wooden floor, i. e. underneath the container. The pipe wall has collapsed and partially melted.

3.1.3 Explanation/assessment case 1

The fire was most probably caused by an unprotected exhaust pipe heating the wooden floor beneath until it ignited. The following five factors must be considered in this context:

(1) In accordance with the approval note, safety rules for agricultural biogas plants must be complied with. The safety rules specify fire protection for a CHP in section 4. 2. 2 'Installation in residential buildings', and should be applied to the relevant conditions in this case: "Walls, supports and ceilings above and below the installation room must have a minimum fire-resistance (F90 A DIN 4102) and must be constructed from non-flammable materials." Section 6 states furthermore: "Additional measures for fire prevention are to be coordinated with the relevant local fire station."

(2) The fire protection ordinance (FeuVO) issued by the federal state of Sachsen-Anhalt, issue date March 27, 2006, validity date March 15, 2006 applies. The ordinance states that "Exhaust systems must be installed at a sufficient distance from components consisting of flammable materials, and must be insulated in such a way that said components will not reach temperatures above 85℃ at nominal loads of the connected combustion apparatus." Furthermore, specifically with regards to exhaust pipes developing with a min. temperature of 400℃ a min. distance of 40 cm between exhaust pipe and flammable building materials must be maintained. The requirements were definitely not complied with in this case.

Some sections of the exhaust pipe were not insulated. This in particular applies to the sections of the exhaust pipe running close to the wooden floor. Temperatures of over 700℃ can be expected to develop in the exhaust pipe.

(3) Of particular interest is the question why the fire started on that particular day. An important factor in this regard is the fact that continuous operation of the biogas plant/CHP coincided for the first time with a prolonged period of hot weather from the middle of April 2009 onwards. The biogas plant was initially commissioned in spring of 2008. It only began to operate at full load starting around autumn of 2008. In autumn and winter 2008 and the beginning of 2009, the wooden floor was kept moist continuously due to the wet weather conditions. The wooden floor only began to dry out during the dry weather spell in the spring of 2009, which culminated in its ultimate spontaneous

combustion.

(4) Once the fire had started, it slowly advanced through the chipboard flooring. Extreme temperatures developed, which melted the insulation on cables, which were installed in close proximity to the heat source. The molten cable insulation then dripped through the hole in the floor which had resulted from the burning of the wood flooring. These drops fell on the PEHD biogas lines running directly underneath the container, which were only marginally covered with earth. Evidence of melting on this pipe implies that biogas was escaping from the pipe, adding more 'fuel' to the fire.

This secondary hazard was not recognised by the operator on the night of the fire. He did, however, shut off the biogas supply to the container. The supply of biogas to the fire was only discovered the following morning, when the smell of biogas became apparent, at which point the line was shut off at the post-fermenter. Considering that the biogas plant was running at full capacity at the time of the fire, approx. 450 m^3 of biogas will have escaped during the course of six hours. It may be a blessing in disguise that the fire did not ignite again.

We must point out here that the smoke alarm and the gas detector were not placed to full technical advantage. The two devices are mounted in an area, where they are in the flow area for fresh air, which means that a fire or gas leak can only be detected too late or not at all.

3.1.4 Conclusions for case 1

The CHP container with integrated CHP was not installed in compliance with relevant regulations. The container and wooden floor was supplied by a system manufacturer, who has since filed for bankruptcy. It is unknown, whether the Swiss-based CHP provider expressed any doubts regarding installing his CHP system in the container provided. We would expect an expert company to have the relevant knowledge and would advise the system provider with the necessary information. From a German legal point of view, the CHP supplier would have had to submit his doubts concerning wooden flooring in writing, and would have subsequently refused to install the CHP unit in said container. It would be important in this context to see the relevant contractual agreements between the Swiss CHP system provider and the German customer, and/or the contract between the Swiss CHP provider and the Swiss system provider.

From a structural viewpoint, it is negligent to run the biogas supply lines in the PEHD directly underneath the wooden flooring. The situation was further exacerbated by the incorrect assessment of the situation by the operator at the time of the fire in terms of where the biogas supply was interrupted. It was purely down to luck that the damage was not significantly higher.

Neither the local fire brigade, nor the relevant local authority undertook a formal inspection of the biogas plant, even though the approval notification included an inspection as a specific requirement. Furthermore, no fire protection regulation was formulated, although that too was explicitly required in the approval notification. In summary, legal regulations were disregarded in several instances. Although in addition to the CHP supplier and the CHP system provider a professional plant operator was involved, not one of the three parties initiated a controlling of any kind.

3.1.5 Lessons learned from case 1

Fire-protection regulations have to be paid attention to:

- The exhaust gas pipe has to keep safety distances or/and has to be insulated;
- To avoid accidents the engineering of a biogas plant should be done by a biogas specialist.

3.2 Case study 2: Overfilling of hydrolysis tank

3.2.1 Plant description case 2

A typical biogas plant with an electrical capacity of 536 kW_{el} for the processing of renewable raw materials in connection with agricultural operations. The biogas plant was commissioned in 2006.

Krieg & Fischer received an E-Mail from the insurance company. They were asked to analyse the damage of a concrete tank. Which tank was not noticeable from the documentation of the insurance company. The insurance company asked for an examination of the documentation available on site the biogas plant. The result was that the data were only rudimental:

No process flow sheet of the biogas plant exists, neither as an overall diagram nor for individual sections. The visualisation on the monitor in the operations room captures the feeding technology only, but not the actual biogas plant.

The tank was initially constructed as an open tank. At a later time, in 2008, a 3 m wall and a lid were added. After enquiry with the tank manufacturer we were informed that the client carried out the addition himself. The tank manufacturer only supplied the casings. The apparently only documentation is a one page document with basic data on the construction steel and concrete used.

The plant operator was unable to provide any further information regarding what documentation was supplied regarding statics and reinforcements during the approval process and later at the time of modification of the tank. No documentation beyond the approval submission form LP 4 HOAI is available. An inspection of the relevant documentation as part of the order could therefore not be carried out. The plant operator was furthermore unable to provide final billing documentation.

An inspection of the approval submission form showed that the static data for the two steel fermenters, the two steel supply tanks the steel maturation tanks were actually submitted. There is no similar documentation available for the two concrete tanks of the biogas plant, the manure pit and the hydrolysis tank. As the operator does not have the relevant documentation, we must assume that no type-tested structural analysis was carried out for the structural addition of 2005. It can therefore also be assumed that the same applies for the height increase and cover of the hydrolysis tank carried out in 2008.

- The damaged tank was not registered by the insurance company;
- There was no documentation of the plant;
- The site layout was taken from permission documents and obsolete because the concrete tank was erected 2 years later;
- There was no process flow sheet;
- The hydrolysis tank had been enhanced and equipped with a cover by the operator;
- There was no structural calculation of the tank;
- There was no safety-regulated documentation.

Will the insurance company pay for a damaged tank they do not know about?

3.2.2 Damage case 2

An accident occurred at the biogas plant in March of 2009. Overfilling of the hydrolysis tank led to a partial destruction of same. The hydrolysis tank was supplied by a pump. Once the maximum fill level was reached, the plant operator was notified via SMS to his mobile phone. As the plant operator did not have the mobile phone on his person at that time, the pump was not—as planned—shut down, but continued running until the cover of the hydrolysis tank was pushed outwards by the flow pressure and was subsequently torn off the wall of the tank.

3.2.3 Explanation/assessment case 2

The pump used for filling the hydrolysis tank is a model typically used (eccentric screw pump). The pump with an electrical power output of 11 kW is generously dimensioned for the purpose of its application here. It is not equipped with either a pressure-side pressure sensor or a dry run protection. It can therefore only

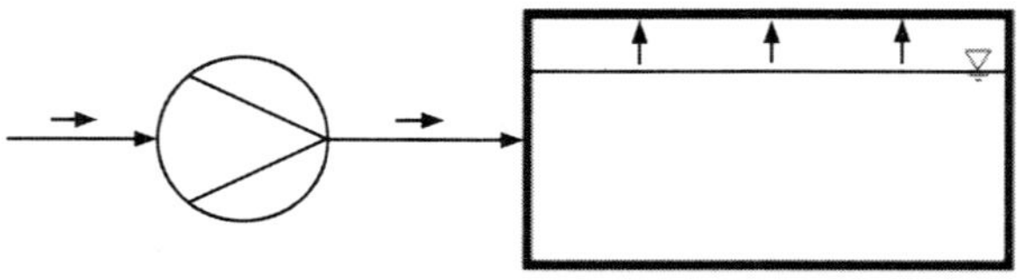

Figure 2 The pump continued working until the cover of the concrete tank was pushed upwards and the cover was torn off the tank wall

be switched off manually.

The fill height of the hydrolysis tank is captured by a so-called load cell. This is a standard approach. The signal 'Maximum fill level' is transmitted via SMS from the load cell to the mobile phone of the operator. Under normal circumstances, the pump is then deactivated manually. According to the operator, this automated alarm chain with SMS notification functioned appropriately, as it also did on the day of the accident. However, he didn't have the mobile phone on his person at that time, and could therefore not react accordingly. The pump was therefore not shut off, and the hydrolysis tank over-filled.

Obviously this kind of system is insufficient to safeguard against over-filling of the hydrolysis tank. We recommend automating the shut-off process for the pump filling the hydrolysis tank.

The construction of the hydrolysis tank—as judged from an external visual inspection—was very clean. The wall was concreted and its upper edge was finished with a joint strip. The joint strip was held in place vertically by way of steel brackets, providing a clean transition for the subsequent concreting of the roof. There were therefore no reinforcements put in place for the transition between cover and wall, there was however somewhat of a steel support, which facilitated a sealing of the transition between wall and cover with a rubber lip (joint strip). When the cover was forced upwards, the joint strip remained on the concrete wall, and the concrete coverage over that area (facing outwards) chipped off.

3.2.4 Conclusion case 2

The only existing documentation available from the operator is the approval submission form and the approval notice. No documentation regarding orders from the various companies, acceptance reports or any other documents on the part of the operating company were available. It can therefore be assumed that no planning was carried out beyond the requirements of the LP 4 of the HOAI (German Fee Structures for Architects and Engineers). Since October 30, 2008, new safety regulations for agricultural biogas plants have been put into place (at the time of construction, the old safety regulations dating from September 5, 2002 were still valid). The new regulations have definitely not been published prior to the commissioning of this biogas plant. It can however be expected from any plant operator to incorporate new regulations into existing processes of the plant after a certain grace period. The new safety regulations offer, in addition to technical requirements for plant safety, a significantly increased number of requirements in terms of documentation and burden of proof for the operator. The current standards at the plant do not fulfil even the most rudimentary standards, and the same applies for agricultural biogas plant regulations. Furthermore, there are no drawings of the pipe network available, no pipework and instruments schematics, no process control engineering, and no safety documentation of any kind. We can therefore assume the lowest possible standards were being applied here.

In conclusion:

The initial approval submission was put together by an engineering firm, which, however, never actually visited the location. Once approval was received, several companies were contracted to carry out the implementation. None of these companies apparently thought it necessary to advise the operator of the necessity of further planning and site supervision. According to statements from the operator, the result was a biogas plant that was initially not fully functional, and which he continued to work on over the next 1.5 years to reach a state in which it 'more or less works'. The companies involved in the project did not fulfil their duty of information towards the client. The unprofessional attitude of the client further exacerbated the problem.

3.2.5 Lessons learned from case 2

- A precise documentation is important;
- Safety devices have to be installed;
- An engineering office specialized in biogas should be consulted;
- The engineering of a biogas plant should be done by a biogas specialist.

3.3 Case study 3: Rupture of a gas holder roof

3.3.1 Plant description case 3

The biogas plant for the utilisation of sustainable raw materials has been in operation since its construction in 2005 and is part of an agricultural operation. The input material is maize silage, liquid manure and grain.

The plant consists of a receiving station and associated input technology, two fermenters and two CHPs. The content of the fermenter is mixed with biogas introduced under pressure. Each fermenter consists of two rings, one inner and one outer ring, whereby the outer ring is subdivided into several chambers. The chambers are used for mixing the input materials, and as storage tanks for almost fully fermented material.

The biogas plant was commissioned with one fermenter and one CHP after the first construction phase was completed in 2006. In summer of 2007, the second fermenter was completed and subsequently commissioned, and the second CHP was connected.

The entire tank is spanned by a double-membrane gas storage roof, which is fastened centrally on a steel support and at the outer edge of the tank rim. The outer membrane, which functions as weather protection, is harnessed to the tip of the central support. The inner membrane of the gas storage tank is stretched over the central support and fixed between two steel plates with a screw connection.

At the time of planning, approval and construction of the biogas plant, the safety regulations for agricultural biogas plants issued in September of 2002 applied. These safety regulations require a continuously active safety device which monitors increases in internal pressure of a gas storage tank installed on top of a fermenter. The safety device is intended to prevent an increase in pressure to amounts that could result in the destruction of the gas storage membrane. Biogas plants are therefore usually equipped with a hydraulic over/underpressure valve, which is installed on the exterior of the tank and is connected directly to the gas storage tank.

A so-called shell deflector measures the fill level of the gas membrane in the damaged fermenter. A hollow steel section of about 2 m length is laid on top of the inner membrane of the double-membrane gas storage roof. A sensor monitors the upward and downward movement of the membrane, which sends a signal to the CHP.

Other safety features, i.e. the emergency flare stack and the over/underpressure device remain active independently. The over/underpressure device installed on the damaged fermenter was set to release biogas, once an overpressure of 3.5 bar was reached. The device was set to open and thus increase pressure, should an underpressure of 1.0 mbar be reached.

3.3.2 Damage case 3

On one day in July of 2009, the biogas plant recorded a deactivation of both CHPs. Fermenter 2 had not issued a signal regarding the fill level of the gas storage tank. After several failed attempts to restart the CHPs, the operator suspected a gas leak in the roof of Fermenter 2. The operator proceeded to shut off the gas carrying pipe connection between the two gas storage tanks of Fermenter 1 and 2 to be able to continue plant operation with Fermenter 1 only. The operator notified the relevant authorities.

Two days later it was found that the gas membrane of the damaged fermenter had torn at the mast, thus leaking biogas. The supplier of the double-membrane roof of the gas storage tank paid a visit to the plant. As the fermenter had not yet been emptied at the time, no exact analysis of the material could be carried out.

Over the course of the following days, Fermenter 2 was evacuated completely, and the centre support, complete with the attached double-membrane roof was lifted from the tank with a crane. A tear in the inner gas storage membrane was discovered at the clamped fixture on the central support. The tear ran tangential and relatively equidistant to the clamping plate on the central support in the area of the single-layer membrane foil cover.

3.3.3 Explanation case 3

After an inspection of the relevant documentation supplied, the following cause for the damage was determined: The over/underpressure safety device is installed on the tank edge of the fermenter. The biogas is supplied to the over/underpressure safety device via a HDPE (high density polyethylene) line DN 100, which reaches about 2.50 m into the tank. The gas line coming from the tank was expanded from DN 100 to DN 150 for the transition to the over/underpressure safety device on the outside of the tank, which is the size required for the installation of the over/underpressure safety device.

A pressure transmitter is mounted at the tank exit point of the DDPE gas line. Pressures inside the gas storage tank are measured here and transmitted to the control center via a control cable. It is of critical importance to examine pressure losses at that point in more detail in order to further assess the actual damage incurred.

These pressure losses are the result of resistances and friction losses on the inner wall of the pipe, which the biogas experiences when flowing through the pipe and formed parts. At low flow rates, the pressure losses are lower, at higher flow rates the pressure loss increases. Every opening and pipe length through which the gas flows, results in pressure losses. It is therefore important to consider angled pipes and other formed parts, including fittings, when assessing pressure losses in pipes.

The following individual components must therefore be analysed before they are exposed to biogas flows:

- The supply openings to the HDPE lines;
- The 90 elbow;
- The HDPE line with a length of approx. 2.50 m and a diameter of 100 mm;
- The two T-parts on the outside between the tank wall and the entry point to the under/overpressure safety device.

The actual pressure losses in the sections listed above were calculated with specialist software. Pressure losses of 2.23 mbar were calculated based on pipe dimensions and the integrated form parts.

The calculations are however based on a smooth and clean surface inside the biogas line. In the area of the biogas line inside the tank, maize silage and substrate sediment is visible. Over the course of 3.5 years in operation, particulate deposits have formed at the entry and inside the gas pipe. It can therefore be assumed that the calculated pressure loss is a conservative estimate, which was in reality surpassed. The following assumption therefore uses a conservative value of 2. 5 mbar as pressure loss resulting from the structural condition upstream of the over/underpressure safety device (*Figure 3*).

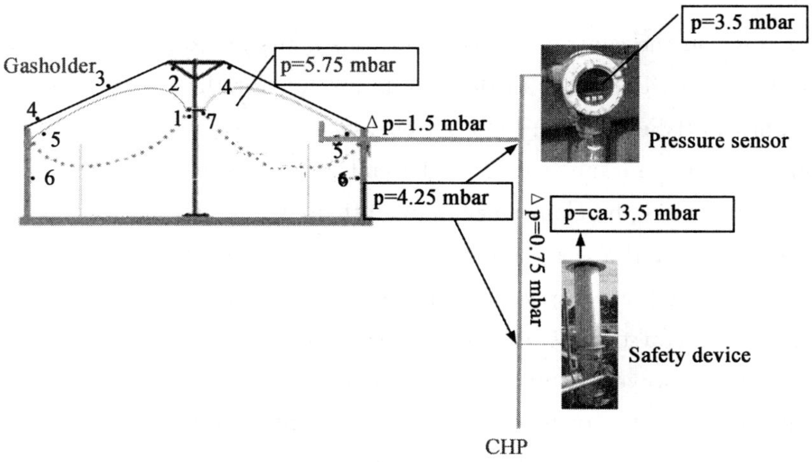

Figure 3 Diagram of pressure losses between fermenter and overpressure safety device

The structure of the over/underpressure safety device results (according to manufacturer data and the installation and operations manual) in a pressure loss of maximum 1 mbar at a gas flow volume of 400 m^3/h. In total, a structure-related pressure loss of 3. 5 mbar upstream of the exit opening of the over/underpressure safety device is to be expected. Due to the location of the measuring point outside the tank, a pressure loss of only 1. 75 mbar can be metrological determined. As a result, the measuring device displays a value that does not reflect the actual pressure underneath the inner membrane.

A further subdivision of the section is required in order to analyse the actual extent of the damage. There are therefore four different pressures to consider:

The pressure inside the gas storage tank (between inner membrane and substrate surface), which represents the stress on the gas membrane, and the pressure which can be metrological determined by the available measuring device. This is the pressure inside the gas storage tank, reduced by the overall pressure loss from the HDPE lines and form parts, i. e. the pressure upstream of the over/underpressure safety device. This is the measured pressure plus pressure loss generated on the two T-parts, and the pressure at the exit point of the over/underpressure safety device. This value includes the pressure loss incurred by the over/underpressure safety device.

We must therefore contemplate the following:

According to manufacturer documentation, the over/underpressure safety device is pre-set to 3. 5 mbar. It can be assumed that this is the measured value. The actual pressure on the under/overpressure safety device would therefore in reality be 4. 25 mbar. Additionally, the pressure loss due to the gas passing through the pipes, plus the entry opening within the tank of about 1. 5 mbar must be considered, which brings the total pressure of approx. 5. 75 mbar affecting the inner membrane.

By calculating the pressure loss of the gas line leading to the over/underpressure safety device, we can therefore prove that the installed pressure gage only seemingly displays the actual pressure affecting the gas membrane. The actual pressures at a displayed pressure of 3. 5 mbar at the pressure gauge will be at least by the calculated amount higher, i. e. at the time the biogas exits the over/underpressure safety device. There will be at least 5. 75 mbar of calculated pressure on the gas storage membrane.

These calculated minimum values can be verified, as the measured pressure values of the damaged fermenter were recorded via the process control technology of the biogas plant. According to those recorded values, we must assume that the calculated pressure underneath the gas membrane was surpassed several times per month.

In the area of the clamping on the central support, obvious signs of excessive stresses were discovered on the clamped screw connections. The holes in the membrane foil for the clamped screw connections on the central support are significantly expanded. The extension of the screw hole openings occurred in radial direction, i. e. the gas tank membrane had to withstand overpressure that was so severe that even the doubled layer of membrane foil at the clamping became affected.

By the time of the generation of the assessment report, it was not possible to extract a specific answer regarding maximum pressure values, for which the inner membrane is designed from either the general contractor or from the supplier of the membrane. Information provided by other manufacturers and from our own experience, we can however assume that a value of 5 mbar may be considered as maximum value. This value was periodically surpassed over the 3. 5 years of plant operation. Based on the pipe construction upstream of the over/underpressure safety device described above, and due to the improper calibration, the gas tank membrane was periodically over-stressed. These conditions finally caused a material failure.

The operator documentation was not updated after the fill level sensor was replaced in spring of 2006. The biogas plant was therefore operated despite the fact that the operator did not have appropriate documentation for service and maintenance purposes. This did not have any influence on the occurrence of the damage, but should be taken into consideration during the repair of the damaged fermenter.

Furthermore, it must be considered in terms of future plant operations that periodic inspections of the functionality of the over/underpressure safety device and a calibration of the pressure gauges upstream of Fermenter 1 and Fermenter 2 must be carried out and the results documented.

The construction of Fermenter 2 and Fermenter 1 is identical. It can be assumed that the pressure conditions in Fermenter 1 are identical to those in Fermenter 2. We must therefore assume that Fermenter 1, which is currently in operation, may suffer similar damage as the one determined for Fermenter 2. Relevant action is required immediately for safety and environmental reasons.

3.3.4 Conclusion case 3

The cause for the damage on the inner membrane of Fermenter 2 of the biogas plant has been determined as a planning/structural deficiency, which has existed since the construction of the fermenter. Specifically, the deficiency is due to an underestimation of the pressure loss caused by the layout of a pipe inside the tank, which supplies the biogas to the over/underpressure safety device. The pressure gauge installed does not account for the pressure underneath the inner membrane, or the pressure at the over/underpressure safety device. In this context, the pipe section between the outer tank wall and the over/underpressure safety device is insufficiently laid out.

It was ascertained that the gas storage tank membrane was periodically overstressed since initial commissioning of the plant, and it is therefore surprising that the membrane held as long as it did. In addition to the findings above we must mention here that even relatively small changes in ambient temperatures and pressures, for example day/night or changes in the weather would have caused even higher stresses on the membrane, when the gas storage tank was full. A remodelling of the structure is therefore required. Reinforcing the membrane will not suffice as a solution. A manufacturing fault in the membrane, which was temporarily considered as the cause for the damage, is now considered improbable in light of the findings regarding the actual pressures occurring in the plant.

This point must be taken into consideration when assessing the status of the identically constructed

Fermenter 1. There too, it might come to a similar failure at any time. We have pointed out the importance of that fact in terms of safety.

In addition to the construction-related deficiency, which ultimately resulted in the failure of the gas storage tank membrane of Fermenter 2, it must be noted that the relevant documentation is incomplete and maintenance of safety-related devices was not carried out.

3.3.5 Lessons Learned from case 3

- The engineer has to understand the system;
- The position of measurements is important;
- In gas pipes reductions of the dimension and bends cause pressure reductions;
- The engineering of a biogas plant should be done by a biogas specialist.

4 Conclusions

Plant safety and the prevention of faults and accidents are important factors in the planning and construction of biogas plants. Insufficient planning, substandard construction or improper operation can lead to faults and cause hazards for people and for the environment. These may result on the one hand from the characteristics of biogas, including the possible generation of explosive gas/air mixtures or the risk of poisoning/suffocation, and on the other hand mechanical risks. In order to successfully prevent accidents and environmental pollution, all components of a biogas plant must comply with specific safety and technical requirements.

The safety of biogas plants has been the focus of attention over the past years, which has resulted in improved and higher quality plant technology, and also in increasingly stricter directives, legal requirements and controls.

Biogas leaks are considered the most dangerous cause of accidents. Leaking biogas can pool in pits and recesses, posing a potentially lethal trap for humans, as it cannot be detected quickly enough to leave the hazard zone in time. Suffocation or poisoning pose a potentially lethal threat. Depending on substrate, biogas may contain dangerous levels of hydrogen sulphide. A concentration of H_2S of only 0.1% can be lethal in minutes, and a concentration of 0.5% in just a few seconds! A loss of consciousness can occur at such concentrations after just a few inhalations.

An explosion of ignitable gas/air mixtures can occur at between 6 and 12% volume of biogas, where there are suitable ignition sources, like electrical systems where sparks can form, where there are balancing currents or static electricity.

The causes for faults and accidents can usually be traced back to insufficient planning. Many planners do not have the required expertise to provide a correct plant layout. For economic reasons, calculations are often based on an overly high stress on the fermenter, and plants are operated on the borderline of their capacity at high risk—also in regards to finances. A further example is the pipework planning. Here, mistakes in the pipe diameters used can lead to problems at a later time if the flow rate is too low and particulate deposits form, or if pipes get clogged due to insufficient diameters. During prolonged rest times of the substrate in the lines, which often occur due to some other fault, gases can form inside the pipes , which can then cause problems at the pumps. A closed-loop pipe segment can burst.

During the construction phase, problems can occur in the tanks. Due to sub-standard work in the concrete construction as low-quality concrete, or unfavourable weather conditions during the concreting, leaks in the tanks can occur, which then have to be sealed at high cost in terms of money and manpower.

References

FeuVOn. (2006): *Feuerverordnung*. Federal state of Sachsen-Anhalt, March 27, 2006.

Gothaer Allgemeine Versicherung AG. (2009): Information given by E-mail.

HOAI. (2010): *Honorarordnung für Architekten und Ingenieure*. (German Fee Structures for Architects and Engineers), Deutscher Taschenbuchverlag, München 2010, pp. 291.

Marsh GmbH. (2009): Information given by E-mail.

Bundesverband der landwirtschaftlichen Berufsgenossenschaften. (2008): *Sicherheitsregeln für Biogasanlagen* (Safety Rules for Biogas Systems), Kassel 10/2008, pp. 75.

从城市固体废弃物中提取生物质原料用于沼气生产

G. Rettenberger

(*RUK* 工程集团，*Stuttgart* 70597)

【摘要】目前常用机械—生物处理工艺分离城市固体废弃物，也可以通过生物干燥工艺利用城市固体废弃物生产垃圾衍生燃料。文章针对两种不同的处理过程展示了不同的应用技术。在生物干燥之前，需要进行机械加工处理，最简单的是配有装料机（如装料吊车）的分类系统，最复杂的是对不同分类废弃物的同时粉碎。在机械处理过程中，低质量的垃圾衍生燃料、金属、外来有害物质被分离出去。通过使用一种多功能筛鼓还可以不进行原料粉碎就能有效地将未受损的外来有害物质进行分离。通过高效率钕磁铁还可以将电子废弃物、喷雾罐或废电池等磁性差的物质进行分离去除。

机械处理完毕后便是生物厌氧处理过程，该过程生产沼气需要去除沙粒。最实用的技术是：在配有水平混合装置的过滤器中，向废弃物上喷水，将选择性粉碎均匀的废弃物加工成有利于后续工艺尺寸及特点的原料。在过滤器末端，将固体废弃物中的水分排干，液体收集后进行机械处理，进入厌氧反应器。厌氧发酵生产沼气所提供的能量要比整个处理系统所需能耗高。通过对含有丰富有机物的液体进行发酵，避免了固体发酵工艺的弊端。

下一步工艺是生物干燥处理。该处理即是通过去水、生物降解过程获得满足后续分选步骤的高质量固体。由于干燥的固体可以利用常见的方法（如筛选或分离进行处理）得到不同的高热值轻物质部分和矿物质部分。该工艺的复杂程度取决于预期垃圾衍生燃料市场趋势进行的成本效益分析，机械分离可以将原料的75%转化为垃圾衍生燃料，剩余物质可以通过填埋或焚烧的方式处理。

机械处理较好的方法是将废弃物在高压条件下进行挤压（压力高达 1 000 bar = 10^8 Pa）。该挤压过程可以将城市固体废弃物或有机废弃物/生物降解废弃物很容易地分离开，并在低成本的条件下得到含有高热值的干物料部分和富含微生物易降解的有机质的液体部分。该加工设备容易与现有的垃圾处理场和沼气工程结合起来。日常的城市废弃物不需要进行预处理（无需分离金属，无需粉碎，无需筛分），且该设备可以实现完全自动化控制。

干物质部分可以进行热处理或其他进一步耗能处理，其不仅包含物理性质稳定材料（如塑料、木头、卡纸和金属等），并含有 25%可降解有机材料，这个比例足以保证生物干燥后能得到高质量的稳定材料。湿物质部分可以用来高效地生产沼气，即将物料挤压后得到的湿物质部分进行升温，分解过程就会自发进行。干法厌氧处理最重要的特点是湿物质部分可以与发酵过程中循环的物质进行很好的混合。

【关键词】城市固体废弃物；机械生物质提取；挤压；机械—生物处理

EXTRACTION OF BIOMASS FROM MUNICIPAL SOLID WASTE FOR BIOGAS GENERATION

G. Rettenberger

Ingenieurgruppe RUK

Contact: Prof. Dr. Gerhard Rettenberger, Ingenieurgruppe RUK, Auf dem Haigst 21, 70597 Stuttgart, Germany, Tel.: +49-71190678-0, E-Mail: info@ruk-online.de

EXECUTIVE SUMMARY

Presented is the state of the art of mechanical-biological treatment systems to separate common municipal

solid waste (MSW) for further anaerobic treatment. In the most cases these are also technologies to produce RDF (refuse derived fuels) from MSW using biological drying systems. For two different process variants, the used technologies are shown.
Before the biological drying step a mechanical treatment is necessary. The mechanical treatment is in the simplest case the sorting with the filling machine (e. g. grab crane) and in the most complex case the sorting of different fractions including shredding. Here RDF with low quality, metals, foreign and hazardous materials will be separated. By using a multifunctional sieving drum and no shredding an effective separation of the undamaged foreign and hazardous material is possible. High-performance neodymium magnets can be used for the removal of weakly magnetic components such as electronic waste, spray cans or batteries.

The anaerobic biological treatment follows after the mechanical treatment. Here biogas is produced and sand has to be separated. The best available technology is the following: In percolators with horizontal mixing gear water is added to the waste by irrigation. The waste is homogenized and shredded selectively in such a way that optimum characteristics for the subsequent processing stages are generated. At the end of the percolator the solid material is drained. The water is collected, mechanical treated and fed into anaerobic reactors. The energy supply from biogas is higher than the needs for the whole treatment process. By fermenting the organic-rich water, the disadvantages known from fermentation plants for solid material are avoided.

The next step is the biological drying. Effect of the biological drying is a water loss, a biodegradation and a solid with good quality characteristics for the following separation step. The dried solids can be treated with common conditioning techniques as sieving or separation. So different high-calorific light-fractions and mineral fractions can be generated. The used complexity is generally the result of a cost-benefit-analysis depending on the anticipated trend of the market for RDF. Product of the mechanical separation is up to 75% RDF. The remaining material is minerals for landfilling and a fraction, which has to be thermally treated.

A good option for the mechanical treatment is the extrusion-process where the waste is compacted with high pressure (up to 1 000 bar = 10^8 Pa). With an extrusion process municipal solid waste or organic waste/biowaste can be separated quite easy and therefore with low costs in a 'dry fraction' with high calorific value and a microbiological degradable organic rich 'wet fraction'. The plant engineering can easily be integrated in existing waste treatment plants and/or biogas plants. Common municipal waste needs no pre-treatment (no separation of metal, no crushing, no sieving). The facility works fully automated.
The dry fraction can be treated with a thermal or after further treatment an energetic process. The dry fraction contains mechanically stable materials (plastics, wood, paperboard and metal) and about 25% of the degradable organic material. This percentage is sufficient to produce stabilized material of very high quality by biological drying.
The wet fraction can be used to produce biogas with high efficiency. The wet fraction leaves the press with a raised temperature. Decomposition processes start spontaneously. Important for the following dry anaerobic treatment is that the wet fraction can be mixed very well with circulated material from the digestion.

Key words: MSW; mechanical biomass extraction; extrusion; mechanical-biological treatment

1 Introduction

MSW is an important source for biomass to be treated in a biogas plant. Normally the typical processes are a part of the so called mechanical-biological treatment (MBT) which is quite common in Germany for treating MSW. In the paper two processes will be described in detail which has been operated in Germany for several years successfully:

- A plant in which the waste is percolated after a mechanical treatment step following by a wet fermentation process for the liquid;
- A plant in which the waste is separated by a press followed by a dry fermentation process.

In both plants the remaining fraction is used as a fuel in incineration plants. Therefore the efficiency of the total process is extremely high.

2 State of the art of Mechanical-Biological Treatment technology

Mechanical Biological Treatment (MBT) used in Germany can be divided in 4 variants: see *Figure 1*.

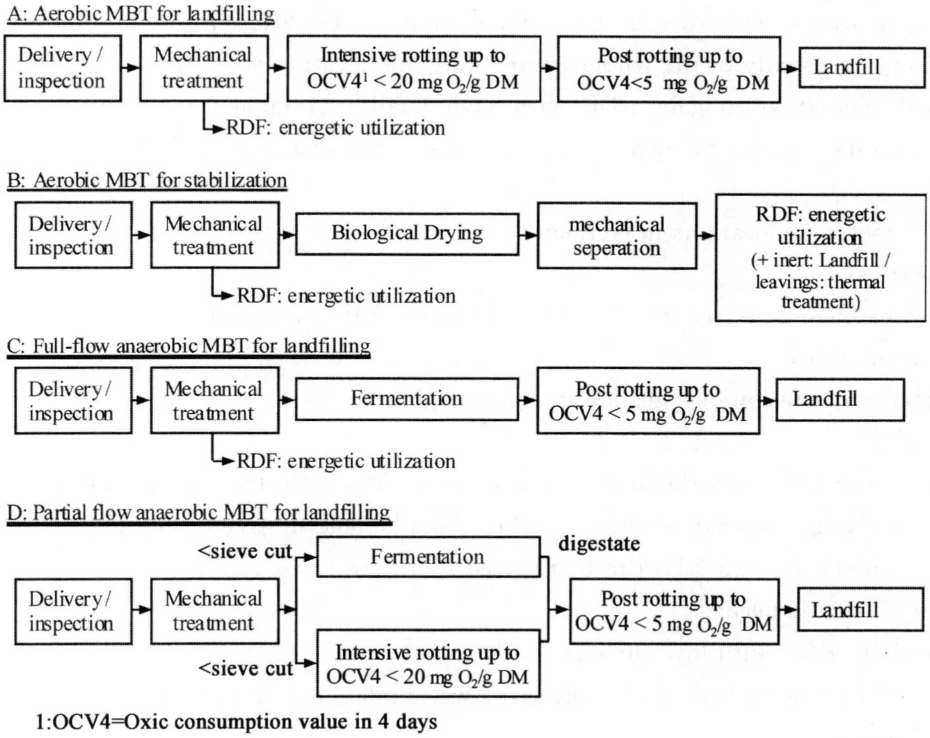

Figure 1 Flow chart of MBT process variants

Aerobic processes are the primarily used systems in Germany, since these technologies are well established and also cost efficient. Generally these processes are not energy self-sufficient which is a disadvantage of the technology. This also applies to variant B because the use of RDF for energy production is more complex than the use of fermentation gas.

In variant C wet fermentation (Water content >90%) and dry fermentation (Water content <70%) are used in the market. Because of problems with sand, mixture and staying time both variants are very complex to operate. An additional problem is the conversion of the digestate from anaerobic to aerobic conditions on the beginning of the post rotting. Because of NH_3-production odor emissions can occur. This problem can be solved through a new variant with a post treatment in form of a wet oxidation (suspension ventilation). Nonetheless the problem of de-watering and the structure of waste for the post rotting exist. Because of the immense technical extra effort Variant B will not be successful on the market in the long-run.

The following processes currently exist on the market:

- Process A: (17 locations in Germany)

Process properties:

- Intensive rotting of a fraction $<\approx 40$ mm (40%-80%) till reaching the requirements of the "Abfallablagerungs - verordnung" (German ordinance for landfilling);
- 15%-50% of the input is sorted out during the first stage as RDF;
- Robust process with established technology;
- Process is not energy autarkic;
- Process leads to RDF with low quality;
- Process B (8 locations in Germany).

Process properties:

- Mechanical treatment with shredding, partly sieving and Fe-discharge;
- Biological drying usually of the whole waste (only one location $<$ 80 mm);
- Mechanical separation to gain RDF with high quality. Complexity of the used combination of constructions depends on the quality requirements of the recipients of RDF.
- Process C

(1) Haase Process (2 locations in Germany)

Process properties:

- Biological treatment of a fraction $<$40 mm (less than 50% of input), combination of:

— Wet fermentation;

—Wet oxidation (aeration of digestate in aqueous phase);

— De-watering;

—Thermical drying through the heat from the generation of electric energy of biogas.

- Product: landfilling material according to the 'Abfallablagerungsverordnung'.
- The odor-problem through NH_3 can be reduced by the wet oxidation;
- Separation of sand is needed;
- Process leads to RDF with low quality;
- Production of RDF with high quality after the wet oxidation is not tested.

(2) Hese process

Process properties:

- Rotting of the fraction $<$5 mm (aprox. 32% of input);
- De-watering;
- Post rotting;
- Fraction $>$5 mm has to be aftertreated to produce RDF.

(3)Process with rotting of a fraction $<$ 40 to 60 mm (6 locations in Germany)

Process properties:

- Rotting of the fraction $<$ 40 to 60 mm (usually less than 50% of the input);
- De-watering;
- Post rotting.

In spite of the comparatively small number of MBT-plants in Germany many different types of constructions exists. This is a consequence of the needed optimization. So the following two processes was developed.

3 Extrusion process

Municipal solid waste and organic waste/biowaste can be separated quite easy and therefore with low costs with an extrusion process (VMPress) in high calorific value 'dry fraction' and microbiological degradable rich organic 'wet fraction'. The dry fraction can be treated with a thermal or after further treatment an energetic process. The wet fraction can be used to produce biogas with high efficiency. The plant engineering can easily be integrated in existing waste treatment plants and/or biogas plants. This offers a cost effective upgrade possibility for the integration of a material separation step.

The waste is compacted with high pressure (up to 1 000 bar = 10^8 Pa) in chambers of hydraulic presses. These are furnished with cylindrical, wear resistant, replaceable and perforated matrixes with defined diameters of the holes (normally 16 or 50 mm).

The core of the facility is a horizontally orientated screw extruder for the generation of high pressures and a rotary drum with a length of about 0. 9 m with a diameter of about 3 m (see *Figure 2*: yellow cover of the drum), in which 3 compression chambers are arranged in an angle of 120° to one another (see scheme in *Figure 2*).

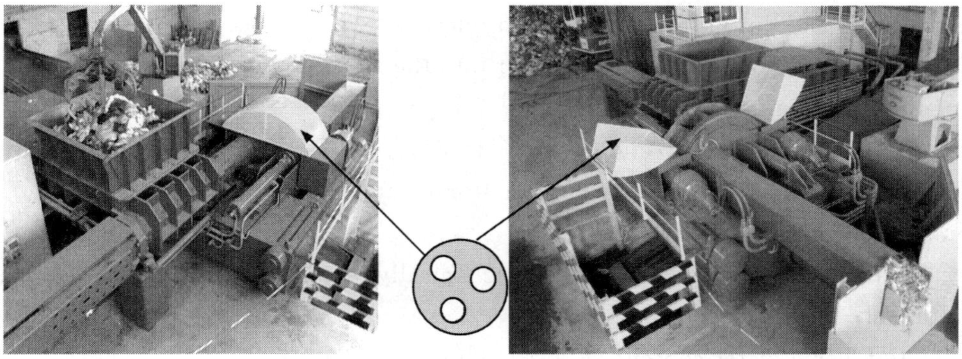

Figure 2 VMPress plant and scheme of the rotary drum in which 3 compression chambers are arranged in an angle of 120° to one another

Because of the rotation of the wheel in turns of 120° each compression chamber passes through the following three phases:

- Filling phase: Filling with waste from the cone which is filled by the included grab crane (see *Figure 2*) using a ram at the bottom of the cone;
- Compression phase: Compaction of the waste with the extruder, whereas the wet fraction is pressed out of the compression chamber through the holes of the matrix;
- Discharge phase: Discharge of the remaining dry fraction from the pressure chamber using a ram.

The whole process cycle proceeds in about 20 sec per filling-, compression- and discharge-phase. The operational sequences are synchronised and controlled. The facility works fully automated. Only for the filling a person is needed.

Common municipal waste or biological waste can be filled into the cone directly after the delivery without any pretreatment (no separation of metal, no crushing, no sieving).

During the compressing procedure a decrease in volume resp. an increase of the density occurs depending on the pressure and the time of compression. The most important variable in the operation of the facility is the pressure. With the rise of the pressure the following factors rise as well:

- The rate of the wet fraction and thus the yield of the biogas;
- The anaerobic microbiological decomposability of the wet fraction;
- The calorific value of the dry fraction.

VMpress-procedure constitutes an economic, quickly realisable, space-saving modular option to produce a fraction with a high heat value (dry fraction) and a fraction which can be treated biologically (wet fraction).

The wet fraction can be characterised in the following way:

- The wet fraction contains about 75% of the degradable organic material;
- The wet fraction contains almost the complete easily degradable organic material, like residuals of food;
- The wet fraction contains small portions of plastics and inert materials;
- Inert substances occur in the wet fraction in a form that they can be separated easily, because e. g. glass is not totally smashed by the compression procedure;
- The wet fraction leaves the press with a raised temperature. Decomposition processes start spontaneously;
- If an anaerobic treatment follows, the wet fraction can be mixed very good with circulated material from the digestion and during the digestion no relevant demixing occurs;
- If an aerobic treatment follows, an input of air into the wet fraction is possible without any problems because of the remaining structure;
- The water content of the wet fraction is 51% to 54%. Thus water doesn't pass out. There is no relevant discharge of fluids during the interim storage and the further processing.

The dry fraction can be characterised as follows:

- The dry fraction essentially consists of mechanically stable materials such as plastics, wood, paperboard and metal;
- The dry fraction has a water content of 30%-40%;
- The dry fraction contains about 25% of the degradable organic material. This percentage of organics is sufficient to produce stabilised material of very high quality by biological drying;
- The density of the dry fraction is about 0.9 to 1 mg/m^3;
- A breaking-up of the compact dry fraction is simply possible. Density after breaking-up: 0.4 mg/m^3;

The standard process flow is shown in *Figure 3*. *Figure 4* shows one of many options.

4 MYT-process

The MYT-process eliminates hazardous and foreign material extremely effectively, and thereafter shreds the waste selectively and homogenizes it in a percolator by circulation and irrigation in a way that optimum characteristics for the subsequent biological drying process and material separation are obtained. Within this process, effluent water accumulates which is rich in organic substances, and which is used for the generation of biogas. Due to the homogeneous drying level, dry matter is generated that can be transformed into above average high-quality fuel which can be stored almost without time-restriction.

The MYT technology consists of the stages shown in *Figure 5* and *Figure 6*. The Sources of the picture in Figure 6 are further pictures are shown in the appendix.

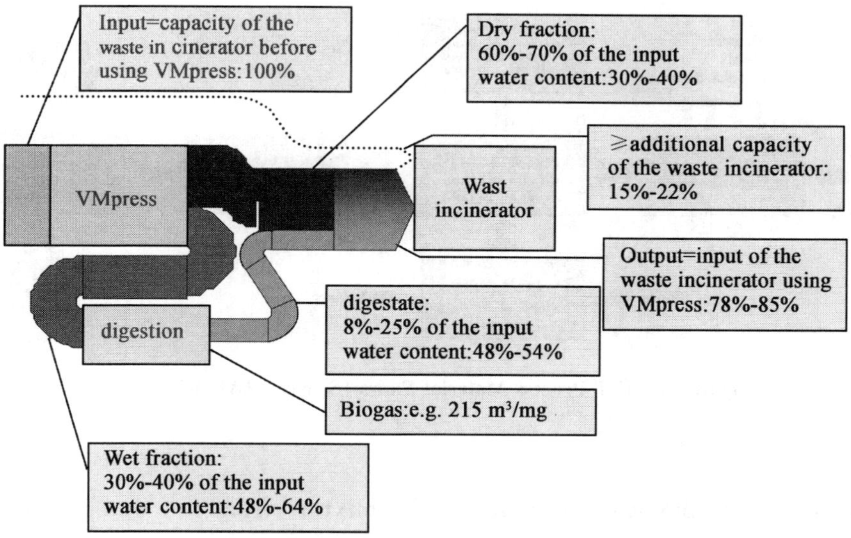

Figure 3 VMPress-Standard process flow

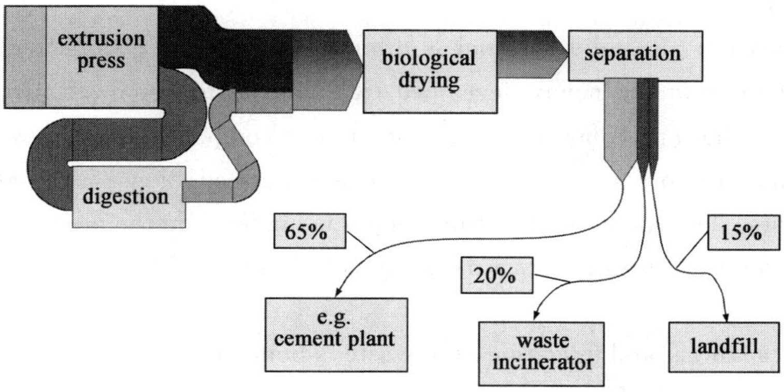

Figure 4 VMPress-Mass flow of one of many possible further options

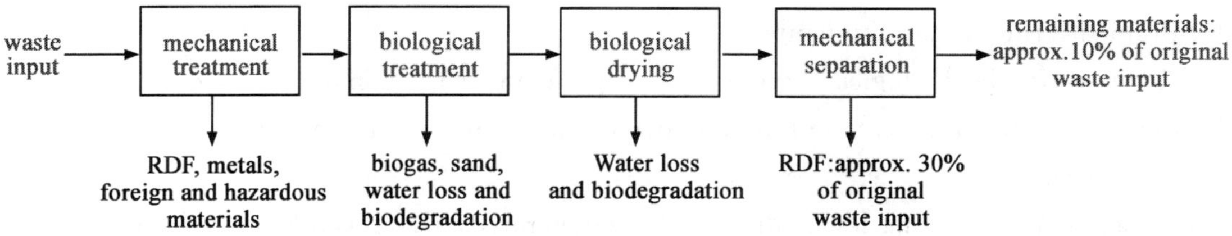

Figure 5 MYT-Process -Material Flows

Mechanical Treatment: In the mechanical treatment stage, a multifunctional sieving drum is used. By using this drum the shredding is no longer necessary, and thus a more effective separation of the undamaged foreign and hazardous material is possible. The effective removal of the foreign and hazardous material at the end of the MYT technology generates very high quality fuels. In the multifunctional sieving drum, waste bags and similar bundles are opened by different tools, and the waste is divided into three fractions. The fine and the middle fraction are individually separated from the foreign and hazardous material and fed into the percolator. The coarse fraction is removed.

For the separation of the hazardous material, amongst others, high-performance neodymium magnets are used for the removal of weakly magnetic components such as, for example, electronic waste, spray cans or

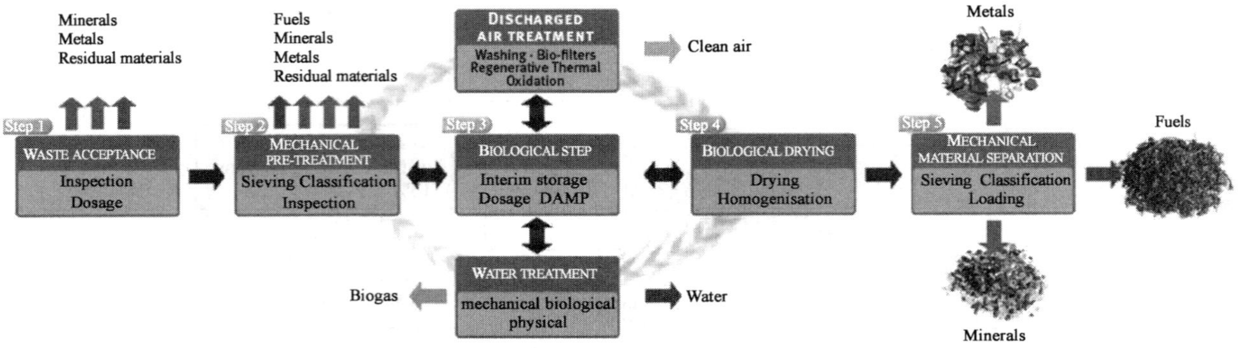

Figure 6 MYT-Process-Material Flows (Source: ZAK Ringsheim)

batteries.

Biological Treatment: In the biological treatment, the mixture from fine and middle fraction remaining after the removal of the hazardous material, is fed into six percolators and circulated continually under the addition of water. The percolator is a concrete half-shell, about 25 m long and 4.5 m wide, with horizontal mixing gear and sieve area.

At the end of the percolator the solid material is drained. The water is collected, and fed via specially adapted conveyor systems into the newly developed treatment stage for mechanical water treatment and subsequently fed into several anaerobic reactors. The organic components of the water are fermented into biogas in the anaerobic reactors (see *Figure 7*). The generated biogas is used for energy and heat generation in gas motors. The energy supply from biogas is higher than the needs of the MBT. A part flow of the run-off water from the anaerobic reactors is further cleaned and largely re-used for the irrigation in the percolator.

By circulation, using the mixer and irrigation, the waste is homogenized and shredded selectively in such a way that optimum characteristics for the subsequent processing stages are generated.

By fermenting the organic-rich water, the disadvantages known from fermentation plants for solid material are avoided (especially the complex operation with relatively low availability, and an odour intensive, and difficulty drained fermentation remnant).

Biological Drying: In the biological drying stage, quick and very homogeneous drying of the material is possible because of the pre-treatment in the percolator. This is supported by a single transfer of the material to be dried.

Due to the homogeneous drying level, dry matter is generated, which can be transformed into a fuel which can be stored almost without time-restriction.

Mechanical Material Separation: The mechanical material separation facilitates an extraordinarily strong separation of the material flows. Fuels of different qualities can be generated by subsequent mixing. Thus, the adaptation to the requirements of the customer of the fuels is possible at all times.

In the mechanical material separation (see *Figure 8*), the dried waste is separated according to its size and weight, in order to remove stones, sand, ceramics, and glass, as well as components with an exceeding pollutant level, or a too low caloric value, and thus obtain a high-quality fuel. To this end, the dried waste is sieved. All components with a size superior to about 25 mm are separated as residual material which has to be incinerated in an incineration plant, since further treatment due to the very low proportion would be too costly.

The main part of the dried waste is smaller than 25 mm, and is separated into several fractions of different

Figure 7 Anaerobic reactors

Figure 8 Mechanical material separation

sizes by further sievings. Each of these fractions is subsequently divided according to its weight, and the separated light components are removed as fuels. The heavy components have a low polluting level, and can be deposited, or used in road constructions.

The generated fuels have very good storage-characteristics, and can be used diversely in industry as an alternative to fossil fuels.

The mechanical material separation has been conceived in such a flexible manner that a quick response is always possible to changed requirements on the fuel market through corresponding quality adaptations, by regulating the material flow within the mechanical separation. Thus, the economic sale of the generated fuels is guaranteed at all times.

In contrast to conventional MBT technologies, the MYT technology, realized in the MBT Kahlenberg has the following ecological advantages:

- High yield of refuse derived fuels;
- Low amount of waste for disposal;
- High marketability of the generated refuse derived fuels;
- Low emission potential of greenhouse gases.

On the MBT Kahlenberg, the objectives of the MYT technology were already achieved after a relatively short operation period, when the operation was still characterised by changes, repairs and adaptations.

The MBT Kahlenberg sets itself very clearly and positively apart from the conventional MBT technologies by its very high technological standard and a virtually dirt-free operation outside the encapsulated aggregates, without higher specific costs, and is thus entirely appropriate as a reference plant.

5 Conclusion

MSW is an important source of biogas plants. To produce the feed out of MSW various technologies are possible. The two technologies described in the paper are now in full operation since several years and are showing best results. This is valid not only for the biogas production but also for the production of feed for the incineration. This means that not only the thermal and electrical energy production is very efficient but also the waste treatment is a nearly total recycling.

Further MYT-Pictures, Sources:
http://zak-ringsheim.de/ (homepage of the MYT-contractor);
http://web317.bwx2.de/media/myt_eng.pdf (brochure in English);
http://web317.bwx2.de/media/myt_cn.pdf (brochure in Chinese).

Figure 9 MVT plant Zweckverband MSW treatmentplant Kahlenberg

References

Rettenberger G., Schneider R. (2006): *A new MBT technology optimizes the output of high-caloric fuels*. waste management world, PennWell Corporation, London (www.waste-management-world.com).

Rettenberger G. (2007): *Experiences with mechanical-biological pretreatment of waste using the press-extrusion-process*. In: Proceedings Sardinia 2007, Eleventh International Waste Management and Landfill Symposium 2007 by CISA, Environmental Sanitary Engineering Center, Italy.

德国生物甲烷的发展

H. Kern

（*Arcanum能源系统有限公司，Unna 59423*）

【摘要】Arcanum 能源已致力于可再生能源市场运行能力方面的工作 10 余年了，并开始涉足沼气与生物甲烷领域。目前的主要工作是致力于德国生物甲烷的发展以及生物甲烷技术开发。生物甲烷作为可再生能源，能应用于许多能源系统中，目前德国和欧洲的生物甲烷市场正在逐步扩大。在欧洲，德国拥有最大的生物甲烷供给能力，从 2006 年起，其供给能力已从 1 000 m^3/h 增加到 30 000 m^3/h。如此巨大的增长完全得益于《联邦政府国家行动计划书》，该计划书中提到在 2020 年和 2030 年，生物甲烷将会替代国家天然气市场消耗总量的 7%和 11.5%。而庞大的生物甲烷市场促进了德国政府对生物甲烷处理方面的投入，使生物甲烷得以实现与气网全面合并，且建有专门为生物甲烷设立的气网。目前有许多种将沼气提纯为生物甲烷的方法，最实用的技术有胺洗法、物理有机气体洗涤法、压力水涤气法和变压吸附法。可见，在“Stadtwerke 沼气供给”商业理念下，Arcanum 能源为沼气并网的市场发展所作的贡献是独一无二的。

【关键词】生物甲烷；沼气提纯；支持政策

BIOMETHANE DEVELOPMENT IN GERMANY

H. Kern

Arcanum Energy Systems GmbH & Co. KG

Contact: Dr. Helmut Kern, Arcanum Energy Systems GmbH & Co. KG, Hertingerstraße 45, Unna 59423, Germany, Tel.: +49-2303-95-232-0, E-Mail: info@arcanum-energy.de

EXECUTIVE SUMMARY

Arcanum Energy is trussing its operational ability on markets of renewable energy sources for more than 10 years. Part and parcel is caisson of biogas and biomethane. The present work delivers insight into biomethane development in Germany and biomethane technologies. The sustainable biomethane is the all rounder of the renewable energies. It can be used for several energy systems. The market of biomethane in Germany and Europe is increasing. Within Europe, Germany has the largest feed-in capacity of biomethane. It's capacity since 2006 has increased from about 1 000 m^3/h up to 30 000 m^3/h. This massive extension derives from the 'National plan of action of the federal government', which says that by 2020 7% and by 2030 11.5% of the market need of the national natural gas consumption should be substituted with biomethane. To bring the biomethane market forward the German government installed special treatments for biomathane. Biomethane has preferred access to the gas grid on all levels. There are special gas-accounting grids only for the biomethane. To upgrade the biogas to biomethane there are several purification techniques that can be used for it. The best available technologies are the amine gas treating, the physical-organic gas treating, the pressurized water gas scrubbing and the pressure swing adsorption. With the business concept 'Biogas Pool für Stadtwerke' the business company Arcanum Energy Systems GmbH & Co. KG contributes an innovative and unique input to the market development of the biogas feed in.

Key words: Biomethane; biogas purification; support policy

1 Introduction

1.1 Background/about Arcanum Energy

Arcanum Energy is an independent enterprise for consulting and services with its registered seat in Unna, Germany. Arcanum Energy is trussing its operational ability on markets of renewable energy sources for more than 10 years.

Part and parcel is caisson of biogas and biogenic natural gas. In the last 2 years, the enterprise was able to enlarge its market position and expertise for trading biogenic natural gas in B2B section and project development. By now Arcanum Energy is the germane leading contractor of economically advantageous and sustainable charge of biogas and biogenic natural gas.

For the future the branch of renewable energy trade in general and biogenic natural gas in particular will increase. Furthermore Arcanum Energy will confront this challenge and cater to the requirements of its customers, whose are among public utility companies, energy providers, contracting companies, methods engineers and agricultural cooperatives.

1.2 Research objectives

The present work delivers insight into biomethane development in Germany and biomethane technologies.

2 Methodology

The concentrate efforts of Arcanum Energy are project development, trading and consulting on renewable energy markets in general and biomethane in particular. Project development under the direction of Dr. Helmut Kern works on feasibility and regional analyses, economic efficiency and technology research and the business concept 'Biogas Pool für Stadtwerke'. The directorate of management under the direction of Vera Schürmann executes biomethane trading, transport and balancing for biomethane as well as marketing and distribution.

3 Market and support policies of biomethane in Germany and Europe

The market of biomethane in Germany and Europe is increasing. Within Europe, Germany has the largest feed-in capacity of biomethane. It's capacity since 2006 has increased from about 1 000 m^3/h up to 30 000 m^3/h. This massive extension derives from the 'National plan of action of the federal government', which says that by 2020 7% and by 2030 11.5% of the market need of the national natural gas consumption should be substituted with biomethane. This means a feed in of 6 billion m^3/a in 2020 and 10 billion m^3/a in 2030. To reach this aim the feed in capacity has to increase more than 22 times until 2020. So there is still a lot to do. For the encouragement of the extension of the renewable energies, which includes the extension of the feed-in capacity, the federal government established a law for the fee of the renewable energies, the so called EEG. It was created in the year 2000 and amended in the years 2004 and 2009. The next amendment will be in 2012. These laws specify the fees for the power creation out of the particular renewable energy source. The fee for the electric power out of biomass is stepped. Applicable for the fee is the average feed-in power (electric) and the date of launching of the unique power creation unit (in most cases a combined heat and power unit, CHP). The following table shows the fees in dependence of the feed-in power and of the applied law.

Table 1 Fees of the different EEG (EEG)

Basis fee	EEG 2000 (ct/kWh$_{el}$)	EEG 2004 (ct/kWh$_{el}$)	EEG 2009 (ct/kWh$_{el}$)
Up to 150 kW	—	11.5	11.67
Up to 500 kW	10.23	9.9	9.18
Up to 5 MW	9.21	8.9	8.25
From 5 MW (EEG 00/04) and accordingly up to 20 MW (EEG 2009)	8.7	8.4	7.79
Degression	1%	1.5%	1%

The degression is calculated to the date of launching and then fix for twenty years plus the year of launching. Besides these fees, there are several bonuses that are added to these fees. There are bonuses for the usage of renewable primary products, technology bonuses for the purification of the biogas and a bonus for the use of the gas in CHP. *Table 2* gives an overview of the different bonuses.

Table 2 EEG-bonuses (EEG)

Basis fee	EEG 2000 (ct/kWh$_{el}$)	EEG 2004 (ct/kWh$_{el}$)	EEG 2009 (ct/kWh$_{el}$)
Up to 150 kW	—	11.5	11.67
Up to 500 kW	10.23	9.9	9.18
Up to 5 MW	9.21	8.9	8.25
From 5 MW (EEG 00/04) and accordingly up to 20 MW (EEG 2009)	8.7	8.4	7.79
Degression	1%	1.5%	1%

The cost for purchasing the biomethane average out at 7.4 to 7.6. To this price the transport-, gas grid usage and on the basis of these fees and bonuses is a profitable running of CHP with biomethane possible.

4 Usage applications of biomethane and market potential

The sustainable biomethane is the all rounder of the renewable energies. It can be used for several energy systems. Today there are three ways of purchasing biomethane.

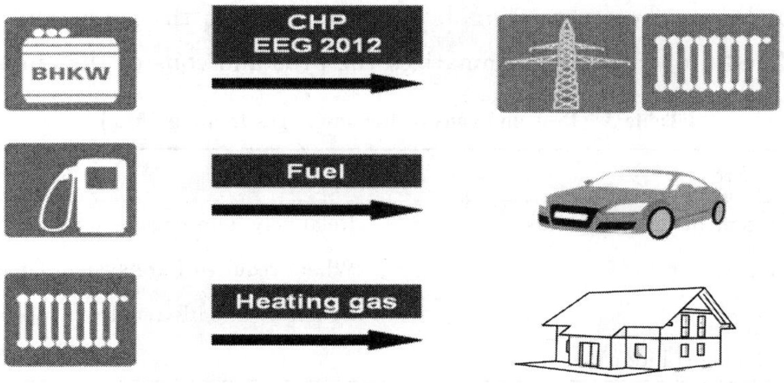

Figure 1 Marketing channels for biomethane

It can be used in CHP's for generating power and heat at the same time. Under the EEG 2009 the use of biomethane in CHP's with an installed electric power of 200-500 kW is most profitable.

Another way of using it is the traffic sector, where it can be used as a fuel in e. g. natural gas cars. The EU targets to have 10% bio fuels in 2020 with biomethane as a great part of it. There are taxation exempts until 2018 if the bio fuel use is within the contingent and there are taxation exempts until 2015 if the bio fuel use is beyond the contingent.

The third way of using biomethane is the heating sector. In this sector is the intermixture of biomethane with natural gas sensible. Regional regulations demand a fraction of 20% biomethane in the heating gas in the case of new buildings or new heating systems. If public buildings are renovated, the fraction of biomathane has to be at least by 30%.

Currently the use of biomethane in CHP's is with 85% the greatest fraction. It is followed with 10% of using it by the private customer. The traffic fuel usage is the smallest part with about 5%. Generally the demand of biomethane is extensively increasing.

5 Purification systems/technologies and profitability

To upgrade the biogas to biomethane there are several purification techniques that can be used for it. The best available technologies are the amine gas treating, the physical-organic gas treating, the pressurized water gas scrubbing and the pressure swing adsorption.

The amine gas treating is a special case of the physical-organic gas treating. The purification process is pressure less. In the first step of the treating the gas is desulfurization. In most cases this is done by an activated coal filter. It decreases the fraction of the hydrogen sulphides below 5 ppm. The desulphurization is followed by the dehydration. After this treatment the carbon dioxide is removed by the amine. The gas flows from the bottom to the top and the amine flows from the top to the bottom. On its way down the amine absorbs the carbon dioxide in a first step physically and in a second step chemically. This is a very effective way of removing the carbon dioxide. After the amine purification are pureness fractions of 99.9% methane possible. When the purified biogas leaves the amine reactor, it is called biomethane. It is now fully saturated of water. This water has to be reduced for a feed-in into the gas grid. A tube bundle heat exchanger reduces the saturation to a level that the rest of the water can be removed by an adsorption. The gas is now purified and can be committed to the gas grid operator. When the purification does not reach the needed gas quality, the iniquitous is burned in a flare system. To run the amine in a circle, it has to be regenerated. The regeneration is done under an absolute pressure of 5 bar and an temperature of about 140℃. Under these conditions the carbon dioxide is separated from the amine and the solvent can recycle to the wash column. The following table summarizes the pros and cons of the amine gas treating.

Table 3 Pros and cons of the amine gas treating (MT)

Pros	Cons
Relatively low power requirement (pressure less)	Relatively high heat requirement (thermal regeneration)
High reachable methane purity (up to 99.9%)	Where required flare system (iniquitous gas flare)
Low methane slip (ca. 0.1%)	Desulfurization is required
Heat extraction possible (amine solution)	

The physical-organic srcubbing gas uses an organic solvent, similar to the amine solvent, to purify the gas. But there are several differences to the amine gas treating. The purification is done under a pressure about 8 bar in the absorber. The absorber separates the methane by physical absorption from the carbon dioxide, most of the hydrogen sulphide, the water, and the other minor components like oxygen and nitrogen. The product gas has to be precision desulfurized and dehydrated after the absorber. When the physical-organic gas scrubbing does not reach the needed gas quality, the gas circulates until it reaches the quality. Due to the physical absorption the regeneration of the organic solvent is easier than the regeneration of the amine. The solvent is regenerated pressure less and without heat. *Table 4* gives an overview of the pros and cons of the physical-organic gas scrubbing.

Table 4 Pros and cons of the physical-organic gas scrubbing (HAASE)

Pros	Cons
Relatively low heat requirement (regeneration)	Relatively high power requirement (gas compression to 8 bar)
No flare system required (recirculation iniquitous gas)	Methane slip about 1%
Heat extraction possible (waste heat compressor)	Methane purity ca. 97%
High guarantee of the plant availability (96%)	Waste air purification required (Vosci-Box®)
	Desulfurization of the product gas is required

The pressurized water gas scrubbing uses, instead of an organic solvent, water. The process works with an absolute pressure equal to pressure of the physical-organic gas scrubbing of about 8 bar and without heat. Unwanted components are removed from the gas stream by physical absorption. In the best case a fraction of 97% methane with a slip of 1% can be reached. After the absorber the methane is desulfurized, but it has to be dehydrated. The pressurized water absorbs the unwanted components. The carbon dioxide and the hydrogen sulphide are separated from the water by desorption on a low pressure level. These components are carried out by air, which has to be desulfurized in a last step before it is led into the atmosphere.

Table 5 Pros and cons of the pressurized water gas scrubbing [Malm]

Pros	Cons
No heat required	Relatively high power requirement (gas compression to 8 bar)
No flare system required (recirculation iniquitous gas)	Methane slip about 1%
Heat extraction possible (waste heat compressor)	Methane purity ca. 97%
High guarantee of the plant availability (95%)	Waste air purification required (RTO)
No separate desulfurization required (RTO)	

Another possible technique to purify the raw biogas is the pressure swing absorption. The raw biogas has to be desulfurized before it is led to the apparatus. In the apparatus is activated coal. When the apparatus is put under pressure of about 4 bar the activated coal adsorbs the carbon dioxide. A sensor registers when the coal is fully loaded. At this point the pressure is drained and the desulfurized raw biogas is led to another apparatus which is under pressure. In the pressure less apparatus the carbon dioxide desorbs from the activated coal and is blown off after purification into the atmosphere. Iniquitous gas is circulated and so there is no flare required. By the pressure swing absorption purities about 96% methane is reached with a methane slip of 3%. The next table shows the pros and cons of the pressure swing absorption.

Table 6 Pros and cons of the pressure swing absorption (Schmack)

Pros	Cons
No heat required	Relatively high power required (gas compression to 4 bar)
No flare system required (recirculation iniquitous gas)	Methane slip about 3%
Heat extraction possible (waste heat compressor)	Methane purity ca. 96%
High guarantee of the plant availability (95%)	Waste air purification required (poor gas flare system)
	Desulfurization of the raw biogas is required

There are some technical preconditions for the operation of a purification plant. These are the gas quality of the raw biogas (e. g. the methane fraction, the hydrogen sulfide concentration, the oxy- and nitrogen concentration etc.), the quality of the bio methane, the gas flow rates, a heat concept, handover pressure between the biogas plant and the purification plant, handover pressure between purification plant and feed in plant and many more.

Of course there are some economical preconditions for an operation of a purification plant, too. A high working load has to be guaranteed and the treaties for the maintenance and the services have to be considered. For the techniques that use chemical gas scrubbing for the purification is the heat import price a very important factor. In case of the physical gas scrubbers the heat is replaced by the price of the power import. Another important factor is the length of the pipeline to connect the plant with the gas grid. Until the length of a kilometer the costs are capped. Beyond one kilometer to a distance of ten kilometers the costs are separated between the plant operator and the gas grid operator in a fixed fraction. If the distance to the gas grid is longer than ten kilometers the plant operator has to pay to addition to the pipeline. *Figure 2* shows how the specific costs of purification can vary for a biomethane production co 350 Nm^3/h.

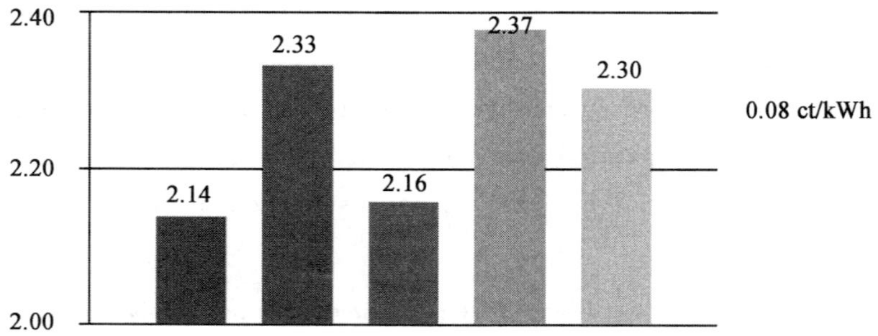

The operating costs of the feed-in plant bears the system operator, except the capital costs of the 250 000 €-part.

Annotation: costs: heat 3.6 cts/kWh_{th}, costs: power 13 cts/kWh_{el}

Source: guiding offer prices, existing offers, pre dimensioning through AES

Figure 2 Specific costs (Investment and operation) gas purification and feed-in with a capacity of 350 Nm^3 biomethane per hour

6 Transport and accounting services

The transportation and accounting of gas through the gas grid is a very complex operation. To get to the starting point of transportation, there are several preconditions that have to be fulfilled. These preconditions are:

- Production of biomethane (location, purification, feed-in) or import of biomethane;
- Connection with the gas grid by a grid connection demand, followed by a grid connection treaty;
- Establishment of a special biogas-accounting grid access/outlet by feed-in, feed-out treaty and delivery frame treaty;
- Declaration of a counting point or rather a gauge;
- Declaration of the feed-in and feed-out capacity;
- Procurement of the proof of origin;
- Biomethane customer has to conclude a biomethane delivery treaty.

A virtual transport of biomethane is only possible, if it can be retraced physically. The simplified process of the physical and virtual gas flow is shown in *Figure 3*.

To bring the biomethane market forward the government installed special treatments for biomathane.

Biomethane has preferred access to the gas grid on all levels. There are special gas-accounting grids only for the biomethane. Feed-in nomination has no bill relevance. The gas is accounted yearly with a flexible frame of 25%, which can be transferred to the following year. Costs for account adjustment are fixed by the gas grid access act.

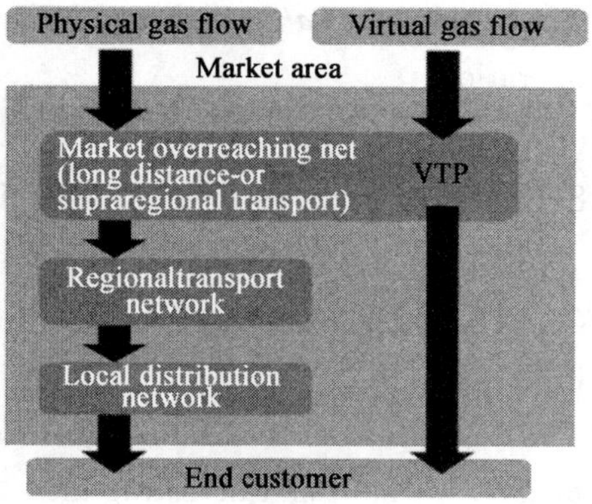

Figure 3 Physical and virtual gas flow from the purification to the customer

7 Profitability of combined heat and power unit (CHP) compared to biomethane

In most cases the biomethane is used in CHP to produce power and heat. Because of that a detailed calculation of the CHP which use biomethane is necessary. The use of biomethane is in competition with the use of natural gas in CHP or in boilers. So a comparative calculation between these options is useful to find out which system is the best for an individual project. The calculations are based on a cost-benefit analysis over a defined period of time. A lot of variables take an influence on the profitability of a CHP. The most important aspects are the electric power, the efficiency factors, the fuel demand, the costs of investment and many more.

The following tables give an overview of a comparative calculation between the usage of biomethane and natural gas in CHP with an installed electric power of 250 kW.

Table 7 Priori statements for CHP calculation (2G)

Specifications	Biomethane	Natural gas
Full load hours (h/a)	8 000	8 000
Electric efficiency factor (%)	37.4	37.4
Thermal efficiency factor (%)	51.5	51.5
Power input (%)	100	50
Heat aging (ct/kWh)	3.50	3.50
Equity (%)	30	30
Interest rate (%)	4.50	4.50
Recovery period (a)	10	10
Price dynamic sampling (%)	0.6	3
Price dynamic sampling heat (Q) (%)	3	3
Specific gas price(ct/kWh)	7.60	3.40
Bonuses	RPP, 2ct/kWh technology, CHP	CHP, avoided net usage fees

Table 8 Characteristic numbers of a CHP calculation

CHP (kW_{el})	Biomethane 250 kW_{el}	Natural gas 250 kW_{el}
Gas price	8.0	3.8
Aging power	22.36	10.62
Aging heat	3.50	3.50
Investment	298 000	298 000
EBT 10 years averaged	14 275	−38 332
ΔEBT	52 607€	

The results in *Table 8* show that the use of biomethane under these circumstances is a very profitable operation.

8 Biogas Pool as a successful investment model

With the business concept 'Biogas Pool für Stadtwerke' the business company Arcanum Energy Systems GmbH & Co. KG contributes an innovative and unique input to the market development of the biogas feed in.

The 'Biogas Pool für Stadtwerke' enables the municipal energy suppliers a long term secured acquisition of biogas, without self investing into the raw biogas production. The investment for the biogas production is done by the farmers, who develop on this way a permanent and secure business field as 'energy hosts'. The farmers cater for the long range substratum supply and drive the biogas equipment for the production of raw biogas. By this means is a clear cutting site between the raw gas production and the Biogas upgrading to Biomethane created, in first instance because the technology of the gas industry is not a core competence of the agriculture.

Farmers invest into the biogas equipment and produce the raw biogas. Municipal energy suppliers invest as a pool together on different sites in Germany. This municipal energy suppliers pool constructs on each site a gas conditioning plant for the raw biogas for editing biomethane, which is fed in the natural gas distribution system for the usage of the customer. The farmers sell the raw biogas to the municipal energy suppliers pool based on well defined and long lasting (generally 20 years) raw biogas supply contracts. The single municipal energy supplier can procure via the pool individual amounts tailored to suit a market need. By this each supplier increases his delivery security compared to contracts with single producers and equipments. Through the cooperation the investment costs decrease, synergistic effects occur and the risks are reduced.

This form of investment in renewable energies advances the own competitiveness and improves the image of a supply company, that is serious about environment and climate protection in terms of a sustainable and local production.

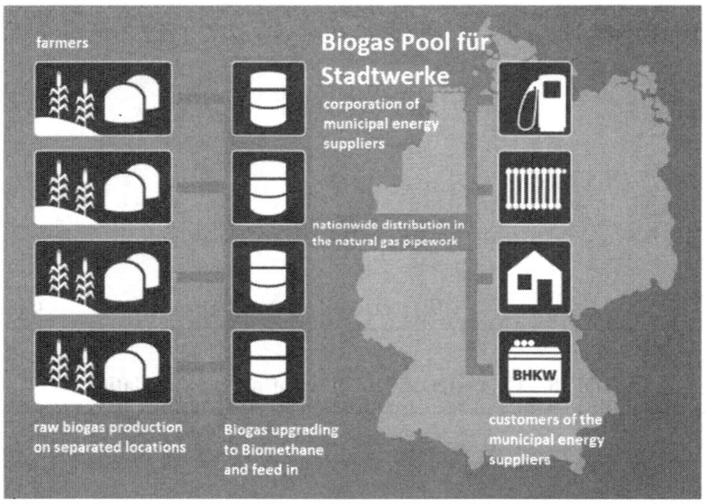

Figure 4 Schematic exposition of the business concept

By the foundation of a 'Biogas Pool für Stadtwerke' several municipal energy suppliers invest together in the conditioning and feed in of biogas, whereby synergies occur, that allow an efficient and less risky usage of biogas. Furthermore this pool effect enables a flexible and individual supply of biogas. A high planning and financing security, inter alia on the basis of well defined and fair raw biogas delivery contracts, is guaranteed by that.

Presently there are three feed in projects were put into Practice ('Biogas Pool 1'). All in all the biogas pool produces 1 400 Nm^3 of biomethane per hour. Besides there are the 'Biogas Pool 2' in the realization phase. The construction has already started. A wise investment in renewable energies with the focus on the acquisition of biogas is realized clever and efficiently that way. Bright prospects for an expanding biogas market based on an innovative value creation concept.

Table 9 Advantages of the business concept

Long term purchase warrant for bio methane
Assurance of the individual demand
High planning and financing security
Use of synergy effects
Investment only in Biogas upgrading to Biomethane and feed in
Well defined and fair raw biogas delivery contracts
Adequate task sharing
Reduction of the default risk by the pool effect
Increasing competitiveness
Wise investment in renewable energies

9 Conclusions

The federal government aspires to an annual biogas feed of 6 billion m^3 until the year 2020 and 10 billion m^3 until 2030, whereby the renewable energy production gets more and more important.

In this regard the production, conditioning and feeding of the biogas into the natural gas pipe work afford a seminal business model for the farmers as well as the energy providers. Furthermore plays the arrangement of an intelligent and efficient value creation chain for the use of biogas by the energy providers today more than ever a central role in your preservation of competitiveness. Investments and participations in biomethane plants are profitable and tailored to a market suit assurance supply of bio methane. The Aracnum Energy Investment model affords a high investment assurance by the EEG fee (20 years), profitable market entry and marketing of a forward-looking and climate protecting gas product. Arcanum energy can avouch a sustainable growth through customer loyalty and competitive capability and an ecological and image aiding positioning of the urban energy suppliers in the liberal energy market.

References

EEG, (2008): *Gesetz zur Neuregelung des Rechts der Erneuerbaren Energien im Strombereich und zu Änderung damit zusammenhägernder Vorschriften*. Bundesgesetzblatt Jahrgang 2008 Teil I Nr. 49, ausgegeben zu Bonn am 31. Oktober 2008.

MT, (2010): *MT BioMethan GmbH*. Internal business information.

HAASE, (2010): *HAASE Energietechnik AG & Co. KG*. Internal business information.

Malm, (2010): *Malmberg Biorerdgastech GmbH*. Internal business information.

Schmack, (2010): *Schmack Carbotech GmbH*. Internal business information.

2G, (2011): *2G Bio-Energietechnik AG*. http://www.2-g.de/module/dateidownload/juli2010_erdgas-deutsch-komplett.pdf.

PROSPECTS AND OVERVIEW OF BIOMETHANE INDUSTRY IN CHINA

CAI Changda, TAN Jing

Hangzhou Energy and Environment Engineering Co., Ltd.

Contact: Mr. CAI Changda, Hangzhou Energy and Environment Engineering Co., Ltd., No. 118 East Fengqi Road, Hangzhou, PR China, Tel.: +86-571-8604-4375, E-Mail: changdacai@163.com

EXECUTIVE SUMMARY

This paper summarizes the strategic significance of developing the bio-methane industry in China. Comparing the current situation of bio-methane generation and use with Europe, the author has identified that the use as vehicle fuel is one of the high value application of biogas and the most promising driver for the development of the Chinese bio-methane industry.

Key words: Bio-methane; high value application; market prospect

中国生物甲烷产业市场前景与概览

蔡昌达，谭婧

（杭州能源环境工程有限公司，杭州 310020）

【摘要】本文概述了发展生物甲烷产业的战略意义，在比较欧洲及国内生物甲烷产业发展现状的基础上，分析得出车用生物甲烷是沼气高值利用的重要方向，并展望了中国生物甲烷产业的市场前景。

【关键词】生物甲烷；高值利用；市场前景

1 引言

生物甲烷具有清洁、高效、安全和可再生四大特征，它的高效制备和高值利用是一种具有代表性的双向清洁过程。开发生物甲烷是国家中长期科学和技术发展规划纲要(2006-2020)重点领域的优先发展主题，生物甲烷产业属于国家战略性新兴产业政策支持的节能环保、生物、新能源以及新能源汽车的等四个领域[1]，国家可再生能源中长期发展规划预计 2020 年建成规模化沼气工程 16 000 座[2]。

2 发展生物甲烷产业的战略意义

2.1 替代化石燃料，改善能源结构

我国已探明的人均煤炭、石油和天然气资源分别为世界平均值的 42.5%、17.1%和 13.2%，人均资源严重不足。石油的对外依存度逐年增大，2008 年已超过 50%，供需矛盾日益突出。但我国有丰富的生物质资源，每年的资源量大约折合 7×10^8 t 标准煤，其中可用于生产生物甲烷的生物质资源量约折合 2.5×10^8 t 标准煤，可转化成约 1.5×10^{11} m^3 的沼气当量，折合天然气 9×10^{10} m^3，按照 2007 年的能源消费总量计算，生物甲烷发展将使中国气体能源消费的比重提高到 8%左右。因此，发展生物甲烷，对于缓解我国能源紧张局面、改善能源结构、减轻对传统化石能源的依赖具有十分重要的意义。

2.2 减少温室气体，应对气候变化

近年来，日益严重的全球变暖问题不断受到世界各国的重视，对于温室气体减排的需求也越来越紧迫。

若有机废弃物得不到有效利用，直接暴露在空气中将产生大量的沼气，而沼气的主要成分是甲烷，其温室气体效应是二氧化碳气体的 21 倍，因此，必须寻求一种妥善处理有机废弃物的方案，减少甲烷排放。以有机废弃物为原料制备生物甲烷是解决此问题的有效途径，它既能控制有机废弃物向空气中排放甲烷，又可生产清洁能源，从而减少化石能源的消耗，实现进一步减排。一座年产 3.65×10^6 m^3 的沼气工程，年温室气体减排量相当于 4.2×10^4 t 二氧化碳当量，由此看出利用有机废弃物生产生物甲烷是应对气候变化的重要举措。

2.3 保护生态环境，推进新农村建设

我国每年产生的有机废弃物大部分都未得到妥善处理和利用，造成了严重的环境污染。通过有机废弃物发酵制备生物甲烷，可以在厌氧消化过程中有效控制病原微生物的扩散和传播，其发酵残余物还可用作农田有机肥，从而减少化肥消耗，改善土壤理化性状。如此既可解决环境污染问题，又能提供能源和肥料，有利于生态系统的良性循环和农业持续发展，为循环经济提供有力支持。

生物甲烷产业将从根本上改变传统的农村废弃物和粪便的利用方式，推进农业废弃物的无序利用向有序利用转变，提高农村环境卫生水平；加强农业废弃物和农村生活废弃物的资源化、能源化利用；推进农村生活用能从低品质向高品质转变；调整农村产业结构，扩大农民就业、增加农民收入、提高农民生活质量、加速社会主义新农村建设。

3 国内外生物甲烷产业发展现状

3.1 欧洲生物甲烷产业发展现状

瑞典是沼气提纯用于车用燃气发展最好的国家，沼气年总产量折合约 1.3×10^8 m^3 天然气当量，占燃气消费总量的 13%。全国有 223 家沼气工厂，主要发酵原料是市政污泥、有机垃圾和农业废弃物。生物甲烷的利用方式主要为车用燃气和管道燃气两方面，全国有 98 个生物甲烷加气站，13 407 辆轿车、369 辆重型车和 760 辆巴士使用生物甲烷，同时瑞典还设计了世界上第一辆沼气火车。瑞典政府计划到 2020 年，50%天然气将由生物甲烷代替，到 2050 年，天然气将完全被生物甲烷代替。

Figure 1 Bio-methane plant in Sweden

图 1 瑞典某生物甲烷项目

瑞士是将生物甲烷用于车用燃气最早的国家，瑞士首都伯尔尼已有 15 000 多辆公交大巴和出租车以生物甲烷为动力，生产沼气的原料主要是市政污泥和有机垃圾。

德国是欧洲生物甲烷发展最快、数量最多的国家，生物甲烷的产量约为 2.07×10^9 m^3 天然气，占德国燃气产量的 14.5%，总消费量的 2.5%。德国传统沼气利用方式主要是热电联产，现已逐渐转向车用燃气和天然气并网，2008 年有 13 家沼气厂提纯 4.2×10^7 m^3 生物甲烷进入天然气管网，2009 年有 30 家沼气厂提纯 1.8×10^8 m^3 生物甲烷进入天然气管网，德国梅克伦堡州居斯特罗市 2009 年建成全球最大的以秸秆为原料的沼气工程，日产 2.5×10^5 m^3 沼气，经提纯用于 50 000 个家庭。

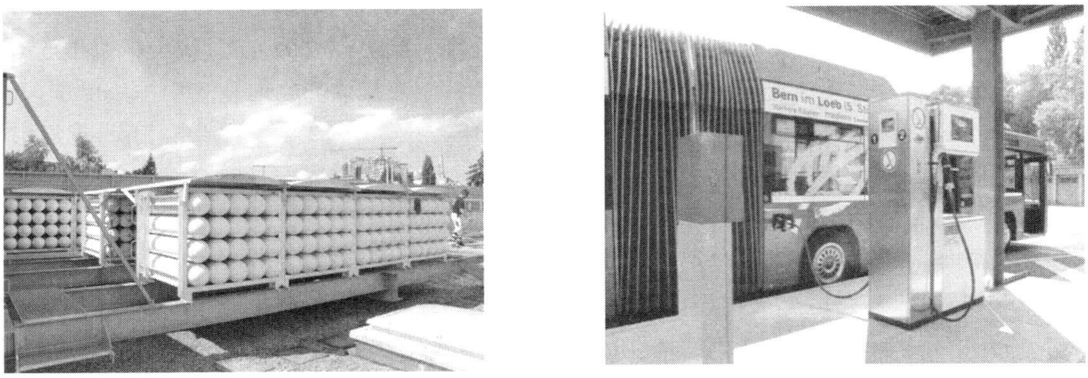

Figure 2　CNG Gas station in Swizerland

图 2　瑞士 CNG 加气站

Figure 3　100 000 m³/d Bio-methane plant in Germany

图 3　德国 100 000 m³/d 生物甲烷项目

3.2　中国生物甲烷产业发展现状

我国生物甲烷产业作为一个新型产业，它的雏形才刚刚展现，具有巨大的发展空间。可再生能源中长期发展规划中指出，到 2020 年，建成大型畜禽养殖场沼气工程 10 000 座、工业有机废水沼气工程 6 000 座，年产沼气约 1.4×10^{10} m³[2]。而我国 2008 年集中型生物燃气产气量才 7.18×10^{8} m³，仅相当于当年天然气消费量的 0.5%。根据各地的具体情况，生物燃气产业呈现热电联供、生产车用压缩天然气、管道燃气、罐装燃料等多种形式。畜禽养殖场规模化沼气工程已有山东民和 3 MW、德青源 2 MW 以及内蒙古澳亚牧场 1 MW 沼气发电工程等许多示范工程，并产生了良好的效益。

Figure 4　Demonstration project for biogas power generation in China

图 4　国内沼气发电示范工程

目前厌氧发酵技术及关键设备国内虽已有工业示范，但用于沼气发电的关键设备主要还是从国外进口，部分国产设备也还有待进一步完善并标准化。

沼气生产车用生物甲烷和管道燃气在国内尚处于起步阶段，目前已在南宁安宁淀粉厂、山东博兴、内蒙古通辽等地建有沼气提纯生物甲烷示范项目。

Figure 5　Demonstration project for biogas upgrading in China

图 5　国内沼气提纯生产生物甲烷示范工程

有关生物甲烷提纯的部分关键技术还需自主开发或国外引进。目前我国适宜发展车用的生物燃气量不足生物燃气总产量的 3%，不足天然气消费量的 0.1%，因此我国急需发展大型沼气工程，加强高效多原料混合厌氧发酵、高品质生物燃气净化、提纯、提质制备车用燃气等方面的研究。

4　车用生物甲烷是沼气高值利用的重要方向

4.1　市场优势

首先，随着化石能源的日趋紧缺，生物甲烷的市场需求日益增大。据有关部门预测，2020 年我国石油需求量达 5.6×10^{8} t，而届时国内产量仅 1.9×10^{8} t，石油进口依存度将达 65%以上。随着汽车数量的日趋增长以及汽车能源结构的不断变化，燃气需求量将逐年增长，从 2000 至 2008 年，燃气消费量平均增速在 16%；预计 2020 年天然气消费量将突破 2×10^{11} m^3，各地都会出现天然气汽车排队加气的状况。

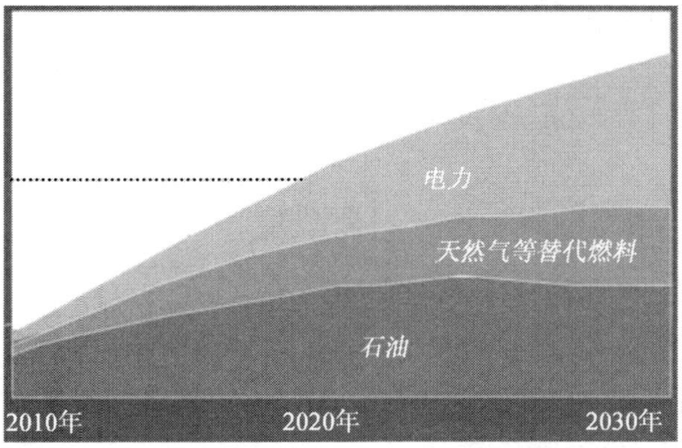

Figure 6　Structure chart of vehicle energy

图 6　汽车能源结构图

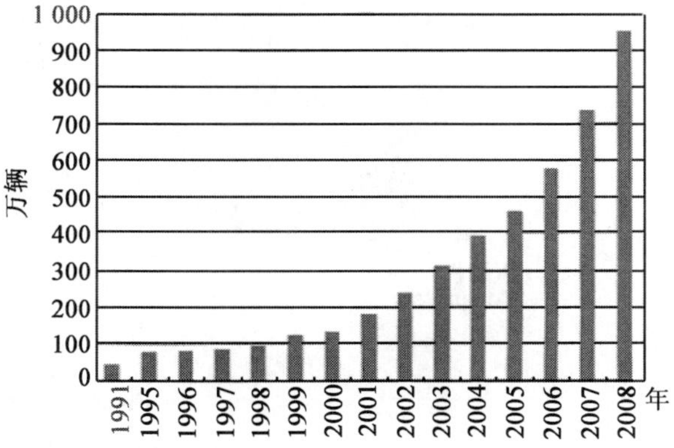

Figure 7　Development chart of natural-gas-powered vehicles worldwide

图 7　世界天然气汽车增长趋势图

其次，生物甲烷是一种最清洁的优质气体能源。沼气经提纯净化后完全可以达到车用燃气的技术标准，燃气质量不存在任何问题，其热值≥31.4 MJ，$CH_4\geq95\%$，$CO_2\leq3\%$，$H_2S\leq10$ mL/L，$O_2\leq0.5\%$；压缩罐装到 250 个大气压的 CNG 贮存罐，输配到各加气站，与石油天然气、CNG 动力汽车通用，推广非常便捷。

再次，由于国家要求各地原则上按车用天然气销售价格与汽油最高零售价比价不得低于 0.75∶1 进行调整。按此调整后，以天然气为燃料的车辆比使用汽油的车辆更具有成本优势，且有利于促进城市环保。与沼气发电相比，随着石油天然气 CNG 的价格日趋上涨，折算成生物甲烷原气价基本在 2.40 元/m^3 以上，收益明显高于沼气发电。

4.2 提纯技术及设备

目前常用的纯化技术有 PSA(变压吸附)、WS(水洗)、MEA(乙醇胺吸收)、DEA(二乙醇胺吸收)等，各技术工艺的比较如表 1 所示：

Table 1 Comparison of different upgrading technologies

表 1 提纯技术比较

纯化技术	PSA	WS	MEA	DEA
预处理	是	否	是	是
操作弹性(%)	±10	50～100	50～100	50～100
甲烷回收率(%)	>96	>97	>99	>99
操作压力(bar)	4～7	4～7	常压	常压
电耗(kWh/Nm^3)	0.25	<0.25	<0.15	<0.12
热量消耗	否	否	160℃	160℃

由于水洗法有工艺简单，无需添加任何化学试剂，运行成本低等优点，目前已在欧洲广泛应用，其工艺流程如图 8 所示：

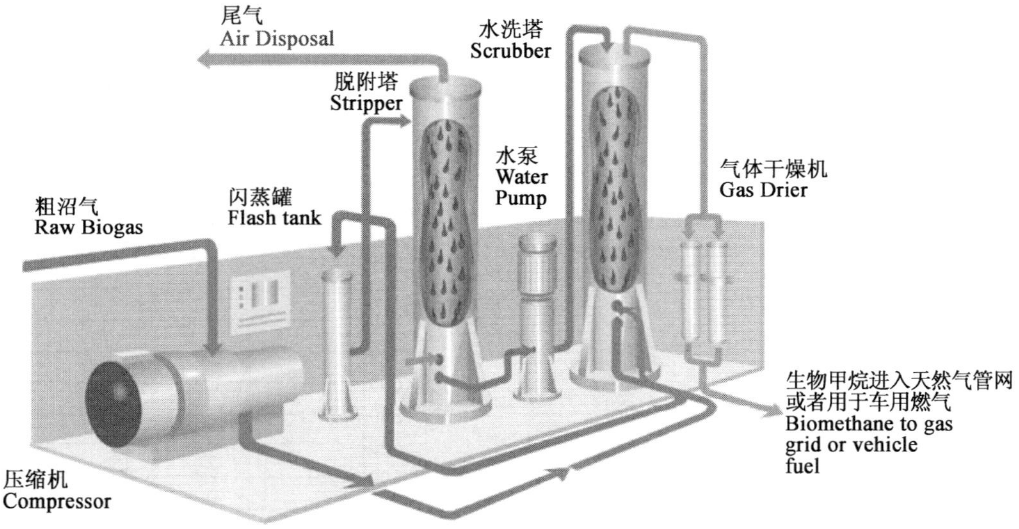

Figure 8 Water scrubbing process

图 8 水洗提纯工艺流程(Greenlane)

沼气经压缩机增压至 8 bar 后进入水洗塔，水由水洗塔顶部喷淋，与沼气逆流接触，吸收沼气中的二氧化碳、硫化氢等杂质气体，净化后的生物甲烷从水洗塔顶部进入气体干燥机进行干燥，然后并入天然气管网或者进入后续压缩系统压缩装罐用于车用燃气。冲洗水由水洗塔底部流出，经过闪蒸塔进入脱附塔，释放溶解的二氧化碳和硫化氢，脱附后的水可循环用于水洗塔。

沼气提纯设备在国外已基本实现了产业化，但国内尚在起步阶段，仍需解决设备制造关键技术，实现设备国产化。

Figure 9 Upgrading equipments in foreign countries

图 9 国外提纯工艺设备

5 中国生物甲烷产业市场发展前景

5.1 可利用资源

目前，我国生物甲烷可利用资源产气潜力为 1.5×10^{11} m^3/a，其中畜禽粪便 7.5×10^{10} m^3/a，1/3 的作物秸秆产气潜力为 6×10^{10} m^3/a，有机垃圾 9×10^{9} m^3/a，农产品加工残余物 6×10^{9} m^3/a。每年的产气潜力总量相当于 9×10^{10} m^3 天然气，是我国 2008 年天然气消费量 8.07×10^{8} m^3 的 1.12 倍。沼气可利用资源潜力大，产业发展前景广阔。

5.2 生物甲烷产业发展方向

从世界范围内综合分析，生物燃气未来的发展趋势主要如下：

由单一原料发酵技术向多元原料共发酵技术发展。生物燃气工程的原料由单一畜禽粪便向作物秸秆、农业及农产品加工废弃物、生活垃圾等多种原料发展，发酵技术也从单一原料向多元原料共发酵技术发展，原料多元化的产业拓展趋势明显。

由低效直燃热利用向高品质生物甲烷产品发展。生物燃气的利用由低效直燃热利用向高效热电联供、高品质民用、车用燃气的方向发展。生物燃气用于热电肥联产发电机组，热电能源利用总效率可达 80%以上，电能并入国家电网，热能用于厌氧系统冬季增温和其他生产用热；生物燃气用于生产车用燃气和管道燃气，甲烷含量达 95%以上，1 m^3 甲烷气相当于 1.2 L 汽油。随着化石能源日益枯竭，汽油价格上涨，生物燃气的利用逐步由热电联产向车用燃气转变。发展沼气提纯技术，生产生物车用燃气势必成为生物甲烷产业的一个重要发展方向。

由单一的终端用途向联产利用方向发展。沼液沼渣多元化利用，生物质的能源利用率和资源利用率更高，污染物排放更少，过程更清洁。如沼液用于各种农作物、果园、饲料地的优质液态有机肥料或无土栽培营养液；沼渣用于生产固态有机复合肥或蔬菜大棚土壤修复剂。

5.3 产业化前景

随着生物甲烷产业市场的日趋成熟，近两年国内将建成日产 2×10^{5} m^3 沼气的生物甲烷项目 2～3 处，日产 1×10^{5} m^3 沼气的生物甲烷项目 3～5 处；日产 3×10^{4} m^3 沼气以上的生物甲烷项目 8～10 处；日产 1×10^{4} m^3 沼气以上的生物甲烷项目 20 余处。这批大型生物甲烷高值利用的标志性示范工程的建成运行，将对提升中国生物甲烷产业的技术水平、自主创新能力和核心竞争力起着积极的推进作用。

预计在未来 3～5 年内，为五个城市 1 000 辆燃气汽车提供生物甲烷，为 500 个新农村提供生物甲烷气，建设 1×10^{6} 亩生态农业示范区。生物甲烷产业的技术提升和稳步发展，对形成工程设计施工、成套化装备制造、沼肥高值利用和技术服务的巨大产业链有重要意义，每年可增加大约 40 000 人就业，促进战略性新兴产业的快速发展。

6 结语

生物甲烷产业的发展不仅对于保护环境、减少温室气体排放有着重要作用，对于改善我国的能源结构也至关重要。由于油价的不断上涨，国内外沼气的利用方式已呈现出从沼气发电向提纯生物甲烷制备车用燃气转变。中国的沼气资源丰富，市场前景广阔，且目前制沼气的相关技术及设备已基本成熟，但沼气纯化生产生物甲烷的技术尚处在起步阶段，因此，沼气纯化技术的研究及设备开发是今后中国生物甲烷产业发展的重点工作。

参考文献

[1] 国家中长期科学和技术发展规划纲要(2006-2020).
[2] 国家可再生能源中长期发展规划. 2007.

STATUS QUO OF BIOGAS PURIFICATION TECHNOLOGY APPLICATION IN CHINA AND ABROAD

ZHAO Ming, LIU Lin

Qingdao Tianren Environment Company Limited

Contact: Mr. ZHAO Ming, Master, Intermediate Engineer, Qingdao Tianren Environment Co., Ltd, 266101 Qingdao, Shandong Province, PR China, Tel.: +86-532-8099-0306, E-Mail: zhaoming@tianren.com

EXECUTIVE SUMMARY

By analyzing the development trend of biogas plants both home and abroad, this paper reaches the conclusion that biogas purification is the main direction of biogas industry development. Meanwhile, this paper introduces the principles of the main purification technologies as well as compares various technical indicators of PSA, Chemical Absorption Method and Pressure Water Scrubbing Method, under consideration of the current development of domestic biogas plants. The overall result of this paper is that the Chemical Absorption Method and PSA should be adopted for biogas purification in China.
Key Words: Biogas; purification; chemical absorption method; pressure swing adsorption method

国内外沼气净化提纯技术应用现状

赵明*,刘林

(青岛天人环境股份有限公司,青岛 266101)

【摘要】本文通过对国内外沼气工程发展趋势进行分析,发现对沼气进行净化提纯是沼气行业发展的主要发展方向。同时文中对主要净化提纯技术,包括变压吸附法(PSA)、化学吸收法、压力水洗法进行了工作原理介绍和各项技术指标对比,结合我国沼气工程发展现状,确定在我国沼气净化提纯技术中应该以化学吸收法和变压吸附法为主要发展方向。
【关键词】沼气;净化提纯;化学吸收法;变压吸附法

1 引言

由于全球化石原料面临枯竭,对生物质燃料的开发和应用已成为各国热点。大型沼气工程的快速发展,也将沼气脱硫脱碳净化作为高附加值生物燃气的需求日益迫切,沼气净化提纯技术得到了不同程度的应用和发展,例如,变压吸附、压力水洗、化学吸收法、膜分离法等一些净化技术在沼气净化提纯技术上被广泛引用,通过净化提纯后的沼气可以取代天然气作为城市管道燃气或车用燃气,目前已经成为沼气技术发展的主要方向。本文结合对国内外沼气净化提纯技术的考察,简单介绍几种目前国际上比较常用的沼气净化提纯技术[1][2]。

* 通讯作者:赵明,男,硕士,中级工程师,青岛天人环境股份有限公司,青岛 266101,zhaoming@tianren.com,0532-80990306

2 沼气发展趋势

随着全球经济的高速增长，能源和环保问题日益突出，在全世界积极推动可再生能源开发和利用的背景下，利用有机废弃物生产沼气的模式已得到广泛推广应用，利用沼气作为生物燃气已经成为今后能源发展趋势，图1为沼气在未来50年内的发展趋势。

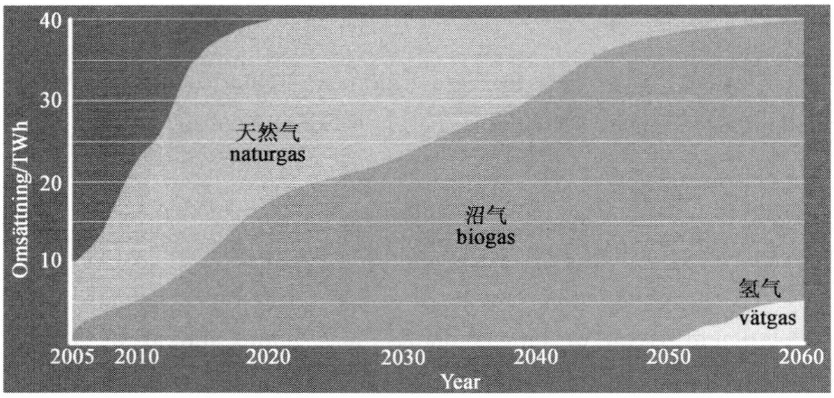

Figure 1 Development Trend of Biogas in the coming 50 Years

图1 沼气在未来50年内的发展趋势

3 沼气净化提纯技术现状

通常沼气工程生产出的生物气为沼气，其含量为：甲烷50%～65%；二氧化碳30%～39%；以及少量的硫化氢、氮气等气体，沼气由于二氧化碳浓度高，在高端应用方面受限，大规模的储存运输并不经济，而德国前些年95%以上的沼气工程生产的沼气均用作热电联产(CHP)，近年来才加快了沼气净化提纯方面的工程建设和应用，近3年就建设了30多家沼气净化提纯燃气工厂。通过对沼气进行净化提纯，去除沼气中的杂质气体(主要是二氧化碳)，将提高净化后的甲烷气体浓度和燃烧热值，即可替代天然气用作城市管道燃气或者用作车用燃气，大大提高沼气工程的经济效益和沼气的深度利用前景。

近年来沼气净化提纯工厂在欧美发达国家发展很快，新建工厂快速增多，并且在处理规模上也日益提高，到2009年全球总共有约100多家沼气净化工厂(从1987年到2009年沼气净化提纯工厂的数量发展过程见图2)。

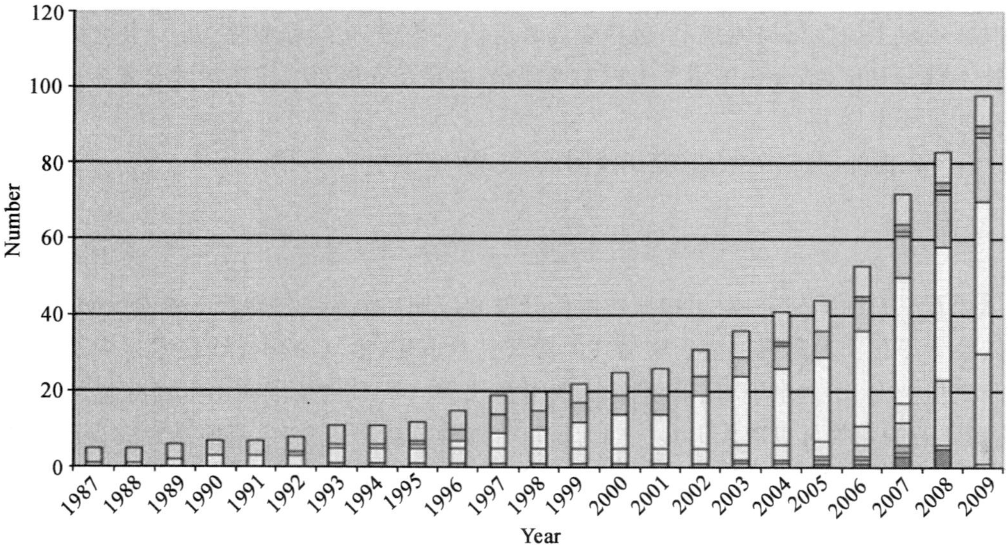

注：摘自 IEA Bioenergy Task 37 - Energy from biogas and landfill gas

Figure 2 Number of Biogas Purification Plants from 1987 to 2009

图2 1987年到2009年沼气净化提纯工厂的数量发展过程

4 沼气净化提纯技术

目前沼气净化提纯工艺有变压吸附、压力水洗、化学吸收法(胺液)、选择分离法、膜分离法等多种方法，其中沼气净化提纯技术发展比较快的国家为德国、加拿大和瑞典，图 3 为欧洲沼气净化使用技术的状况。通过多种方法的使用情况可以看出：变压吸附、压力水洗和化学吸收法这三种方法在沼气净化提纯方面使用的比较多。

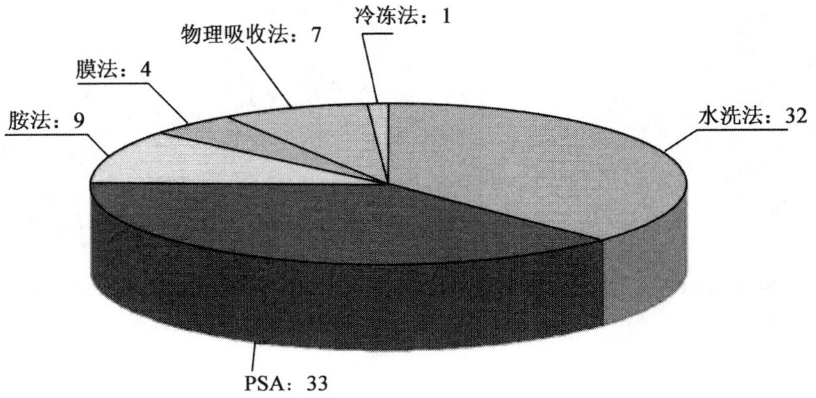

Figure 3 Main Purification Technology Application in Europe

图 3 欧洲沼气净化使用技术的状况

4.1 变压吸附法

变压吸附(Pressure Swing Adsorption 简称 PSA)是在加压条件下，使二氧化碳吸附在活性炭、分子筛等吸附剂表面实现气体分离脱除二氧化碳，图 4 是一种新型气体吸附分离技术，它有如下优点：①产品纯度高；②一般可在室温和不高的压力下工作，床层再生时不用加热，节能经济；③设备简单，操作、维护简便；④连续循环操作，可完全达到自动化；⑤无废水产生，环保性能好[3]。因此，当这种新技术问世后，就受到各国工业界的关注，竞相开发和研究，发展迅速，并日益成熟。图 5 和图 6 分别为加拿大瑞必科公司开发的 PSA 沼气净化提纯设备及瑞典变压吸附沼气净化提纯工程实例。

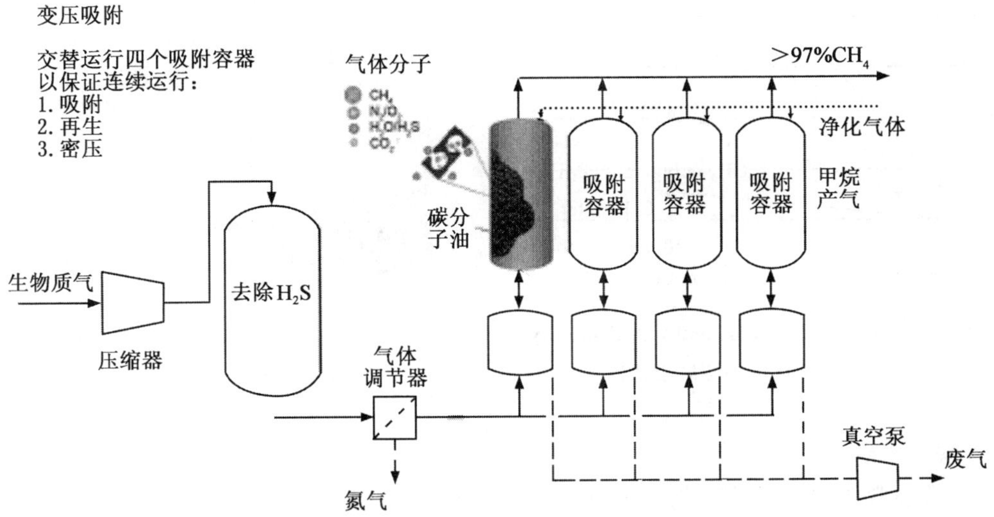

Figure 4 PSA Work Principle Flowchart

图 4 变压吸附法工作流程示意图

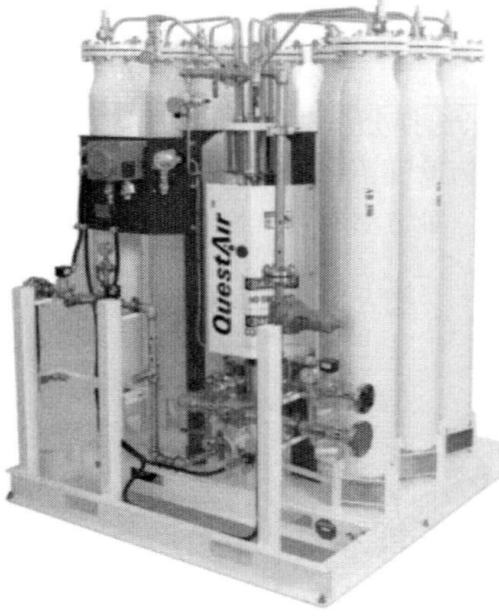

Figure 5 PSA Device

图 5 变压吸附设备

Figure 6 PSA Engineering Project in Sweden

图 6 瑞典变压吸附沼气提纯工程实例

4.2 压力水洗法

该法脱除二氧化碳是物理吸收过程，根据沼气中主要组分甲烷和二氧化碳在水中的溶解度不同而实现气体分离，在整个过程温度变化不大的条件下，甲烷在水中的溶解度随压力变化不大，二氧化碳和硫化氢在较高压力下水中的溶解度大，在较低压力情况下因二氧化碳和硫化氢在水中的溶解度小而释放出来，从而实现对二氧化碳气体分离完成沼气净化提纯，工作原理图见图 7。图 8 为丹麦沼气压力水洗发净化脱碳工程实例。

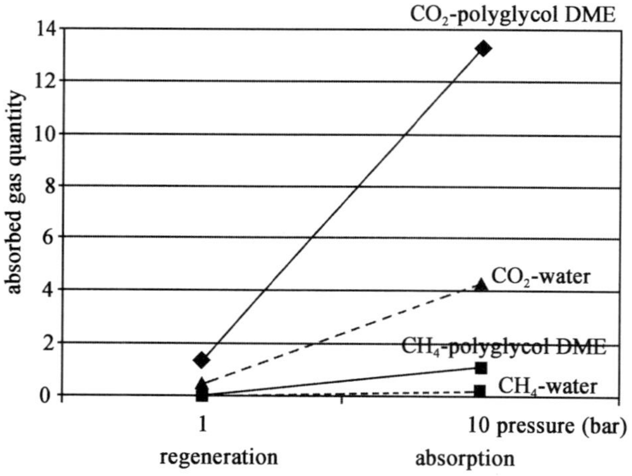

Figure 7 The Effect of Pressure on CH_4 and CO_2 Solubility

图 7 压力对甲烷与二氧化碳的溶解度影响

Figure 8 Pressure Rinsing Project in Denmark

图 8 丹麦压力水洗法工程实例

4.3 化学吸收法

二氧化碳在常温常压下极易溶于化学吸收液（贫液）中形成富液，富液在高温下，二氧化碳又很容易被解析出来，从而实现与二氧化碳的分离达到沼气净化的目的。化学吸收法对二氧化碳和甲烷气体的选择性很强，溶液与二氧化碳等酸性气体反应迅速，但与甲烷气体不发生化学反应。该法具有甲烷气体纯度高、损失率小等特点。工艺流程图如图 9。图 10 和图 11 为化学吸收法净化提纯沼气工程实例。

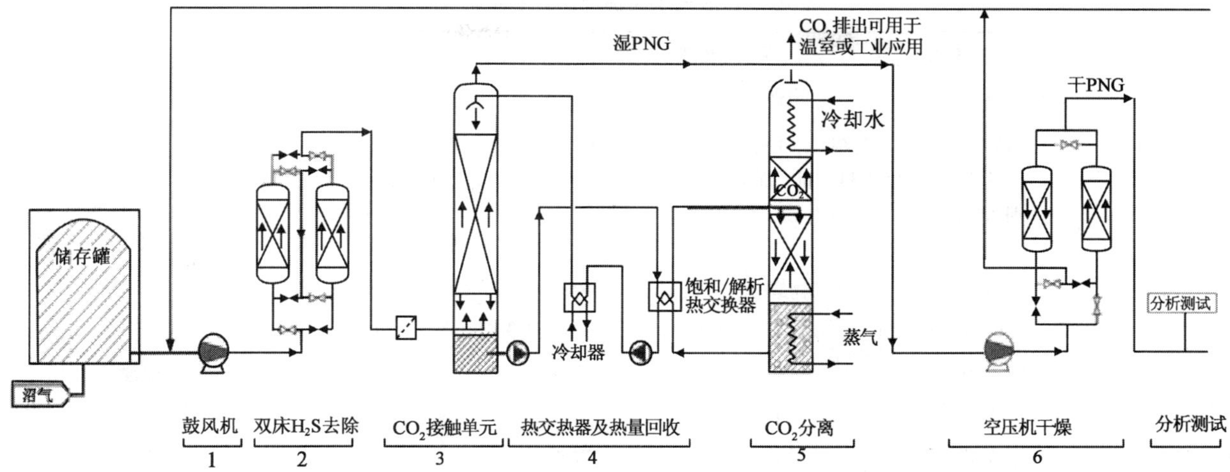

Figure 9　PSA Technology Flowchart

图 9　化学吸附法工艺流程图

Figure 10　Demo Project of Sino-Danish Bio-energy Plant

图 10　安阳中丹生物能源工程实例

Figure 11　Case Project of RATHERNOW in Germany

图 11　德国 RATHERNOW 沼气工厂实例

4.4　膜分离法

膜分离法利用渗透膜对不同气体的选择性能差异，将甲烷和二氧化碳等气体分离，即在 0.36 MPa 下二氧化碳气体不能通过渗透膜，而甲烷能通过，使甲烷的纯度达到 96%左右，实现甲烷和二氧化碳气体的分离[4]。由于膜分离效果不好，分离成本高，因此目前应用不多。

4.5　主要几种技术的工艺比较

表 1 为目前国际上使用较多的几种净化提纯工艺的参数对比，通过工艺参数的比对可以发现，每种工艺都有其优缺点，根据工程现场的具体情况，可以选择满足其条件的净化提纯工艺，通常来说变压吸附、压力水洗和化学吸收法是三种比较成熟和稳定的技术，目前在国外都得到了较为广泛的应用，在国内已开展相关研发应用和探索。

Table 1 Comparison of Processing Parameters of Biogas Purification

表 1 沼气净化提纯工艺参数对比

	变压吸附法	压力水洗法	化学吸收法		
简称	PSA	DWW	MEA 或 DEA	LP Cooab	Genosorb
工作原理	CO_2 在高压下吸附于吸附剂，而 CH_4 则不能。	CO_2 在高压下溶于水，而 CH_4 则不能。	CO_2 在常压下溶于胺液，而 CH_4 则不能。	化学剂 Cooab 在一定的压力温度条件下吸收 CO_2	化学有机溶剂在一定的压力温度的条件下有吸收 CO_2，H_2S 的功能
工作方式	物理法	物理法	物理＋化学法	物理＋化学法	物理＋化学法
工作压力	0.6～0.8 MPa	0.8～1.2 MPa	0～0.1 MPa	0.6～0.8 MPa	0.6～0.8 MPa
工作介质	吸附剂	水	胺液	化学剂 Cooab	化学有机溶剂
工作热能	不需要	不需要	大量需要(160℃)	大量需要	需要(55～88℃)
化学吸收剂	不需要	不需要	需要	需要	需要
工作用水	不需要	大量需要	需要	不需要	不需要
CH_4 含量	＞96％	＞97％	＞99％	＞99％	＞93％
CH_4 损失率	3％～8％	0.05％～6％	小于 0.1％	小于 0.1％	9％～18％
技术设备系统	复杂	复杂	相对简单	相对简单	复杂
自身耗电量 kWh/m^3 Untreated gas	0.24～0.29	0.22～0.4	0.06～0.14	0.15	0.511
自身耗热量 kWh/m^3 Untreated gas	0.1	0.1	0.3～0.7	0.58	0.07
设备自身耗能	低	低	低	低	高
沼气提纯规模	大型	大型	中型	小型	大型
设备一次性投资	高	中等	较低	较低	高
系统运行费用	高	中等	较低	较低	高
系统运行可靠性	高	高	较高	较高	高
实际工程经验	许多	许多	较多	较少	较少

主要几种技术比对分析：

(1)从提纯后甲烷的纯度对比分析可以看出：化学吸收法比变压吸附法和压力水洗法提纯后的甲烷纯度高。

(2)从甲烷损失率对比分析可以得出：变压吸附法以及 Genosorb 法对甲烷的损失率较大，压力水洗法以及化学吸收法(MEA 和 LP Cooab)甲烷损失率较小。

(3)从系统运行可靠性对比分析得出：变压吸附法和压力水洗法运行可靠性高于化学吸收法。

(4)从适合沼气提纯规模分析：变压吸附法和压力水洗法适合于大型沼气工程，而化学吸收法适合于中小型沼气工程。

5 结束语

沼气净化脱碳技术已经发展了十几年，在世界上已有 100 多处工程应用实例，主要集中在德国、丹麦、瑞典、加拿大这些国家，形成了一系列非常成熟的技术。同时，国内也开始了沼气净化提纯技术的研究与应用，其中变压吸附法和化学吸收法开始应用于实际工程中，例如安阳中丹生物质能源公司大型车用沼气工程选择的就是化学吸收法，实现了沼气净化提纯后作为车用燃气使用。因此，通过对各项工艺技术指标的对比以及我国沼气工程发展现状分析，沼气净化提纯技术主要以化学吸收法和变压吸附法为主要发展方向。

参考文献

[1] 倪慎军. 政策支持 法律保障 德国沼气工程发电成气候——关于德国沼气发电技术应用的考察报告[J]. 河南农业，2005(2).
[2] 史立山. 瑞典、丹麦、德国和意大利生物质能开发利用考察报告[J]. 阳光能源，2005(12)：53-55.
[3] 叶恩东，李贵华，王丽娟. 浅析工业尾气中二氧化碳的吸收及综合利用[J]. 攀枝花科技与信息，2001(2)：30-31.
[4] 沈光林，李新培. 膜法分离二氧化碳/甲烷的理论研究[J]. 膜科学与技术，1994(12)：29-33.

EMISSIONS FROM BIOGAS PURIFICATION TECHNOLOGIES IN COMPARISION TO OTHER BIOGAS USES

J. Clemens

gewitra (Ingenieurgesellschaft für Wissenstransfer)

Contact: Dr. Joachim Clemens, Managing Director, gewitra GmbH, Mülheimer Straße 26, Bld. 159, 53840 Troisdorf, Germany, Tel.: +49-228-266-7882, Fax: +49-3221-10899-9721, E-Mail: clemens@gewitra.de, www.gewitra.de

EXECUTIVE SUMMARY

Biogas produced in biogas plants can be used on site or can be transported either as biogas or purified 'Biomethane'. In any case, emissions during treatment and use of the gas occur. For electricity production on site, emissions via stack gas are most relevant. If gas is purified, the different gas purification technologies may emit up to 5% of the produced methane. A gas treatment - usually some kind of thermal oxidation- can be used to minimize methane emissions.

Key words: Bio-methane; purification; CH_4 emissions; leakage

1 Introduction

Biogas plants produce biogas from organic waste and residues. The gas is used in different ways such as

- Electricity and/or heat production e.g. in a combined heat and power generators (CHP);
- Biogas in micro gas grids for direct use in households;
- Biomethane after purification (upgrading to >95% CH_4).

Biogas technology is regarded as a way to produce renewable and environmentally friendly energy. However during the fermentation process and the use of biogas methane can be emitted into the atmosphere. Methane is an effective greenhouse gas with a carbon dioxide equivalent of 25.

For example, if the biogas process emits more than 575 g CO_{2eq}/kWh (23 g CH_4/kWh), the production process is 'as bad' as the overall electricity mix of Germany that includes all different kinds of energy sources, especially coal (*Figure 1*, *Figure 2*).

2 Leakages from the biogas process

Under normal operation emissions from digesters and the storage of fermented digestate occur. Recently the CH_4 emissions from biogas plants fed with agricultural wastes and energy crops were determined. The emissions were in a range of 0.2%-14% of the produced CH_4 with a mean of 3.9% (Liebetrau et al. 2010). Emissions occurred from the non covered storage from digested material, from wholes in the gas storage foils and at the connection between fermenter wall and gas storage.

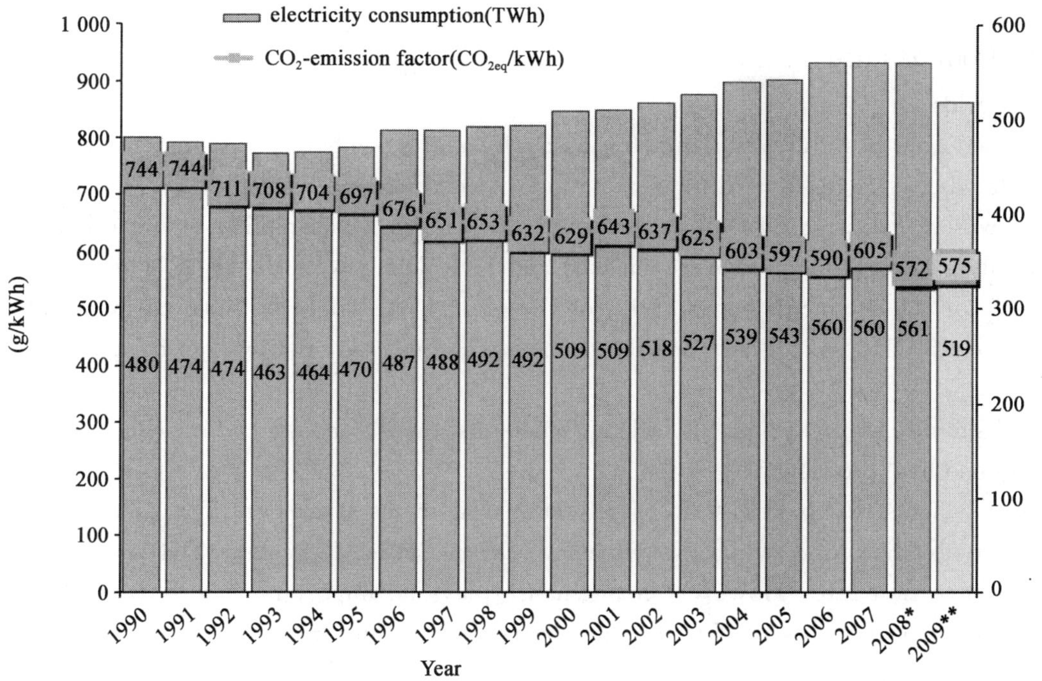

Figure 1 Electricity consumption and ist CO_2 emission factor in Germany (Umweltbundesamt, 2009)

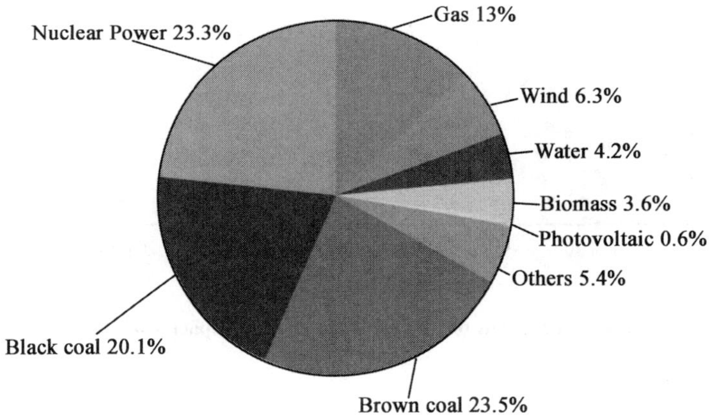

Figure 2 Energy sources for electricity production in Germany (http: // de. wikipedia. org/wiki/ Datei: Strommix-D-2008. png)

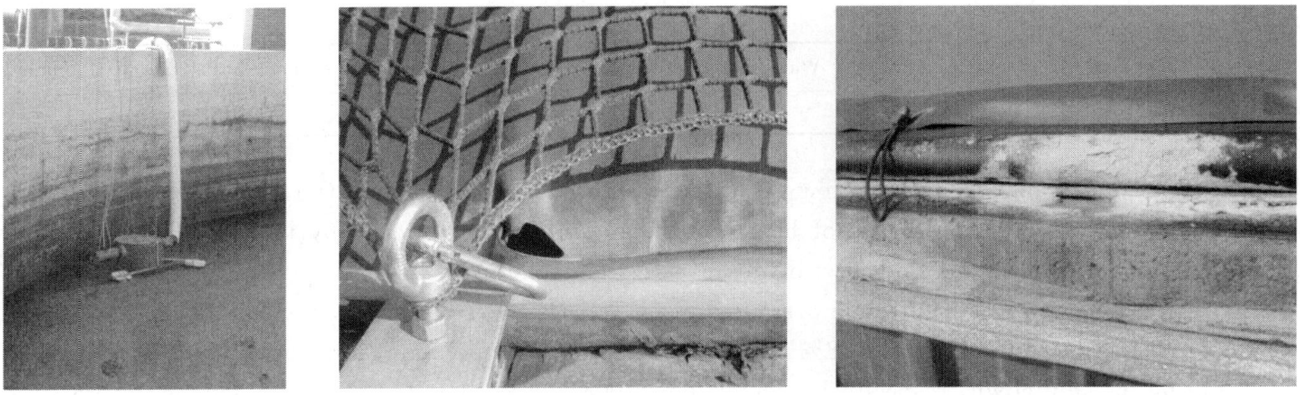

Figure 3 Analysis of CH_4 emissions from digestate, whole in a gas storage foil, leakage between digester wall and foil

3 Leakages from gas use

3.1 Micro gas grids

Biogas can be used after desulfurization directly in small gas grids. The gas is stored and then supplied to e. g. households where the gas is used for heating and/or cooking.

Such a system was investigated by Cuhls et al. , 2010 in China. 245 sampling points (flanges, screws, manometers, valves, grease nipples) were checked and 13 leakages were detected. The leakages emitted more than 0. 6% of the produced biogas. Nine of the 13 leakages could be repaired immediately by the operator. The overall CH_4 leakage was only 0. 34%. In another case, the leakage from the gas grid was more than 3%.

Micro gas grids should be checked on gas leakages and subsequently maintained in regular intervals.

3.2 Electricity and heat production

CHPs and other systems that oxidize CH_4 usually emit CH_4 in the stack gas. The share of emitted CH_4 depends on the combustion process and the efficiency of e. g. CHPs. In general, CHPs emit more gas at a higher electric conversion rate. However, maintenance plays an important role as well. From ongoing analysis on German biogas plants the CH_4 leakages were analyzed in a range from almost 0- >10 g CH_4/kWh (*Figure 4*).

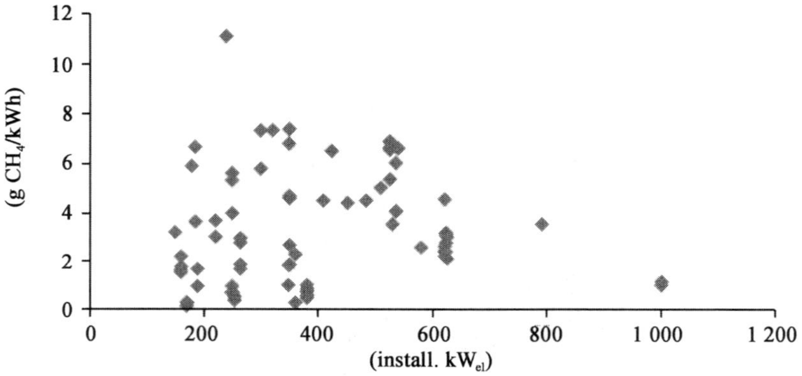

Figure 4 CH_4 in the stack gas dependent on the installed electric capacity of the CHP (unpublished data)

3.3 Biogas purification

There are different technologies to purify biogas. According to the information of the manufacturers, the CH_4 loss is between 0. 1%-5%.

Table 1 CH_4 leakage from different gas purification technologies according to the information of the producers (Urban et al. 2008)

	Pressure Swing Absorption	Water Scrubber	Genosorb, Seloxol	Mono ethanol amine	Diethyl amine
Leakage (%)	2-5	<1	2-4	<0. 1	<0. 1

Independent studies on CH_4 loss are few: One study showed that the current German threshold of 0. 5% CH_4 loss (GasNzV, 2010) because of biogas purification was not matched by the investigated plants, however amine scrubbers were not part of the investigation (DVGW).

In another study the emissions from a pressure swing Absorber were determined. The overall losses were 2. 6% of the produced CH_4. About 88% occurred during normal operation, 2% during start up and about 10% during the shut down, respectively (Baum et al. 2008). According to own investigations at a PSA the emissions were between 2%-6% of the produced CH_4.

For PSA, the emissions could be reduced by longer cycles and more absorbers installed in a row. Another possibility for all purification units may be the oxidation of emitted CH_4 (burning). At rather high concentrations of CH_4 (PSA), the exhaust gas may be burnt without additional energy input (Flux Flare). These technologies require 6%-7% CH_4 in the exhaust gas. At lower CH_4 concentrations, the exhaust air treatment requires additional energy. In such a case, Biogas may be added to t he exhaust gas and then Flux Flares or regenerative thermal oxidation can be used to oxidize the CH_4 to CO_2.

Currently a threshold for CH_4 emissions from biogas upgrading facilities is discussed. In Germany, the CH_4 emissions have to be lower than 0.2% of the produced CH_4 from 2012 on (currently the threshold is 0.5%, GasNzV, 2010). In Switzerland the threshold is 5% of the produced CH_4. A threshold should consider all different kinds of biogas use. According to own investigations in Germany, biogas plants emit around 3.9% of the produced CH_4 without considering the biogas use. One major emission source is the open storage of digestate. The emissions from biogas use are up to 3% for micro gas grids (Cuhls et al., 2010), up to 4% from CHPs and up to 5%-6% from biogas purification. As purification plants treat usually high amounts of biogas the impact of CH_4 emissions from purification units may be high. However, thousands of CHPs in Germany also contribute to the CH_4 emissions. If the low German thresholds are in place, the green house gas emissions from biogas purification plants are significantly lower as compared to the other biogas uses. But for smaller biogas plants a biogas purification is not realistic as costs are too high.

4 Conclusions

Every use of biogas goes along with CH_4 emissions. Independent from the technology, it is important to maintain the biogas usage devices (check-repair-check philosophy) to avoid higher CH_4 emissions.

The CH_4 leakage has to be minimized because of its environmental impact as green house gas. For unavoidable emissions, a post treatment is required if CH_4 emissions are too high. An overall threshold for the emissions from biogas use may be not realistic as the level of technology of the biogas facilities varies. For bigger units-such as biogas purification- a lower threshold may be useful.

References

Cuhls C., Clemens J., XU Pan, GUO Jianbin, DONG Renjie, (2010): *Monitoring and reduction of methane emissions from biogas plants*. Biogas Engineering and Application, Eds: DONG Renjie, Raninger B., Volume 1, pp. 318-322.

Liebetrau J, Clemens J, Cuhls C, Hafermann C., Friehe J, Weiland P., Daniel-Gromke J, (2010): *Methane emissions from biogas-producing facilities within the agricultural sector*. Engineering in Life Sciences, 10(6), pp. 595-599.

Urban W., Lohmann H., Girod K, (2009): *Beseitigung technischer, rechtlicher und ökonomischer Hemmnisse bei der Einspeisung biogener Gase in das Erdgasnetz zur Reduzierung klimarelevanter Emissionen durch Aufbau und Anwendung einer georeferenzierten Datenbank*. Band 4, Oberhausen, http://www.biogaseinspeisung.de/download/2008_UMSICHT_Technologien_und_Kosten_der_Biogasaufbereitung_und_Einspeisung_in_das_Erdgasnetz.pdf.

Baum S. Baier U., Johannes Judex J., Biollaz S., (2008): *Methanverluste bei der Biogasaufbereitung*. Studie im Auftrag des Bundesamtes für Energie (BFE), Wädenswil, November 2008, ZHAW Zürcher Hochschule für Angewandte Wissenschaften, IBT Institut für Biotechnologie, Fachstelle Umweltbiotechnologie, http://www.umweltbiotech.zhaw.ch/fileadmin/user_upload/life_sciences/_Institute_und_Zentren/ubt/Dateien/Methanverluste_bei_der_Biogasaufbereitung.pdf.

DVGW. DVGW Technologie-Report Nr. 1/10, http://www.dvgw.de/angebote-leistungen/forschung/technologie-report/aufbereitung/ausgabe-110/.

Gasnetzzugangsverordnung vom., (2010): *Verordnung über den Zugang zu Gasversorgungsnetzen* (*Gasnetzzugangsverordnung - GasNZV*).

沼气提纯与沼气其他利用途径的甲烷排放比较

J. Clemens

(*gewitra* 公司,波恩 53840)

【摘要】沼气工程生产的沼气可供工程点直接使用也可以以沼气或甲烷的形式输送到工程点外使用。无论使用哪种方式,沼气在净化和使用过程中都会有甲烷泄漏。当沼气用于工程点内发电时,最主要甲烷泄漏的来源是尾气。沼气提纯时,不同的提纯技术使甲烷最高排放量可达到总产生量的 5%。通常可用热氧化气体处理技术来降低甲烷排放。

【关键词】生物甲烷;提纯;甲烷排放;泄漏

1 引言

沼气工程利用有机废弃物生产的沼气可以用于以下用途:

- 发电或加热,例如利用热电联产技术(CHP);
- 沼气进入小型管网,直接供家庭使用;
- 提纯后生产生物甲烷(甲烷浓度大于 95%)。

沼气技术是生产可再生能源和环境友好能源的途径之一。然而,在厌氧发酵和沼气使用过程中,甲烷可能会被排放到大气当中,且甲烷的温室气体效应相当于同体积二氧化碳的 25 倍。

例如沼气生产利用过程中排放超过 575 g CO_2 当量/kWh(23 g CH_4/kWh),则其排放水平和德国利用其他常规能源(比如煤炭)生产电能的温室气体排放水平一样"糟糕"。(图 1,图 2)。

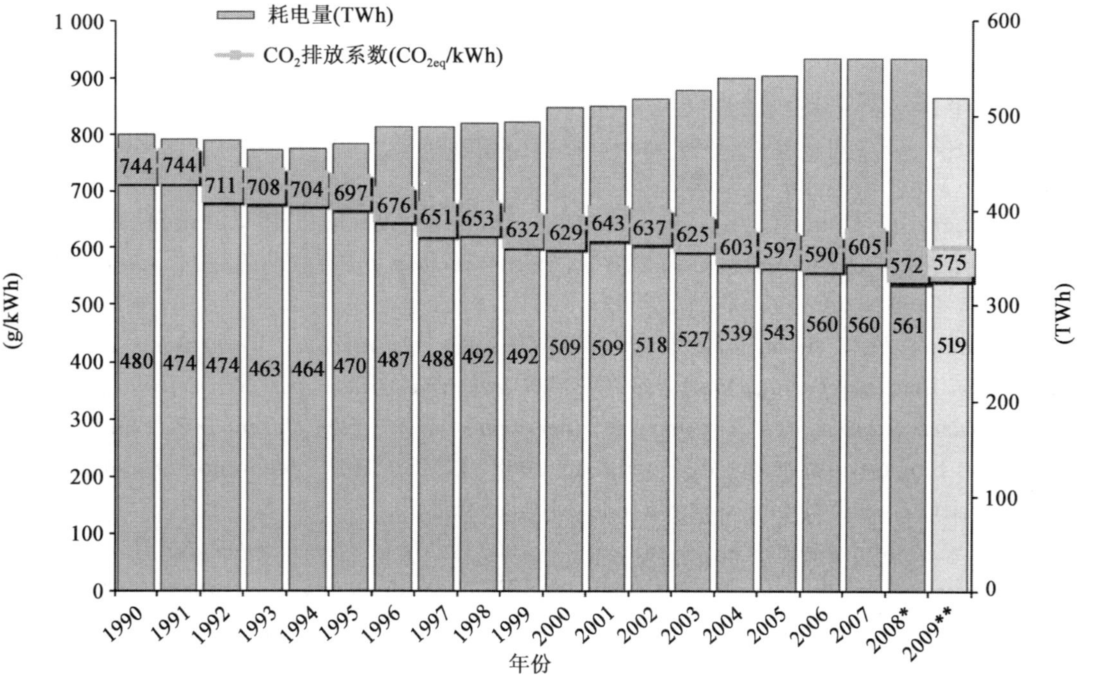

图 1 德国电力消耗和 CO_2 排放系数 (Umweltbundesamt, 2009)

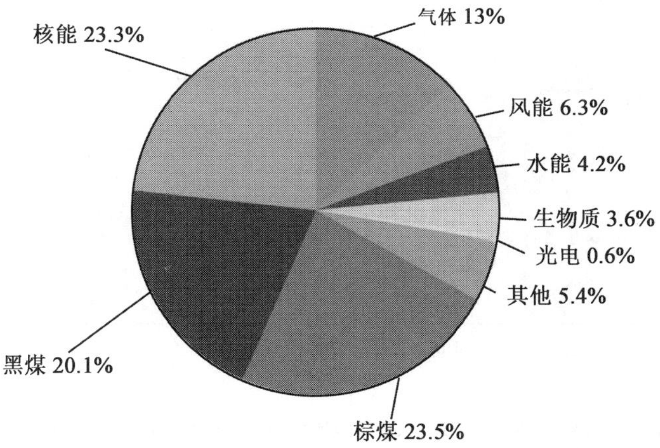

图 2　德国用于电力生产的能源（http://de.wikipedia.org/wiki/Datei:Strommix-D-2008.png）

2　沼气生产过程中的泄漏

沼气工程正常运行时，甲烷可能从发酵罐和沼液沼渣储罐内排出。最近 Liebetrau 等人测定了以农业废弃物和能源植物为原料的沼气工程甲烷泄漏量，其比例为所生产甲烷的 0.2%～14%，平均为 3.9%(Liebetrau et al. 2010)。甲烷泄漏发生在未密封的沼渣沼液储存装置、沼气储膜的破洞以及发酵罐与储气膜的连接处。

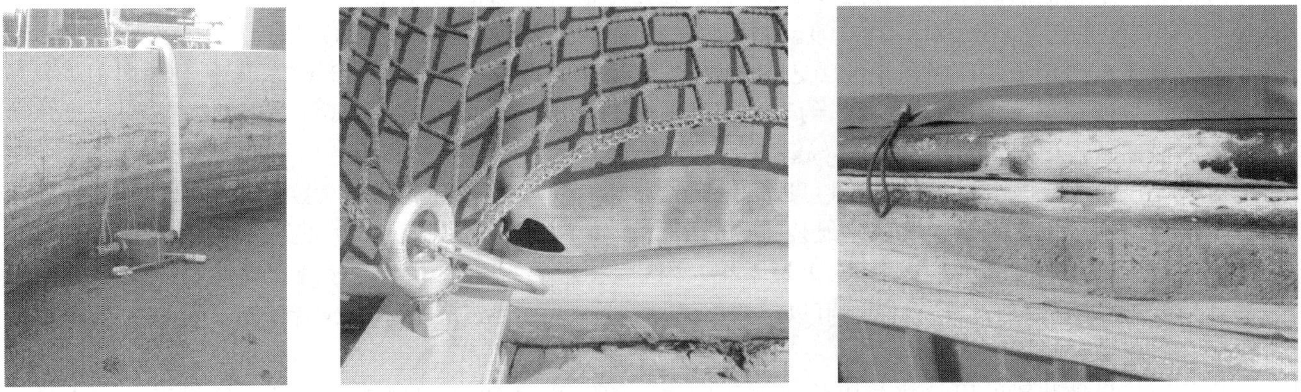

图 3　沼渣沼液储存装置，储气膜的破洞以及发酵罐与储气膜的连接处发生的甲烷泄漏分析

3　沼气利用过程中的泄漏

3.1　小型沼气管网

沼气在脱硫后可直接进入小型沼气管网，储存起来后直接供家庭用于加热或做饭。

Cuhls 等人于 2010 年在中国调查了一个小型沼气管网，分别检查了 245 个采样点(包括法兰、螺口、压力表、阀门、油脂嘴)，发现了 13 处泄漏，其泄漏量超过了沼气产量的 0.6%。这 13 处漏点中有 9 处可由操作工人立即修复，修复之后甲烷泄漏量减少到 0.34%。在另外一处小型沼气管网系统中，管网中泄漏量甚至超过了 3%。

所以，小型沼气管网应当定期检查泄漏并进行维护。

3.2　热电联产

热电联产(CHP)技术和其他燃烧甲烷的系统通常在烟气中会排放甲烷，排放比例取决于燃烧工艺和热电联产的效率。通常甲烷排放量会随着热电联产发电效率的增加而增加，但系统维护也起到重要作用。从现在正在进行的德国沼气工程甲烷泄漏来看，其范围为 0～>10 g CH_4/kWh(图 4)。

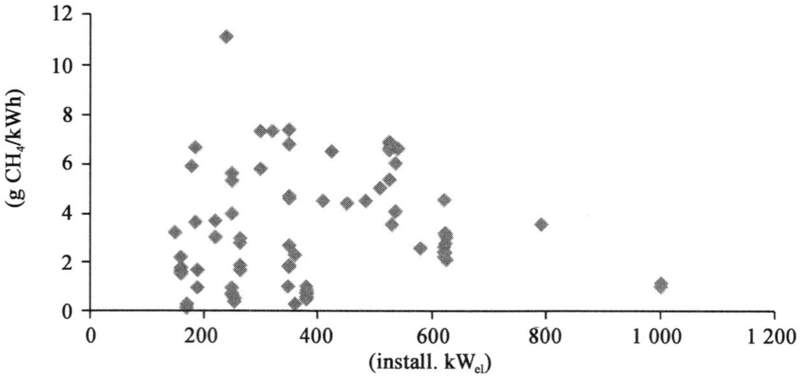

图 4　热电联产装机容量与烟气中甲烷排放量的关系(未发表数据)

3.3　沼气提纯

沼气提纯有多种技术,按照供应商提供的信息,其甲烷损失量为 0.1%~5%。

表 1　供应商提供的不同沼气提纯技术甲烷泄漏量(Urban et al. 2008)

	变压吸附	水洗	乙二醇二甲醚 Selexol 法	乙醇胺	二乙胺
泄漏量(%)	2~5	<1	2~4	<0.1	<0.1

目前关于甲烷泄漏的专门研究还较少:一项研究显示被调查的沼气提纯厂不能达到目前德国甲烷排放总量 0.5%以下的要求(GasNzV, 2010)。而且胺洗提纯法还没有在这次调查的范围内(DVGW)。

另一项研究测定了变压吸附提纯工艺的排放量:甲烷泄漏量占沼气产量的 2.6%,其中 88%的损失发生在正常运行阶段,2%发生在启动阶段,约 10%发生在停机阶段(Baum et al., 2008)。根据调查显示,变压吸附提纯的泄漏量占甲烷产量的 2%~6%。

对于变压吸附工艺,减少其泄漏量可以通过延长循环时间和增加吸附塔串联数量来实现。另一种适用于减少沼气提纯中甲烷排放的方法是燃烧排放的甲烷。如果排放气体中甲烷含量高(如变压吸附),可以直接燃烧尾气而不需要额外的能源输入(火炬系统)。这些技术要求尾气中甲烷含量达到 6%~7%。如果甲烷浓度较低,则需要额外的能耗,这时可以把沼气添加到尾气中,通过火炬系统或者蓄热式热氧化器把甲烷氧化成二氧化碳。

目前沼气提纯装置的甲烷排放限值正在研究中。德国从 2012 年起甲烷泄漏量必须低于总产生量的 0.2%(目前是 0.5% ,GasNzV,2010),瑞士的甲烷排放量不得超过甲烷生产总量的 5%。甲烷泄漏量限值的确定应考虑不同的沼气利用方式。按照在德国的调查,在不考虑利用方式的情况下,沼气工程平均有 3.9%的甲烷泄漏,其主要来源是未封闭的沼液沼渣储存设施。小型沼气管网的甲烷泄漏约为 3%(Cuhls et al., 2010),用于热电联产时约为 4%,而用于沼气提纯时则高达 5%~6%。由于沼气提纯厂处理的沼气量通常较大,甲烷泄漏总量也可能也更高。德国数千个热电联产厂也产生了大量的甲烷气体排放。如果德国降低甲烷泄漏实行标准,沼气提纯厂的甲烷泄漏量将大大低于沼气的其他用途。但对于小规模沼气厂而言,提纯成本过高,采用这种处理方法是不现实的。

4　结论

每种沼气利用方法都会造成甲烷泄漏。不管采用哪种技术,维护沼气利用装置是十分重要的(检查—维修—检查的理念),用以避免更高的甲烷排放。由于其对温室效应的重大影响,甲烷泄漏应该尽力避免,对于不可避免的甲烷泄漏,应增加后处理单元。由于各个沼气工程的技术和水平不同,采用统一的甲烷排放限值并不现实,但对于大型沼气工程而言,如沼气提纯厂,设定一个较低的甲烷排放值是可行的。

德国温室气体减排灯塔项目:高碑店 MBT 处理场

F. Kölsch[1], K. Fricke[1], B. Huang[2], M. Ginter[2]

(1. *Braunschweig 工业大学*, *Braunschweig 38106*, *德国*; 2. *AWN Umwelt GmbH*, *Buchen 74722 德国*)

【摘要】 2006 年,布伦瑞克工业大学废弃物资源管理学院为中国河北省高碑店市设计了市政固体废弃物(MSW)的生物处理厂。一家中德合资公司承建了该项目,德国复兴信贷(KFW)发展银行提供了经济支持。该项目的目标是减少垃圾填埋场有机物降解过程中产生的甲烷排放量。填埋场产生的甲烷温室效应比二氧化碳强 21 倍。通过生物处理后废弃物有机成分比较稳定,从而明显减少了最终处理后的降解。咨询方建议设计处理规模为 40 000 t/a 的机械生物处理厂(MBT),并将处理厂的产物用做甲烷氧化层。在可行性研究时对系统设计进行了调查,在中试研究时对废弃物进行了研究和处理。除了技术研究以外,也探讨了该项目的社会效益和经济效益两方面。

2009 年 6 月,德国联邦环境部批准了项目开发方德国 AWN Umwelt GmbH (Buchen) 公司的资金要求,并将此项目作为德国政府国际气候行动资助的温室气体减排灯塔项目。Pyöry 环境公司为该项目提供设计,并于 2009 年 7 月开始建设施工,预计 2010 年 4 月投入运行。该项目减排量可达 8~25 t CO_2 当量/a,但由于项目是由碳交易费用资助,故不能申请减排认证。

MBT 垃圾处理工程包括机械处理和生物处理两个步骤。机械处理的目的是为了分离出价值高的物品(纸,合成物),同时可以增加有机成分。废弃物被分为粗、中、细三部分,把其中的中、粗两部分运到手工分拣站。根据废弃物分析可以预测分拣后的中等部分主要由较纯的有机物组成,可作为堆肥。细部分在最终处理之前将经过生物处理,之后也可以用作填埋场覆土层改良剂,即作为甲烷氧化层。粗部可以被压碎重新回到进料口,也可以直接运到填埋场。

生物处理步骤采用好氧、主动通风稳定工艺,其处理场地部分露天,部分有顶棚。三角形的料堆会经常翻转,每年的处理能力为 40 000 t。如果采用两班制,处理能力会大幅增加。生物处理后将产生稳定的生物质,预计其呼吸率低于 5 mg O_2/kg,能够满足德国非活性填埋场的入场标准。如果可能,这些稳定生物质还可以作为甲烷氧化层(特殊的填埋场覆土),土壤改良剂或是农用肥料。处理厂的经济效益可以通过出售稳定生物质或者相关的 CDM 项目(填埋场覆土)收益来增加。为了生产高质量的稳定生物质,废弃物进料需进行筛选和控制,以减少污染物和有害物质掺入有机废弃物中。另外还可以采用收集清洁有机废弃物的方法。在可行性调查阶段发现,当地居民在价格合适时会向处理厂提供有机废弃物,费用约为 6 €/t。

【关键词】 机械生物处理;甲烷氧化

GERMAN GHG MITIGATION LIGHTHOUSE PROJECT MBT GAOBEIDIAN*

F. Kölsch[1], K. Fricke[1], B. Huang[2], M. Ginter[2]

1. *Technical University Braunschweig*; 2. *AWN Umwelt GmbH*

Contact: Dr. Florian Kölsch, Technical University Braunschweig, Beethovenstrasse 51 a, Braunschweig 38106, Germany, Tel.: +49-531-391-3969, E-mail: F.Koelsch@tu-bs.de;
Dr. Mathias Ginter, AWN Umwelt GmbH, Sansenhecken 1, Buchen 74722, Germany

EXECUTIVE SUMMARY

In 2006, the Technical University Braunschweig (Dept. Waste and Resources Management) designed a

* This paper was initially presented at the Orbit 2009 CHINA. 本文已发表于 Orbit 2009 CHINA。

biological treatment plant for municipal solid waste (MSW) for the City of Gaobeidian (Hebei province/China). A German-Chinese joint venture had contracted the project; KFW development bank was providing financial assistance. The project identified mitigation measures reducing the methane emissions from decay of organic waste components at landfills. The greenhouse effect of methane, which is generated at landfills, is 21 times higher than the effect of CO_2. Due to biological treatment, organic components are being stabilized, which reduces decay after final disposal significantly. The consultants proposed a MBT plant designed for a capacity of 40 000 t/a. The MBT output should be utilized as a methane oxidation layer. This facility design was investigated in a feasibility study. The residual waste had been analyzed and treated in a pilot plant. In addition to the technical appreciation, the project's social and financial components were considered.

In June 2009, the German Federal Ministry of Environment granted a fund to the project developer, German company AWN Umwelt GmbH (Buchen) to establish the facility as a lighthouse project for GHG mitigation funded by the government's international climate initiative. Pyöry environment delivered the engineering. In July 2009 the construction work launched and the facility is expected to get in operation in April 2010. Emission reductions of approximately 8 000-25 000 t CO_{2eq}/a are expected, but will not be certified since the project is financed from carbon trade overheads.

The MBT plant consists of a mechanical and a biological treatment step. The mechanical treatment aims on segregating of valuable goods (paper, synthetics) and on enrichment of the organic fraction. The waste stream is split into three fractions (fine, medium, coarse). Middle and coarse fraction are being directed to hand assorting stations. Based on the waste analysis it is expected, that the medium fraction after assorting consists mainly of organic, which are slightly clean. This portion may be converted into compost. The fine fraction will undergo biological treatment as well prior to final disposal. This fraction may be also useful as amendment for landfill cover, where it acts as methane oxidation layer. The coarse fraction will be either crushed and fed to the waste stream again or directly disposed at the landfill site.

The biological treatment will be carried out as an aerobic, actively ventilated stabilization. The process will be carried out partly under roof and partly in open air areas. The triangular windrow heaps will be frequently turned. The annual capacity of the plant amounts to 40 000 t and can be significantly increased by two shift operation. The biological treatment will generate a stabilized biomass with an expected respiration activity less than 5 mg O_2/kg. Thus, the material meets the German standards for disposal at non reactive landfill sites. The stabilized biomass (SB) can also be utilized for methane oxidation layer (special landfill cover), as soil amendment, or as fertilizer in agriculture, if possible. The economic benefits of the plant can be increased due to additional revenues from selling the SB and from related CDM-projects (landfill cover). To produce a high quality SB the waste flux management needs to be refined and permanently controlled in order to minimize pollutants and harming matters in the organic waste fraction. Additional measures can be implemented such as acquisition programs for clean biowaste. During the feasibility study it was found that residents drop biowaste at the facility when reimbursed properly. The costs for acquiring biowaste amount to 6 €/t.

Key words: Mechanical biological treatment; methane oxidation.

1 Introduction

1.1 Background

Experts estimate that 8% to 12% of greenhouse gas (GHG) emissions in developing and threshold countries are due to waste management activities. The major sources are methane emissions from disposal of untreated municipal solid waste, which in those countries contains a large portion of degradable organics. Biological waste treatment is an effective technology to stabilize organics and to minimize decay processes in landfills. The MBT plant in Gaobeidian (Hebei province, PR China) is a pilot waste management project to demonstrate this technology in China.

The project is administered by AWN Umwelt GmbH, a German waste management company partly public owned by AWN (Abfallwirtschaftsgesellschaft Neckar-Odenwald-Kreis) based in Buchen. AWN Umwelt has already established a joint venture for the construction of a sewage treatment plant in Gaobeidian The project was recently finalized and started operation. In the summer 2006 AWN Umwelt and the Municipality of Gaobeidian decided to prepare a similar project for the improvement of the waste disposal in Gaobeidian. The treatment facility is under construction and will open in April 2010.

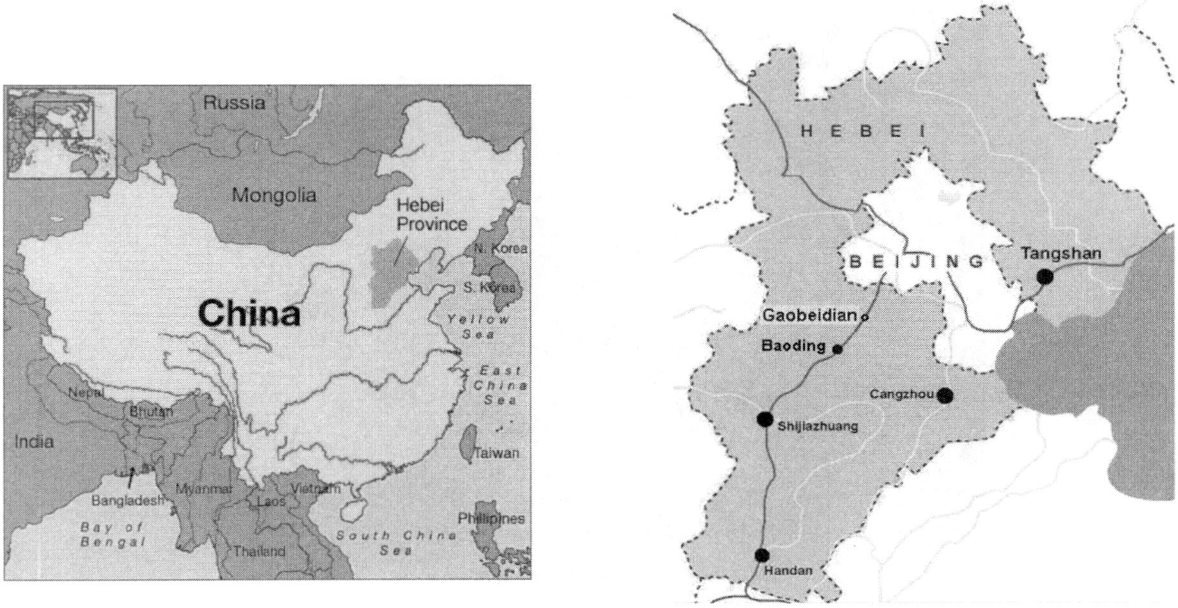

Figure 1 Location of Gaobeidian

1.2 Objectives

The project aims on reducing emissions from waste disposal activities, particularly emissions of methane to reduce the GHG release to the atmosphere by means of biological treatment. With an annual treatment capacity of 40 000 t municipal solid waste the total emission reduction is expected to amount to 80 000-250 000 t CO_2 equivalents during a 10 years period.

2 Methodology

In a first step the technical, economic and legal conditions were investigated in a feasibility study funded by the PPP facility of the KFW development bank, an endowment fund for the support of public private partnership projects. The technical and organizational execution of the study was assigned to the department of Waste and Resource Management of the Technical University Braunschweig. The study was prepared between September 2006 and February 2008. Based on the study AWN Umwelt applied for funding from the International Climate Initiative (IKI) of the German Ministry of Environment, Nature Conservation and Nuclear Safety. In June 2009 the German Government approved the grant. The implementation of the project was launched in July 2009. Currently, the approval process is carried out and the detailed engineering design is prepared. Site clearance will start soon. The construction is expected to be finalized by April 2010.

3 Technical investigations

During the feasibility study several technical investigations were carried out by experts from Technical University Braunschweig, AWN Umwelt and Pöyry Environment Witzenhausen. Data on waste generation and waste composition were examined. The department of waste management of Gaobeidian collects annually approximately 50 000-60 000 t municipal solid waste. The collection is carried out by different collection systems, predominantly by underground drop-off containers and multi-chamber containers that are located at the roadside. The waste disposal site is located near to the city in an exploited clay pit. The disposal site ha no technical barriers and is operated on the poorest technical level as a dumpsite. Dozens of waste pickers are irregularly active on the site (*Figure 2*).

Figure 2 Waste pickers

The composition of the waste was determined by means of hand assorting a 250 kg waste sample. The sample was screened by 40 mm and 10 mm sieves. The sieve overflows were sorted by hand into groups of materials. The material that passed the 10 mm filter has been analysed in the laboratory. *Table 1* shows the waste composition distinguished by material groups, which constitutes approximately 55% of the waste weight.

Table 1 Waste composition

Material group	Portion (weight%)
Organic	28
Plastic foils	8.5
Coal	8.5
Stones	5
Paper	3
Textiles	1.5
<10 mm	45.5

Plastic bottles, metals, rag paper, glass bottles and wood were not found. Regarding to a biological waste treatment there is a particular interest for the portion of the organic group as well as possible disturbing materials. In the fraction < 10 mm the organic fraction is 16%, measured as loss on ignition. Thus the entire organic portion amounts to approximately 35% weight, significantly less than the common value for developing countries (usually: 50%-70% weight). Disturbing materials were not found in the waste. The high ash portion, which originates from burned out coal elements from simple cook places, showed up to be slight in pollutants.

The suitability of the potential biological treatment procedures under the given boundary conditions (local situation, climate, waste composition) was examined in three different pilot plants:

- Passively aerated heaps (so called chimney effect system);
- Actively aerated windrow heaps for MSW;
- Actively aerated composting windrows for biowaste.

Figure 3 Actively aerated windrow for MSW

The results of the biochemical stability of the fraction <40 mm before, during, and after the actively aerated windrow treatment are listed in *Table 2*. The respiration activity of the input material measured as AT4 (respiration during 4 days) amounts to 25 mg O_2/kg, the loss on ignition lay to 18%, the TOC at 8%. *Table 2* illustrates the development of the parameters over the treatment period. The respiration activity dropped significantly down to 1.9 mg O_2/kg after two weeks of treatment reflecting the comprehensive stabilization of the organics. The output material meets the German standards for waste disposal.

Table 2 Evaluation of the biochemical stability—actively aerated window for MSW

	Input	after 14 days (active)	after 6 weeks (active)
	13.10.06	2.11.06	13.12.06
AT_4 (mg O_2/kg)	25.2	1.9	0.91
oDS (weight%)	18.1	14.2	14.3
TOC (weight%)	8.2	5.1	5.2
TOC in the eluate (mg/L)	639	160	270

The results were not fully satisfying. Though the quality of output material was welcomed, the total mass of converted organics was small due to low input concentration. For the large scale application it was anticipated to increase the portion of organic in the input material by either segregating non-organic waste or by separation of organic waste before treatment. Since a separate source collection is not feasible in Gaobeidian a test was carried out for hand assorting of organic materials. A biowaste acquisition campaign was launched at the dumpsite. During a period of 5 weeks biowaste was accepted against payment and has been placed immediately on the composting heap. Larger quantities of biowaste were delivered by both waste pickers and residents at a rate of 6 €/t. Thus there is an economical bench mark for the separate source collection for biowaste available.

Figure 4 Delivery of biowaste by local resident

Based on the technical investigations a pre-design was prepared for a mechanical-biological treatment plant (MBT) by Pöyry Environment Witzenhausen.

4 Design

The MBT plant consists of a mechanical and a biological treatment step. The mechanical treatment aims on segregating of valuable goods (paper, synthetics) and on enrichment of the organic fraction. The waste stream is split into three fractions (fine, medium, coarse). Middle and coarse fraction are being directed to hand assorting stations. Based on the waste analysis it is expected, that the medium fraction after assorting consists mainly of organic, which are slightly clean. This portion may be converted into compost. The fine fraction will undergo biological treatment as well prior to final disposal. This fraction may be also useful as amendment for landfill cover, where it acts as methane oxidation layer. The coarse fraction will be either crushed and fed to the waste stream again or directly disposed at the landfill site. The design of the delivery and pretreatment plant is shown in *Figure 5*.

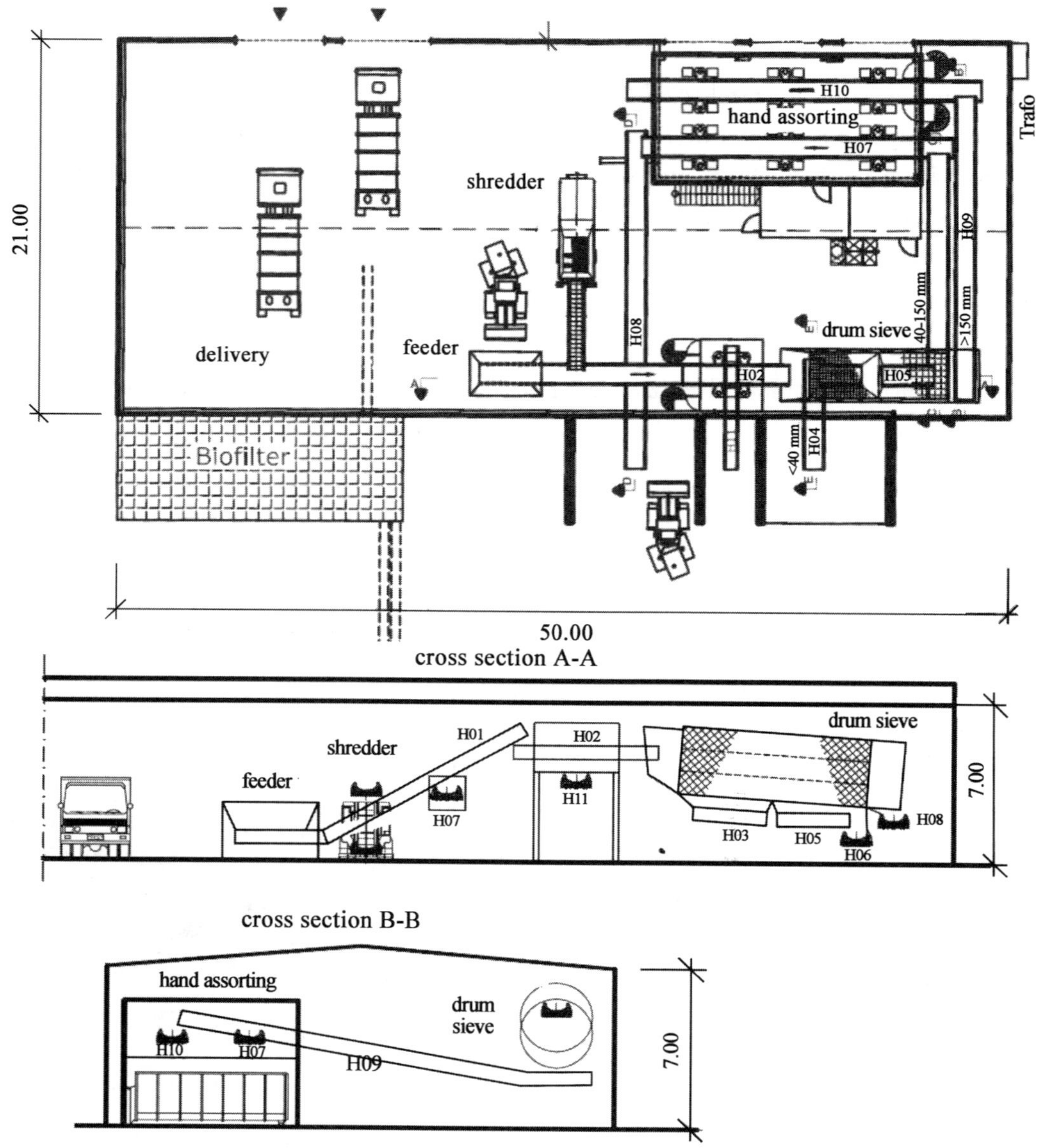

Figure 5 Delivery and mechanical treatment

The biological treatment will be carried out as an aerobic, actively ventilated stabilization. The process will be carried out partly under roof and partly in open air areas. The triangular windrow heaps will be frequently turned by means of mobile equipment as shown by example in *Figure 6*.
The output material will be disposed at the adjacent landfill, which is under construction. On a longer view the economic benefits of the plant will be increased particularly by means of reutilization of the stabilized biomass for methane oxidation layer (special landfill cover), as soil amendment, or as fertilizer in agriculture, if possible. To achieve this goal the waste flux management needs to be refined in order to minimize pollutants and harming matters in the organic waste fraction.
The annual treatment capacity amounts to 40 000 t. The capacity can be significantly increased by Introducing two working shifts. The location still provides contingency area for an expansion of the biological treatment.

Figure 6 Windrow heap under roof with revolving machine (example)

5 Climate mitigation relevance

The generation of methane in the landfill will be dramatically reduced by disposing stabilized biomass instead of untreated waste. The emission reductions were calculated using the methodologies of UNFCCC, which are relevant for CDM projects. Since the facility has been financed by national revenues from carbon trade a registration as CDM project is not permitted. Nevertheless, the Federal Ministry of Environment requests the verification of the emission reductions in a similar procedure. Thus, the calculation of the baselines and the monitoring of project emissions is required.
The methodology AMS Ⅲ. F (Avoidance of methane production from decay of biomass through composting) serves as the base of the baseline study and the monitoring concept for the Gaobeidian MBT project. For calculations an appropriate separate tool is available, the 'Tool to determine methane emissions avoided from dumping waste at a solid waste disposal site'.
In the greenhouse gas balance of the whole project, the emissions which result from the project itself must be considered as a negative influence. Therefore all CO_2 emissions of motor vehicles and electric machines counts by power or fuel consumption, which is reduces the CH_4-emissions. The calculation algorithms for both scenarios—baseline and project—are, as far as possible, given by the existing methodologies and tools. For the calculation of the methane emissions by the disposed waste, a biological decay of the deposited waste is modelled over several years by a layer model. With this model the highest methane

emissions in a layer emerge during the first years and show in the subsequent years a regressive behaviour. *Table 3* shows the provisional calculated emission for the project Goabeidian during one period of 10 years. According to that, the projected amount of emission certificates results in 83 812 t CO_2 equivalentes for the project period of 2010-2019. The calculation does not include any of the planned measures to increase the organic fraction prior to the biotreatment. These gain calculations represent only a prognosis on the basis of the emission reductions which are expected. The height of real income can diverge strongly. It depends on the plant capacity, the waste composition, the resource input, and on the commercial course of the certificates. In fact, the amount of emission reductions may even become three times higher resulting in approximately 250 000 t CO_2 equivalents reductions.

Regarding the monitoring process, all data that were used for the determination of the emissions, have to be determined ex-post, if not constant. Based on the monitored data the real emissions during the lifetime of the project will be recorded. To these data belong e. g. the power or fuel consumption, as well as the waste composition, which has to be examined several times in the year. *Table 3* provides the breakdown of expected emission reductions over the first ten years of operation.

Table 3 Emission calculations for the MBT Gaobeidian

Year	Project emissions (t CO_{2eq})	baseline emissions (t CO_{2eq})	Emission reductions (t CO_{2eq})
2010	497	2 339	1 842
2011	537	4 376	3 839
2012	577	6 156	5 579
2013	616	7 714	7 097
2014	655	9 081	8 426
2015	694	10 284	9 590
2016	732	11 345	10 613
2017	770	12 283	11 513
2018	807	13 114	12 306
2019	845	13 852	13 007
Total	6 729	90 541	83 812

6 State of execution

On 3rd of July 2009 the construction of the MBT was officially launched in a ceremony by laying the first stone. After the clearing of the construction site the local counterpart had turned the area into a festival zone with stage, flowers, dragons, balloons and red carpets. On a 3×15 m^2 sign the future plant was visualized. German and Chinese stakeholders form the municipality, province and national government, project developers and donor joined the event. Several hundred spectators from the region gathered as audience. Finally fire works crowned the opening.

The execution of the project proceeds quickly. In accordance with the Chinese regulations a feasibility study and an environmental impact assessment is carried out and will be finalized by the end of October. Meanwhile Pöyry Environment in cooperation with CRAES prepares the detailed engineering design. The official approval is expected by November. Site clearance and construction of access and supply infrastructure shall be finalized in advance. The construction of the key parts of the MBT will start right after the official approval.

Figure 7 Opening ceremony for the MBT Gobeidian

Acknowledgements

The project developers appreciate the funding from the International Climate Initiative by German Federal Ministry of Environment, Nature Conservation and Nuclear Safety. The International Climate Initiative is financing climate protection projects in developing and newly industrialising countries and in transition countries in Central and Eastern Europe since 2008. Through this, the Federal Environment Ministry is making an effective contribution to emission reductions and adaptation to climate change. This new form of environmental cooperation complements the Government's existing development cooperation. Funding of 120 million€/a is available for the International Climate Initiative from the revenues of the sale of emission allowances. A decision by the German Bundestag forms the basis of this worldwide investment in climate protection. When selecting projects, the Federal Environment Ministry attaches great importance to the development of innovative and multipliable approaches that impact beyond the individual project itself and are transferable. Through targeted cooperation with partner countries the Climate Initiative provides important momentum for negotiations on an international climate agreement for the post-2012 period. One focus of the International Climate Initiative lies in the areas of promoting a climate-friendly economy. In the field of a climate-friendly economy, the goal is to support partner countries in establishing a climate-friendly economic structure that prevents climate-damaging greenhouse gas emissions where possible. This support covers areas such as increasing energy efficiency, expanding renewable energies, reducing environmentally harmful greenhouse gases and investment-related measures, know-how transfer and policy advice in the partner country.

References

Kölsch F., Wiesbaum A., Heidelmann N., Fricke K., Ginter M., (2007): *CDM Projekt MBA für Gaobeidian (VR China)*. Müll und Abfall 39, Heft 8, Berlin: E. Schmidt.

Kölsch F., Ginter M., Fricke K., Müller W., (2009): *Grundsteinlegung des Leuchtturmprojecktes MBA Gaobeidian (VR China)*. Müll und Abfall 41, Heft 9, Berlin: E. Schmidt http://www.bmu-klimaschutzinitiative.de/de/aktuelles.

Bahr T., Fricke K., Schulte B., Kölsch F., (2010): *Center for Research, Education and Demonstration in Waste Management (CReED)*. ASA (Hg.): MBA-Technologie-Schaltstelle für Stoffströme und Energieeffizienz. Internationale 8. ASA-Abfalltage. Weimar: ORBIT e. V., S. pp. 303-319.

A CASE STUDY ON THE CARBON FOOTPRINT OF A BIOGAS PLANT IN BEIJING

ZHOU Jie, WU Shubiao, GUO Jianbin, DONG Renjie, PANG Changle, CHEN Li

China Agricultural University

Contact: CHEN Li, Vice Professor, China Agricultural University, College of Engineering, Qinghua East Road No. 17, Haidian District, 100083 Beijing, PR China, Tel.: +86-10-6273-7885, E-Mail: chenli329@yahoo.cn

EXECUTIVE SUMMARY

Based on the clean development mechanism (UNFCCC CDM) and a life cycle inventory analysis, this paper emphasises on the concept of carbon footprint and estimates the net carbon footprint and emission reductions (ER) of a biogas plant (functional unit) with a fermenter volume of 1 100 m^3. The annual carbon footprint of the biogas plant operation (electricity, fuel consumption, etc.) was calculated to be about 2 835.32 t CO_{2equiv}. Further we have to consider the negative carbon footprint (respectively ER) from the biogas production which amounts annually to 2 914.23 t CO_{2equiv}. The net balance is −78.91 t CO_{2equiv}/a and the total emission reduction (ER) of this project during its lifetime of 20 years amounts to 1 578.20 t $CO_{2\ equiv}$. In total the biogas project has a high reduction benefit value and it's effective to promote greenhouse gases emission reduction.

Key words: Biogas project; carbon footprint; greenhouse gases; emission reduction

---

沼气工程碳足迹案例分析

周捷，吴树彪，郭建斌，董仁杰，庞昌乐，陈理*

（中国农业大学，北京 100083）

【摘要】基于清洁发展机制以及生命周期清单分析，本文引入“碳足迹”概念并估算了一座规模为 1 100 m^3 发酵罐容积的沼气厂（功能单位）碳足迹净值及该项目的减排量。通过对年产气量的碳足迹分析，保守估计沼气工程运行（电力、燃料消耗等）碳足迹为 2 835.32 t CO_2 当量，同时考虑到由于沼气生产而减少的碳足迹（即减排量）约为 2 914.23 t CO_2 当量，故沼气工程功能单位的碳足迹净值为−78.91 t CO_2 当量，且该项目共可减排 1 578.20 t CO_2 当量。因此该沼气工程的减排效益值较高，可有效促进温室气体减排的进行。

【关键词】沼气工程；碳足迹；温室气体；减排

1 引言

2004 年，农业活动甲烷年排放量为 1.7196×10^7 t，动物粪便甲烷年排放量约为 8.67×10^5 t，以猪粪便管理系统甲烷排放为主，占 61%[1]。对畜禽粪便管理系统的温室气体进行评估可有效促进农业活动的减排

* 通信作者：陈理，女，硕士，副教授，中国农业大学工学工学院能源工程与低碳技术研究室，北京市海淀区清华东路 17 号 100083，chenli329@yahoo.cn，010−62737885

策略的制定。

碳足迹评估是目前用来形象准确衡量温室气体排放，抵抗气候变化制定有效合理减排策略的途径之一。碳足迹是指产品或服务在整个生命周期内（从原材料的获取，到生产、使用和废弃的处理）温室气体排放量[2~3]，碳足迹评估标准方法为PAS2050(Publically Available Specification)，即《商品和服务生命周期温室气体排放评估规范》[4~7]，计算过程主要是绘制流程图；界定系统边界；收集数据；计算碳足迹；检验结果等。此外，如果项目的某一碳排放量小于其总的碳足迹1%、人类活动所导致的碳排放以及购买产品而产生的交通碳排放则排除在边界之外。

沼气工程可以减少粪畜禽便等产生的甲烷排放，对其进行碳足迹评价具有十分重要的现实意义。因此，本文通过对北京农村沼气工程运行现状调研，基于清洁发展机制(Clean Development Mechanism，CDM)以及生命周期清单分析（Life Cycle Inventory，LCI)，引入"碳足迹"，并对北京农村大中型沼气工程功能单位的碳足迹进行了评估分析，估算其碳足迹净值以及该项目减排量。

2 北京沼气工程调研概况

中国农业大学能源工程与低碳技术研究室分别于2010年夏季以及2011年1月通过实地取样调查，获得了北京郊区农村84个大中型沼气工程调研数据，其类型大多为能源生态型，调研显示在冬季正常运行的沼气工程点有47个，停用的16个，已废弃4个，已建或正建9个，其他8个，如图1所示。冬季运行沼气过程中全采用燃煤加温的有38个，占80.85%；仅太阳能加温的3个，燃煤及太阳能加温的4个，暖气与燃气加温的1个，其他2个，详见图2。北京沼气工程点产气多为产区自用以及农户炊用，无发电自用或上网情况。在运行的沼气工程中原料需购买的有24个占51.06%，原料自供的21个占44.68%，其他2个。

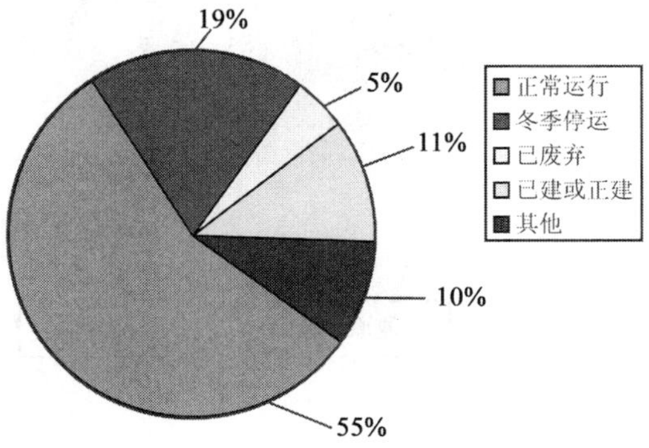

Figure 1 The operation of biogas projects

图1 沼气工程运行情况

可以看出，原料需购买且为能源生态型的沼气工程是被调研工程的主要类型，本文特选取该种类型的工程点进行碳足迹分析，对其温室气体减排量进行估算。

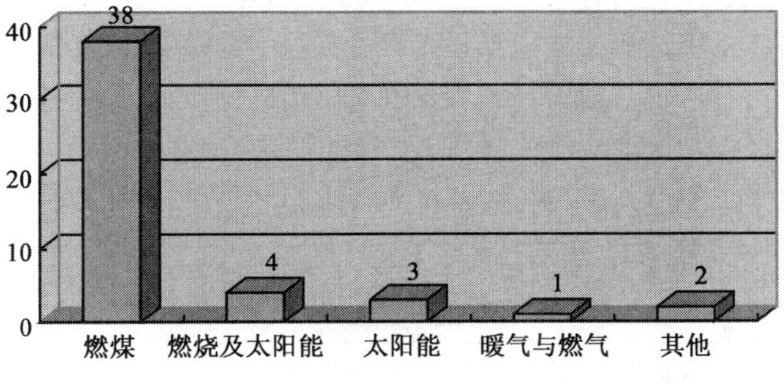

Figure 2 Winter heating modes

图2 冬季加温方式

3 沼气工程碳足迹案例分析——D村沼气工程

3.1 流程图

本文选取北京市房山区石楼镇D村能源生态型沼气工程进行分析评价，该工程的原料需购买、冬季采用燃煤加温，如图3所示，其工艺流程如图4所示。该工程为独立运行的沼气炊用工程，发酵罐容积550×2 m^3，储气柜容积260×2 m^3，规划日产气量1 000 m^3，冬季实际日产气量400 m^3，中温单相发酵，发酵工艺为升流式固体反应器(Upflow Solid Reactor,USR);购买原料猪粪,日进料 7～8 m^3;供气户数 600～700 户,果树、部分温室大棚等足够消纳沼液沼渣。此外,该沼气工程冬季燃煤加温 6 个月,年耗煤约 100 t,年耗电约 50 000 kWh。

Figure 3 Biogas overview

图 3 沼气工程概况

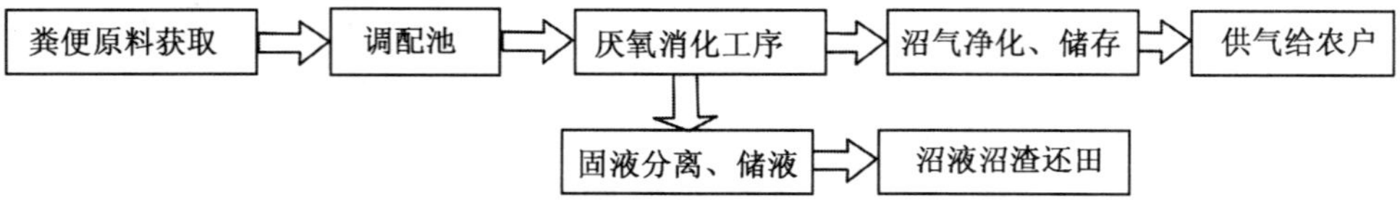

Figure 4 Flow chart of eco-energy biogas

图 4 沼气工程能源生态型流程图

3.2 系统边界

界定系统边界是分析评价过程最重要的一步,需要划分提出阶段、提出假设、确定清单分析数据需求以及限制条件等。该案例系统边界不包括产品的生产和维护,如建筑物、机器等;也不包括用量小于 1%的化学品的生产,如氢氧化钠等。系统边界主要包括以下几个阶段:原料获取、发酵产气、"三沼"输出、沼气使用以及沼液沼渣处理阶段,如图 5 所示。假设原料粪便在 24 h 之内投入发酵罐中,沼气全被农户利用,甲烷含量 60%,沼液沼渣还田使用合理,保证不产生甲烷;工程可以至少运行 20 年。功能单位为数据清单中需要量化的沼气量,设定为沼气工程年产气量。

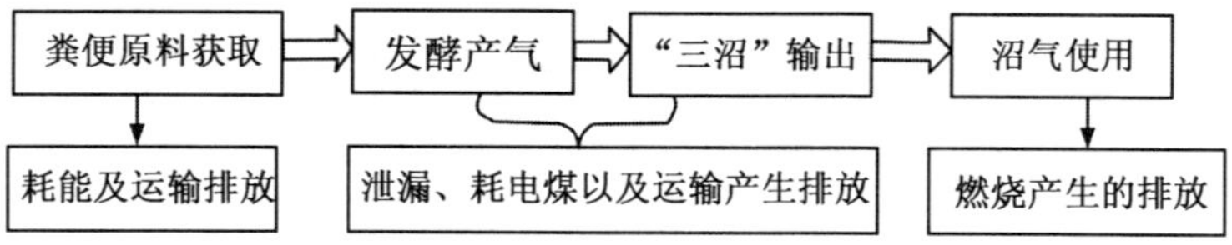

Figure 5 System boundary

图 5 系统边界

3.3 数据选取及分析

无沼气工程时,畜禽粪便引起的温室气体排放主要为甲烷[8]。由于运输距离短且获取耗能小,属于小于 1%的部分不计入边界中,故此部分不做考虑;发酵以及输出过程中涉及泄漏排放,选取方法学 AMS III. D[9] 对物理泄漏规定的默认值 10%。此外较多排放主要是耗电煤以及使用沼气时的温室气体排放。

3.4 计算碳足迹及检验结果

依据下述公式(1)、(2)计算碳足迹,公式中 $CF_{总}$ 为总碳足迹;$CF_{耗电}$、$CF_{耗煤}$、$CF_{泄漏}$、$CF_{沼气燃烧}$分别代表耗电、耗煤、沼气泄漏以及沼气燃烧产生的碳足迹；CF_i 代表各碳足迹；AD_i 代表活动水平；EF_i 代表各个活动的排放因子。

$$CF_{总}=CF_{耗电}+CF_{耗煤}+CF_{泄漏}+CF_{沼气燃烧} \quad (1)$$

$$CF_i=AD_i \times EF_i \quad (2)$$

北京电网排放因子为 0.992 8 t CO_2 当量/MWh[10]，则 $CF_{耗电}$ 为 49.64 t CO_2 当量；原煤二氧化碳排放因子为 1.98 kg/kg[11]，则 $CF_{耗煤}$ 为 198 t CO_2 当量；通过夏季正常产气量总和与冬季产量之和得年产沼气量为 256 200 m^3，温度为 20℃和一个大气压下，CH_4 的密度 0.67 kg/m^3；京都议定书规定，全球现默认值为 21[12]，物理泄漏规定的默认值 10%[9]，则 $CF_{泄漏}$ 为 2 162.84 t CO_2 当量；$CF_{沼气燃烧}$ 为 424.84 t CO_2 当量；故得其总的碳足迹值 $CF_{总}$ 为 2 835.32 t CO_2 当量。

估算因建设沼气工程，而每年减少的碳足迹。每立方米沼气可代替 0.714 kg 标煤[13]，标煤的二氧化碳排放因子为 2.45 kg CO_2 当量/kg[14]，故本案例每年相对应可节约 164.634 t 标煤，相对减少温室气体排放 403.35 t CO_2 当量。冬季每日进料 8 m^3 猪粪，设粪便的含水量为 80%，根据转换系数[15]得沼气工程年处理猪粪为 2 388.93 t，标准煤折换系数为 0.429 t/t[16]，进而可得处理原料猪粪减少温室气体 2 510.88 t CO_2 当量。故年减少碳足迹共 2 914.23 t CO_2 当量。

4 结论

该案例年产沼气量为 256 200 m^3，相应节约 164.634 t 标煤，相对减少温室气体排放 403.35 t CO_2 当量。通过对功能单位年产气量的碳足迹分析，保守估计沼气工程运行碳足迹为 2 835.32 t CO_2 当量，因建设沼气工程而减少的碳足迹约为 2 914.23 t CO_2 当量，故沼气工程的碳足迹净值为 −78.91 t CO_2 当量，所以该沼气工程功能单位减排温室气体 78.91 t CO_2 当量，该项目共可减排 1 578.20 t CO_2 当量。

在本案例分析中，虽然假设原料粪便在 24 h 之内投入发酵罐以及合理使用沼液沼渣还田并保证不产生甲烷等，但仍可反应出北京大中型沼气工程整体是促进温室气体减排的，而且减排效益较好，可有效促进温室气体减排的进行。

项目及基金支持

1.“十一五”国家科技支撑计划“规模化沼气工程沼液、沼渣高值利用技术研究与工程示范”课题(2008BADC4B00)；

2.北京市教育委员会科学研究与科研基地建设项目“规模化养殖场粪污双向清洁处理与利用研究”。

参考文献

[1]《中华人民共和国气候变化初始国家信息通报》[M].北京：中国计划出版社，2004:15-20.

[2] 王微，林剑，崔胜辉，等. 碳足迹分析方法研究综述[J]. 环境科学与技术，2010，33 (07):71-78.

[3] 罗云阔，周亮梅，朱美英. 碳足迹解析[J]. 江西农业大学学报 (社会科学版)，2010，9(2):123-127.

[4] T. Wiedmann，J. Minx. A Definition of “Carbon Footprint”. ISA (UK)，Research Report 07-01 [EB/OL]. www.isa-research.co.uk/docs/IS A -UK-Report-07-01-carbon-footprint z.pdf，2007.

[5] Wiedmann T.，Minx J. A definition of carbon footprint [J]. SA Research & Consulting，2007: 9.

[6] Carbon Trust. Carbon footprint measurement methodology[R]，2007.

[7] 中国标准化研究院. 商品和服务在生命周期内的温室气体排放评价规范(PAS 2050:2008)及使用指南(中文版)[S]. 2009.06.04.

[8] 段茂盛，王革华. 畜禽养殖场沼气工程的温室气体减排效益及利用清洁发展机制(CDM)的影响分析[J]. 太阳能学报，2003，24(3):386-389.

[9] UNFCCC. CDM Methodologies: Small Scale [EB/ OL]. http://cdm. unfccc. int/methodologies/PAmethodologjes/approved. html. 2009-04 -08. AMS III. D: "Methane recovery in animal manure management systems" (Version 14), 2009.

[10] 中国清洁发展机制网. 关于确定 2008 中国区域电网基准线排放因子的公告[EB/OL]. http://edna. ceehina. gov. en/web/News Info. asp? Newsld=3239, 2008.

[11] 董红敏，李玉娥，陶秀萍，等. 中国农业源温室气体排放与减排技术对策[J]. 农业工程学报，2008，24(10):269-273.

[12] IPCC. 2006 IPCC Guidelines for National Greenhouse Gas Inventories. Volume 4-Agriculture, Livestock [M]. Prepared by the IPCC national greenhouse gas Inventories program. Kanagawa, Japan. IGES, Japan, 2006.

[13] 中华人民共和国国家发展和改革委员会. 可再生能源发展中长期规划[S]. 2007，9.

[14] 石元春. 决胜生物质[M]. 北京:中国农业大学出版社，2011:429.

[15] G. Riskowski，张伟力. Swine Manure Management [EB/OL]. http://www. asaimchina. org/technical/hog/41002. html.

[16] 石元春. 决胜生物质[M]. 北京:中国农业大学出版社，2011:298.

COMPARATIVE OVERVIEW OF EEG VERSIONS 2000-2012: EXAMPLE BIOGAS

J. Daniel-Gromke, J. Giersdorf

Deutsches BiomasseForschungsZentrum (DBFZ)

Contacts: Dipl.-Umweltwiss. Jaqueline Daniel-Gromke, Dipl. Pol. Jens Giersdorf, Department Biochemical Conversion, Deutsches BiomasseForschungsZentrum (DBFZ), Torgauer Straße 116, 04347 Leipzig, Germany, E-Mail: info@dbfz.de, www.dbfz.de

EXECUTIVE SUMMARY

The German Renewable Energy Source Act (EEG) from 2000 and its amendments in 2004, 2009, and 2012 was analysed in regard to renewable energy from biogas. A comparison of the respective impacts on the biogas market development and the generated effects and even problems lays the foundation to further analyse the necessity to adapt the biogas policy in order to steer the biogas sector development in Germany in a sustainable direction.

Key words: German renewable energy law; biomass energy

1 Renewable Energy Source Act (EEG) (2000)

- Guaranteed priority feed in framework ;
- Guaranteed electricity feed in tariffs (FiT);
- Planning guarantee: feed in conditions for 20 years;
- Degression of feed-in-tariffs for biogas plants 1% p. a., depends on starting year of operation of biogas plant;
- Technology-open remuneration of bioenergy should force the competition of concepts;
- Small plant sizes have higher specific investment costs than large ones → Basic FiT according to different plant sizes (≤500 kW_{el}, ≤ 5 MW_{el}, >5 MW_{el});
- Difference costs between electricity from conventional and renewable resources paid by consumers, not by government (since 2009 regulated in AusglMechV 2009), difference costs amounted to 2.047 €-ct/kWh in 2010.

2 First Amendment of the Renewable Energy Source Act (EEG) (2004)

2.1 New regulations

- Introduction of new plant size of ≤ 150 kW_{el} to promote decentralized concepts;
- Annual degression of FiT increased from 1% p. a. to 1.5% p. a.;
- Introduction of several boni, difference between basic payments (plant size) and additional payments (bonus): Cultivated biomass bonus (e. g. energy crops, manure), technology bonus (upgrading to biomethane since it is more expensive, dry fermentation) and CHP bonus (waste heat utilization → higher efficiency, highest reduction of GHG emissions) in combination with given options for heat utilization (positive list).

2.2 Impacts

- Increase of average total payments (basic tariff and boni);
- Average installed capacity of biogas plants increased considerably.

3 Second Amendment of the Renewable Energy Source Act (EEG)

3.1 New regulations

- Increase of basic tariff for biogas plant ≤ 150 kW_{el} and of some boni (energy crops, CHP/heat utilization);
- Decrease of basic tariff for biogas plants > 150 kW_{el};
- Introduction of the manure bonus and the emission control bonus;
- Dry fermentation not considered innovative technology anymore (no further special support in comparison to wet fermentation necessary anymore);
- Bonus for the utilization of biowaste;
- Additional bonus for the utilization of manure (mass limit: at least 30% of manure);
- Annual degression of FiT reduced from 1.5% p.a. to 1% p.a.;
- Introduction of degression of bonus payments;
- Progress report about development of EEG every four years as basis for amendments.

3.2 Impacts

- Start of operation of new biogas plants increased from 350 in 2008 to 900 new biogas plants in 2009;
- Around 50 biogas plants with biogas upgrading to biomethane operating in December 2010 (biomethane plants benefit in general from cultivated biomass, CHP and technology bonus; FiT requirement: 100% waste heat utilization, emission and energy demand limits, covering heat demand with renewable energy;
- Increase of average payments for biogas plants compared to 2004*.

3.3 Existing problems

- Remuneration system too complex due to further differentiation of bonus system → costs for documentation and auditing of used substrates;
- Manure bonus not so effective because of the mass limit of manure →support of artificial concepts, windfall gains in the case of plants using manure, 'manure tourism';
- Coupling of manure and cultivated biomass (energy crops) → further concentration of maize production in regions with intensive livestock farming and an excess of nutrients;
- Discussion about the increase of maize production in Germany due to biogas production→problem has to be solved within regional-based regulations;
- Additional bonus of innovative concepts or technologies are not often used;
- Reduction of emissions (Formaldehyde of the CHP unit) → incentive to use (thermal) post-combustion systems not really effective, standards for technologies (emission control) should be regulated in specific laws and compliance with these regulations should be the pre-condition for remuneration via Feed-in-tariffs.

---

* Biogas plant 150 kW_{el} with Cultivated Biomass- and CHP-Bonus and 35% manure (incl. formaldehyde max. value) from 18.07 (2004) to 25.77 (2009)€ -ct/kWh, biogas plant 500 kW_{el} with Cultivated Biomass- and CHP-Bonus (incl. formaldehyde max. value) from 17.03 (2004) to 20.03 (2009) €-ct/kWh.

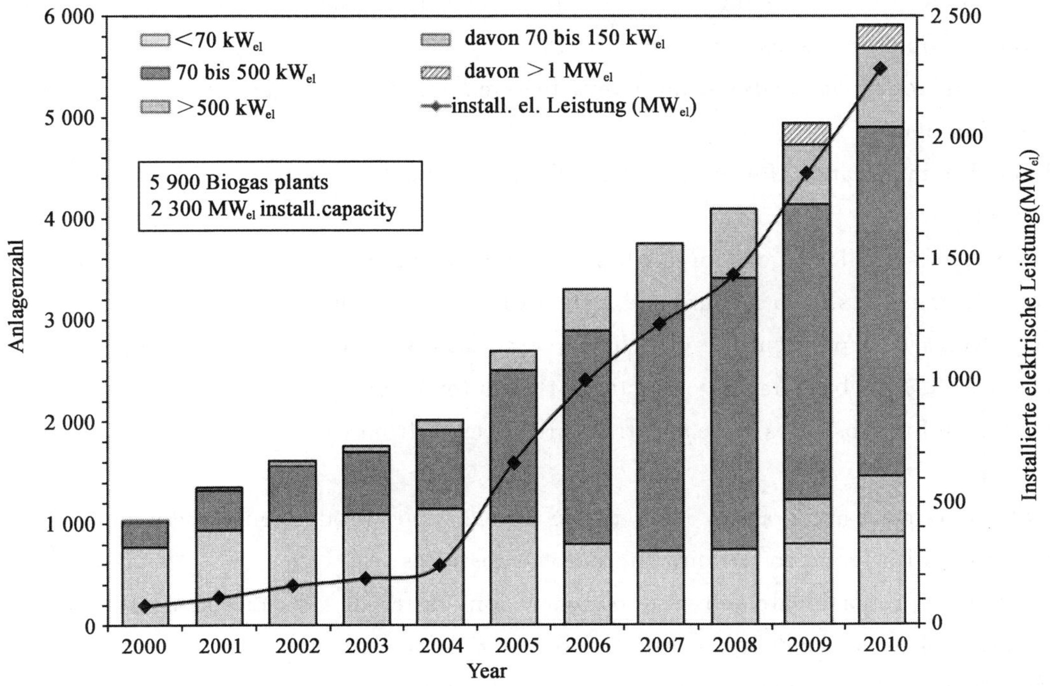

Figure 1 Development of number of biogas plants and installed electrical capacity in Germany 2000-2010

Table 1 Feed-in-tariff (Biogas) system according to EEG 2009

Installed capacity	Basic tariff	Boni					
		Cultivated biomass①	Manure②	Landscape conservation material③	Emission reduction④	Technology⑤	CHP⑥
≤150 kW$_{el}$	11.67	+7.00	+4.00	+2.00	+1.00	+2.00	+3.00
≤500 kW$_{el}$	9.18	+7.00	+1.00	+2.00	+1.00	+2.00	+3.00
≤5 MW$_{el}$	8.25	+4.00	—	—	—	+2.00	+3.00
≤20 MW$_{el}$	7.79	—	—	—	—	—	+3.00

4 Third Amendment of the Renewable Energy Source Act (EEG) (2012)

4.1 New Regulations and Expected Impact

- Simplified tariff system with 4 categories for plant sizes and 2 substrate categories → more

① Cultivated biomass, manure and vegetal by-products as substrates allowed. Raw material log with data about type, amount and origin of biomass substrates necessary for documentation of used substrates. If biogas plant needs immission control licencing → gas-proof cover of digestate storage and gas consumer installation for breakdown necessary. Used by 88% of all biogas plants in 2010.

② Coupled to cultivated biomass bonus. Only granted to plants < 500 kW$_{el}$ in order to avoid "manure tourism". Share of manure must exceed 30% (mass), Used by 70% of all biogas plants in 2010.

③ Material from landscape conservations (streets, parks, etc.), > 50% of material from landscape conservations (mass FM).

④ Introduced in 2009, only for biogas plants. < 60 mg/m^3 (40 mg/m^3) formaldehyde emissions. Only for plants licenced via federal immission control act (BimSchG).

⑤ Innovative technologies: Fuel cell, Gas turbines, ORC, stirling motors (dry fermentation technology not anymore).

Biogas upgrading if max. 0.5% CH_4 emissions and max. 0.5 kWh/ Nm^3 raw biogas energy demand. Max. 700 Nm^3/h capacity for biogas upgrading: < 350 Nm^3/h: 2.00 €-ct /kWh, < 700 Nm^3/h: 1.00 €-ct/kWh

⑥ Positive list with several utilization forms for exhaust heat (heating buildings, feed into heating grid, process heat, heating of poultry, livestock buildings, process heat for digestate treatment used as fertilizer). Used by 73% of all biogas plants in 2010.

transparency (specific costs for substrates do not depend on plant sizes, bonus payments regarding substrates should not depend on plant size);

- Separate tariff for biogas plants using biowaste →mobilization of organic waste and residues for new and existing plants;
- Additional remuneration for biomethane plants, increase of tariff → promotion of biomethane production;
- Introduction of special category for small biogas plants using at least 80% manure (mass) →reduction of methane emissions from manure and groundwater protection;
- Average decrease of payments of 10%-15%, especially for small plants and increase of degression from 1% to 2% for basic tariff→reduction of costs for biogas production for consumers;
- No degression for substrates since substrate prices result from world market prices, no cost reduction potential;
- Limit of use of maize and grain to 60% (mass) for electricity production →reduction of competition of energy crops with food production, alternative substrates;
- Introduction of proportionate substrate remuneration, different substrate categories can be mixed → mobilization of use of landscape conservation material for example;
- Introduction of minimum requirements: Every biogas plant either has to use 60% of exhaust heat or at least use 60% of manure or market directly produced electricity → Additional benefit beyond electricity production;
- Introduction of a flexibility bonus for the market-oriented production of biogas;
- Support of direct marketing for the produced electricity (market bonus); up to 2014 biogas plants with an installed capacity more than 750 kW_{el} has to market their produced electricity directly;

Table 2 Feed-in-tariff system (Biogas) according to EEG 2012

	Feed-in-tariffs					
	Biogas plants (excl. biowaste)				Biowaste biogas plants①	Small manure plants
Installed capacity (kW_{el})	Basic tariff	Substrate tariff Ⅰ	Substrate tariff Ⅱ②	Biogas upgrading/ Biomethane bonus (€-ct/kW_{el})		
≤75	14.3					25③
≤150		6	8	≤ 700 Nm^3/h:3	16	
≤ 500	12.3			≤1 000 Nm^3/h:2		
≤ 750	11	5	8/6④	≤1 400 Nm^3/h:1		
≤ 5 000	11	4			14	
≤ 20 000	6	—		—		

References

Deutscher Bundestag, (2012): *Gesetz für den Vorrang Erneuerbarer Energien (Erneuerbare-Energien-*

①Only for biogas plants using specific biowaste and post rotting process for solid fermentation residues. The stabilized fermentation residues have to be used materially (fertilizer). The tariff is only to be combined with the bonus for biomethane production.

②Only for selected, environmental friendly substrates according to definition.

③Special category for manure plants until 75 kW installed capacity, not to be combined with other tariffs.

④Electricity from manure > 500 kW, 6 €-ct/kWh

Gesetz - EEG). Konsolidierte (unverbindliche) Fassung des Gesetzestextes in der ab 1. Januar 2012 geltenden Fassung; http://www.erneuerbare-energien.de/files/pdfs/allgemein/application/pdf/eeg_2012_bf.pdf.

Bundesministerium für Umwelt, Naturschutz und Reaktorsicherheit, (2011): *Erfahrungsbericht 2011 zum Erneuerbare-Energien-Gesetz (EEG-Erfahrungsbericht) gemäß § 65 EEG vorzulegen dem Deutschen Bundestag durch die Bundesregierung*. Stand 3.5.2011; http://erneuerbare-energien.de/files/pdfs/allgemein/application/pdf/eeg_erfahrungsbericht_2011_entwurf.pdf.

Deutscher Bundestag, (2008): *Erneuerbare-Energien-Gesetz vom* 25. Oktober 2008 (BGBl. I S. 2074), das durch Artikel 1 des Gesetzes vom 28. Juli 2011 (BGBl. I S. 1634) geändert worden ist; http://www.gesetze-im-internet.de/bundesrecht/eeg_2009/gesamt.pdf.

Bundesgesetzblatt, (2004): *Gesetz zur Neuregelung des Rechts der Erneuerbaren Energien im Strombereich*. vom 21. Juli 2004: http://www.bgbl.de/Xaver/start.xav? startbk=Bundesanzeiger_BGBl&bk=Bundesanzeiger_GBGI&start=//*[@attr_id=%27bgbl104s1918.pdf%27].

Bundesgesetzblatt, (2000): *Gesetz für den Vorrang Erneuerbarer Energien*. Datum: 29. März 2000; http://www.gesetze-im-internet.de/bundesrecht/eeg/gesamt.pdf

德国可再生能源法(EEG)2000—2012年版本比较:以沼气为例

J. Daniel-Gromke, J. Giersdorf

(*莱比锡德国生物质研究中心,Leipzig 04347*)

【摘要】本文基于沼气这一可再生能源分析了2000年颁布的德国可再生能源法,以及在2004年、2009年和2012年颁布的德国可再生能源法修订案。通过比较各个版本对沼气市场发展方面的影响和由于补贴方式不同所产生的问题,进一步分析适应沼气行业政策的需求,进而使德国沼气行业向可持续发展的方向迈进。

【关键词】德国可再生能源法;生物质能

1. 可再生能源法(EEG)2000年版

- 保证优先上网;
- 保证入网电价补贴(FiT);
- 规划保证:入网条件20年;
- 沼气工程入网补贴每年递减1%,按沼气工程运行起始年算;
- 生物能源的技术开发补贴应该引入促进竞争理念;
- 小型沼气工程的单位容积投资高于大型沼气工程,所以基础入网电价补贴(FiT)数额取决于沼气工程的规模(≤ 500 kW,≤ 5 MW,> 5 MW);
- 传统电力和可再生能源之间的成本差价应该由消费者来支付而不是由政府支付(2009年补偿机制法(AusglMechV)规定),2010年的差价是2.047 €-ct/kWh)。

2. 可再生能源法第一次修订(2004年)

2.1 新规定

- 引入对≤ 150 kW的沼气工程的补贴,用以促进分散式布局的概念;

- 入网电价补贴从每年递减1%提高到每年递减1.5%；
- 引入多种补贴机制，根据沼气厂规模制定基本补贴并提供额外补贴：生物质种植补贴（例如能源作物，粪污），技术补贴（甲烷提纯补贴（因为投入高），干发酵补贴），热电联产与给定的热利用方式（清单）结合补贴（废热利用→更高效的温室气体减排）。

2.2 影响

- 补贴的平均总支付额度增加（基本补贴加各种额外补贴）；
- 沼气工程平均装机容量大幅增加。

3.可再生能源法第二次修订(2009年)

3.1 新规定

- 对于≤150 kW的沼气工程增加基本补贴和一些额外补贴（能源作物，热电联产/热利用）；
- 降低>150 kW沼气工程的基本补贴；
- 引入了粪污补贴和排放控制补贴；
- 干法发酵不再是创新技术（和湿法发酵相比，不再给予特殊的支持）；
- 生物废弃物利用补贴；
- 对利用粪污给予额外补贴（数量限制：至少利用30%粪污）；
- 入网电价补贴从每年递减1.5%减到每年递减1%；
- 引入补贴支付递减制度；
- 每四年撰写可再生能源法发展进展报告并作为法案修订的依据。

3.2 影响

- 新运行的沼气工程数量从2008年的350个增加到2009年的900个；
- 截止到2010年12月，大约有50个沼气提纯厂在运行（沼气提纯厂的利润一般来自于生物质种植、热电联产和技术补贴；入网电价补贴要求：废热100%得到利用，限制排放和能源需求，热需求由可再生能源来提供）；
- 相较于2004年①，沼气工程的平均补贴支付增加了。

3.3 存在问题

- 补贴系统进一步细化，造成补贴系统过于复杂→归档和原料应用情况的核查需要费用；
- 由于粪污使用有底线要求，所以粪污补贴效率不高→一些使用粪污原料的沼气工程利用该规定的漏洞赚取政府补贴；
- 粪污与种植的生物质（能源作物）结合→进一步在畜牧业集中区以及土地营养过剩地区大量种植玉米；
- 由于沼气生产造成德国玉米种植量增加的讨论→这个问题必须利用地区法规解决；
- 创新理念或技术的补贴利用的很少；
- 减少排放（热电联产机组的甲醛问题）→对于使用（热）后燃烧系统的补贴不是很有效，技术标准（排放控制）应在具体法律中规定，上网补贴的发放应该以遵守这些法规为前提。

---

①150 kW的沼气工程含有生物质种植补贴，热电联产补贴以及35%牛粪补贴（包括甲醛减排最大值）其补贴从18.07 €ct(2004)提高到25.77 €ct(2009)，500 kW的沼气工程含有生物质种植补贴，热电联产补贴以及35%牛粪补贴（包括甲醛减排最大值）其补贴从17.03 €ct(2004)提高到20.03 €ct(2009)。

表 1　2009 年德国可再生能源法(沼气)上网补贴系统

装机容量	基础补贴	额外补贴					
		生物质种植①	粪污②	景观保护原料③	减排量④	技术⑤	热电联产⑥
≤150 kW_{el}	11.67	＋7.00	＋4.00	＋2.00	＋1.00	＋2.00	＋3.00
≤500 kW_{el}	9.18	＋7.00	＋1.00	＋2.00	＋1.00	＋2.00	＋3.00
≤5 MW_{el}	8.25	＋4.00	—	—	—	＋2.00	＋3.00
≤20 MW_{el}	7.79	—	—	—	—	—	＋3.00

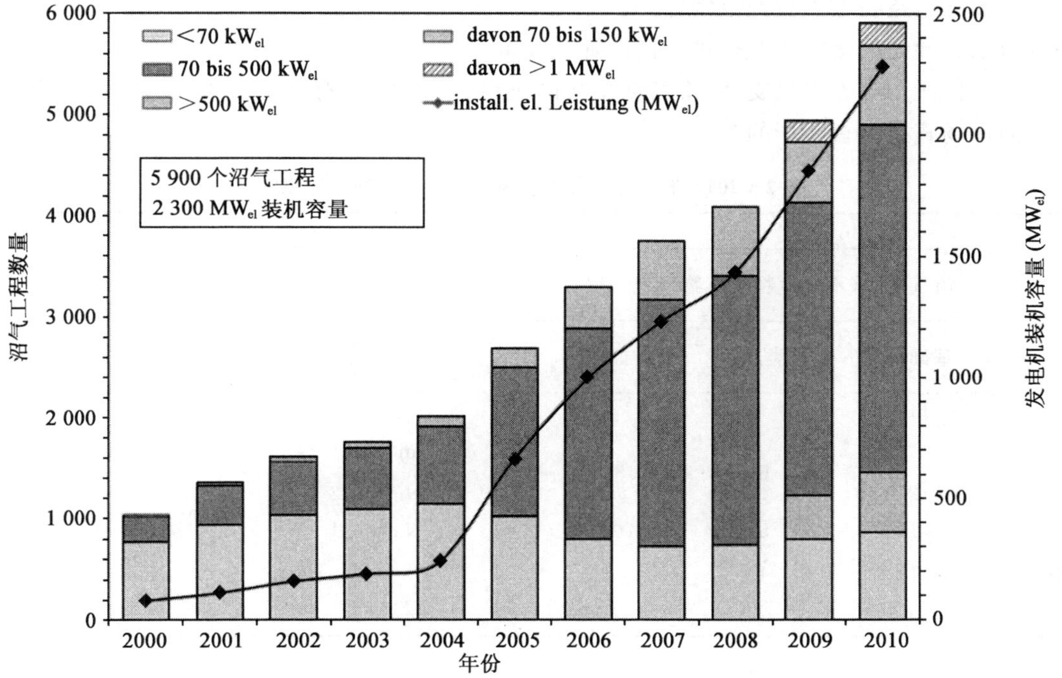

图 1　德国 2000-2010 年沼气工程与发电装机容量的发展

4. 可再生能源法第三次修订(2012 年)

4.1　新规定与预期影响

- 按规模将补贴体系简化为四类，按原料分为两类→更透明(原料支出与沼气工程规模无关，所以对于原料的补贴不应与沼气工程规模挂钩)；
- 将使用生物废弃物的沼气工程补贴单独列出→促进新的和已建的沼气工程使用有机废弃物；
- 对于甲烷提纯厂给予额外补贴→促进生物甲烷生产；
- 对于应用了至少 80％粪污(数量上)原料的小规模沼气工程给予特殊补贴→为降低粪污的甲烷排放

①种植的生物质，粪污以及植物副产品可以作为发酵原料。原料的类型和数量等数据应归档。如果沼气工程需要引入控制执照→那么沼液储池应被覆盖，安装沼气摧毁装置以应对紧急情况。88％的沼气工程在 2010 年获得该项补贴。

②与生物质种植补贴相结合。仅支付给< 500 kW 的沼气厂，为避免钻粪污补贴的漏洞，粪污的比例必须超过 30％(数量比)，2010 年 70％的沼气工程获得该项补贴。

③园林剪枝原料(行道树、公园等)，需使用> 50％的园林剪枝原料(鲜原料重量比)。

④2009 年引入的沼气工程是仅对于甲醛排放< 60 mg/m³(40 mg/m³)的沼气工程和通过联邦准入控制法案(BImSchG)获得许可的沼气工程。

⑤创新技术：燃料电池，气体涡轮机，ORC，Stirling 发动机(干法发酵技术不再支持)。如果甲烷排放最大 0.5％且能源需求最大 0.5 kWh/Nm³，进行沼气提纯。对沼气提纯最大支持 700 Nm³/h 规模：< 350 Nm/h，2 €-ct /kWh，< 500 Nm³/h，1 €-ct /kWh

⑥废热积极利用形式清单(加热建筑物，进入热网，畜禽舍加热，用于沼渣处理生产废料)2010 年 73％沼气工程获得该补贴。

及保护地下水；

- 平均降低 10%～15%的补贴支付额，尤其对于小型沼气工程，基本补贴额年递减从 1%增加到 2%→对于消费者而言，沼气生产成本降低；
- 因为原料价格受世界市场的影响，不存在降低的趋势，所以原料补贴不递减；
- 玉米和谷物的使用限制在 60%→降低能源作物与食品和其他原料的竞争；
- 不同原料清单内的原料可以混合使用，根据原料使用比例确定补贴发放比例→例如鼓励使用园林剪枝废物；
- 底线规则的引入：每个沼气工程必须使用 60%的废热或者至少使用 60%粪污或者直接出售生产的电能→额外收益在电力生产范围之外；
- 为以市场为导向的沼气生产提供灵活的补贴政策；
- 为直接出售所产电力提供支持（市场补贴）：最晚到 2014 年，所有装机容量超过 750 kW 的沼气工程必须向市场直接出售所产的电能。

表 2　2012 年德国可再生能源法（沼气）上网补贴系统

	上网补贴					
	沼气工程（不含生物废弃物沼气工程）				生物废弃物沼气工程①	小型粪污沼气工程
装机容量（kW_{el}）	基础补贴	Ⅰ类原料	Ⅱ类原料②	沼气提纯/生物甲烷补贴（€-ct/kW_{el}）		
≤75	14.3					25③
≤150		6	8	≤ 700 Nm^3/h：3	16	
≤ 500	12.3			≤1 000 Nm^3/h：2		
≤ 750	11	5	8/6④	≤1 400 Nm^3/h：1		
≤ 5 000	11	4			14	
≤ 20 000	6	—		—		

①仅用于明确使用生物废弃物的沼气工程，对于发酵后产物进行堆肥处理。稳定处理后的发酵剩余物必须被原料化应用（肥料）。补贴仅与生物甲烷生产补贴结合使用。

②仅对于根据定义选出的，环境友好的原料。

③针对粪污沼气工程且最小装机容量小于 75 kW 的，不与其他补贴合用。

④＞ 500 kW 的粪污发电厂，6 €-ct/kWh。

ECONOMIC ANALYSIS OF DRY FERMENTATION OF BIODEGRADABLE MUNICIPAL WASTE WITH GARAGE TYPE TECHNOLOGY

LU Hongyan[1], YAN Jiong[1], P. Heck[2], M. Blim[2], M. Speck[3], F. Baur[3]

1. Sichuan University; 2. Institute for Applied Material Flow Management (IfaS), Trier University of Applied Sciences; 3. Institute for Future Energy Systems (IZES)

Contact: Dr. LU Hongyan, Lecturer, Chengdu University, Yihuan Lu Nanyiduan No. 24, 610065 Chengdu, Sichuan Province, PR China, Mobile: +86-15388101387, E-Mail: redbird66@gmail.com;
Prof. Dr. Peter Heck, Institute for Applied Material Flow Management (IfaS), 55768 Hoppstaedten, Germany, Tel.: +49-6782-17, E-Mail: p.heck@umwelt-campus.de;
Mike Speck, Institute for Future Energy Systems (IZES), 66115 Saarbruecken, Germany, E-Mail: speck@izes.de

EXECUTIVE SUMMARY

The bio-decayable organic component is about 60% of municipal solid waste in China. Energy and material recovering from source-separated organic waste contributes to the reduction of green house gas emissions, environmental protection, saving land resource and creating jobs. It therefore helps to promote a sustainable growth of Chinese economy. This article aims to make an economic analysis for such projects. It takes an example for treating 30 000 t/a of source-separated organic waste in China, based on the garage dry fermentation technology. The calculating results show when organic fertilizer price and electricity price to power grid are 20 €/t and 0.06 €/kWh, the disposal fee for recycling organic waste should be not less than 5.16 €/t, in order to get this example project a profit and loss balance or a surplus.

Key words: Municipal solid waste; biogas; dry fermentation; organic waste; economic analysis; waste disposal fee

车库式干法发酵处理城市有机垃圾的经济分析

卢红雁[1*]，颜炯[1]，P. Heck[2]，M. Blim[2]，M. Speck[3]，F. Baur[3]

（1.四川大学，成都 610065；2.特里尔应用技术大学，德国物质流管理研究所 IfaS，Hoppstaedten 55768；3.德国未来能源研究所 IZES，Saarbruecken 66115）

【摘要】中国城市固体废弃物中有机质生物可降解部分约占 60%。有机垃圾的源头单独收集和资源化利用能减少温室气体排放，保护环境，节约土地资源和创造就业，有利于中国经济的可持续增长。为此，本文分析年处理 3×10^4 t 城市有机生活垃圾干法厌氧发酵示范项目的经济可行性。根据示范项目的投资、运行成本及发电和有机肥收益，计算出在不同电价和有机肥价格水平下，为了保证项目盈亏平衡所需要的垃圾处理费。经济分析表明有机肥价格对示范项目的经济可行性影响较大。当有机肥价格 20 €/t，并网电价 0.06 €/kWh 时，有机垃圾处理费要达到 5.16 €/t，示范项目才实现盈亏平衡。

【关键词】城市固体废弃物；沼气；干法发酵；有机垃圾；经济分析；垃圾处理费

* 通信作者：卢红雁，女，博士，讲师，四川大学，成都市一环路南一段 24 号 610065，redbird66@gmail.com，15388101387

1 背景

中国经济快速增长的同时也伴随着城市化进程的加剧和城市人口的增加。由此也使得中国城市生活垃圾管理面临日益严峻的挑战，如垃圾量增加快、无害化处置率低、新建填埋场和焚烧场选址难和资源化利用率低等问题。2009 年中国 600 多个城市共清运的城市生活垃圾总量约为 1.57×10^8 t[1]。如图 1 所示，尽管中国城市生活垃圾的无害处理率从 2006 年的 53%提高到 2009 年的 71%，但是处置方法仍然以填埋为主（56.6%），此外，焚烧 12.9%、堆肥 1.1%。我国城市生活垃圾中的有机垃圾量接近 60%[2]，而北京地区的厨余垃圾占垃圾总量的 63%[3]。有机垃圾进行填埋处理不仅占用大量土地，而且也是巨大的资源潜力浪费。每吨有机生活垃圾的生产沼气的潜力是 80～120 m^3 沼气，沼气中甲烷气体的比例可达 58%～65%[4]。

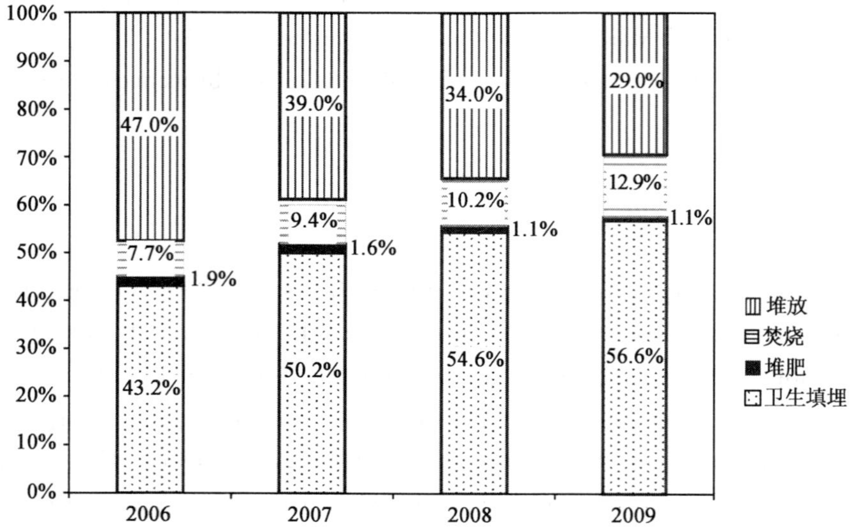

数据来源：中国统计年鉴 2010

Figure 1 Disposal methods of municipal solid waste in China

图 1 中国城市生活垃圾的处置方法

鉴于中国城市生活垃圾中大量有机物资源没有得到有效利用的现状，德国政府和中国政府在中德科技合作框架下执行了中国城镇、农业和养殖业有机垃圾资源化研究的课题，课题执行期从 2008 到 2012 年①。本文根据课题的研究结果，分析在中国建立有机垃圾厌氧处理示范项目的经济可行性。示范项目采用车库式干法发酵工艺，每年处理 30 000 t 源头单独收集的有机垃圾（包括果蔬批发市场的有机垃圾、中药厂药渣、餐厨垃圾和绿化垃圾等）。有机垃圾干法厌氧发酵的工艺类型很多，如瑞士 KOMPOGAS 公司的卧式连续发酵工艺，比利时 OWS-DRANCO 的立式连续发酵工艺和德国 BEKON 的车库式发酵工艺。它们各有优缺点，需要根据项目的实际需要选择工艺。本文主要对 BEKON 公司车库式干法发酵工艺在中国的应用进行经济分析。车库式干法发酵优点是核心设施本土化生产比例较大，因此投资相对较低。此外，该工艺过程产生的废水量较少。但它占地较大、人工需求量较高、异味控制的措施更严格。该工艺还要求发酵原料有一定的空隙率且含水率不宜过高。

2 计算方法

有机垃圾干法厌氧发酵项目的经济分析与传统的投资性项目不同，垃圾资源化利用是外部性很高的市政项目。市政垃圾项目的重要经济指标不是投资回报率，而是每吨垃圾处理费该是多少才能确保垃圾项目

①网址 www.organicresidues.org

的可持续运行。因此，本文以车库式干法发酵年处理 30 000 t 源头单独收集的有机垃圾(包括果蔬批发市场的有机垃圾、中药厂药渣、餐厨垃圾和绿化垃圾等)为例，根据中国的市场情况，计算在不同的电价和有机肥价格水平下，为了保证示范项目盈亏平衡所需要的垃圾处理费水平。为了确定该项目的并网电价和有机肥价格与垃圾处理费的关系，经济可行性分析的投资及运行成本是基于德国 Beckon 公司提供的参数，再结合本土化建造和运行的实际情况折算而来。经济分析的货币单位是€①，计算公式如下：

每吨垃圾处理费＝(年成本－年收益)/年处理量

总成本和总收益所包括的具体内容见表 1、表 2 和表 3。年成本的可变性不大，但是年收益受上网电价和有机肥市场影响很大。因此，为了确定该项目的并网电价和有机肥价格与垃圾处理费的关系，文本分两种情况讨论示范项目实现盈亏平衡时垃圾处理费的征收水平。第一种情况是电价不变，有机肥价格变化对垃圾处理费的影响。第二种情况是有机肥价格不变，而电价变化对垃圾处理费的影响。

另外，根据清洁发展机制(CDM)的标准计算方法，本示范项目每年的二氧化碳减排量超过 7 000 t。但是由于 CDM 项目在未来执行办法的不确定性，因此，在经济分析时没有考虑二氧化碳在清洁发展机制框架内碳交易收益。

3 车库式干法发酵的流程和主要设施

概述：车库式干法发酵工艺来自德国 BEKON 能源技术有限责任公司。铲车将有机垃圾送入发酵仓，关闭气密仓门。发酵期间不进行搅拌，不添加填发酵物。从发酵物中渗透下来的渗滤液由排水设施抽出，收集在一个容器中，再喷洒回发酵物中使之保持湿润。发酵过程的温度保持在 34～37℃之间，在仓壁和仓底有供暖设施。产生的沼气通过热电联产机组转化为电能与热能。为了保证热电联产机组的沼气使用量，在干燥仓储备中有多个发酵仓交替工作。在完成发酵过程后，发酵仓会被清空并重新填充。发酵过的发酵物被加工成有机肥产品。

收运区：工厂入口配备机动车地磅，对有机垃圾、发酵残余物和有机肥称重、登记和监控。建立一个完整的内部和外部的物流数量和费用结算表(垃圾处理量、处理费、肥料等)。秤重和结算程序用计算机系统自动化管理。

储料仓：能容纳平均一星期的有机垃圾量。保证设备遇到故障时也能继续接收垃圾，或者接纳的垃圾数量较少或没有运送垃圾时(例如节假日)，也能使设备持续运行并保质保量。为了防止异味，储料仓为密封设计。储料仓废气利用低压收集并经过生物过滤网后外排。储料仓的大部分埋在地下，只有一小部分露在外边，方便填料。

物料混合区：待处理的有机垃圾与已发酵过的物料进行混合，筛选出体积较大的干扰物质。已发酵物与未发酵物的比例大约在 1∶1。

发酵仓：混合后物料将作为待发酵原料送往车库式发酵仓内，密封仓门。确保在任何时间与任何情况下，杜绝气体泄漏以及由于外界空气同沼气混合而导致爆炸的危险。发酵仓内有供热和温度调节装置，渗滤液从上向下喷淋，湿润发酵原料，渗滤液在底部收集后运到渗滤液储存罐。发酵仓的建材是密封性很好的水泥。发酵仓的设计要适合铲车与垃圾车行驶。多个发酵仓并排而建，交替运行。在装填和清空发酵仓时，需要采取特殊的安全措施来避免甲烷和空气混合而导致的爆炸危险。此外，运行通风设备对发酵仓进行换气通风，使铲车司机不会受到发酵仓气味干扰。在发酵仓中位于发酵物上方的空间作为储存沼气的地方。经过输气控制管道与增压设备将沼气输送至热电联产机组。热电联产机组对沼气实时转换成电和热，这样整个工艺设备里就不需要额外的沼气收集和储存装置。

自控系统：各发酵仓的主要控制参数可以进行逐一设置，如渗滤液循环控制、沼气输送、换气和温度等。通过控制参数的监控可以实现整个工序的最优化并保证发酵仓实现最大的工作效率。

①示范项目经济分析的货币单位是欧元，欧元与人民币的兑换比率按 1∶10 计算。

堆肥区：发酵后的生物质由铲车运出，清空发酵仓。不必对所剩物质进行固液分离，取而代之的是直接把已发酵的生物质堆肥，加工成有机肥。堆肥工艺结束后的有机肥产品进行分类和评级，即可进入市场销售。有机肥产品仓储区需要能够储存1个月的产品量。

用地需要：年处理 3×10^4 t 有机生活垃圾的干法发酵示范项目，需要12个发酵仓。总占地19 000～20 000 m^2，其中储料仓约占地640 m^2、好氧堆肥区约占地8 000 m^2。整个工厂布局中还包括办公室、更衣室、休息室和停车场等区域。

4 示范项目的输入和产出

示范项目拟处理 3×10^4 t 源头单独收集的有机生活垃圾、园林废弃物和果蔬市场的有机垃圾。车库式干法发酵过程中因为没有搅拌，所以发酵原料能容忍一定的杂质。源头单独收集的有机垃圾估计约有2%的干扰物质需要由铲车或人工拣出。项目运行还需消耗自来水2 000 t/a和柴油 1.5×10^4 L/a。

厌氧发酵与好氧堆肥两个工艺将43%的垃圾重量转化为沼气、水蒸气和渗滤液等。55%的垃圾重量留存在有机肥产品里。3×10^4 t 源头单独收集的有机生活垃圾、园林和果蔬市场的有机垃圾产生沼气约 3.19×10^6 m^3，相当于 1.914×10^7 kWh 的能量。示范项目所选热电联产机组的电能转换效率是36%，热能转换效率44%，能量损失20%。沼气在热电联产机组中被转化为电力与热能。部分电能可供自用，其余的将输送至电网以供使用。产生的热能中只有一部分用于发酵仓保温，大部分的热能可供给热能用户。扣除项目自身电能和热能消耗，项目产生净电能为 5.51×10^6 kWh，净热能 6.82×10^6 kWh。此外，3×10^4 t 有机垃圾可生产约 1.54×10^4 t 有机肥产品，废水1 000 t/a。

沼气除了可以在热电联产机组中使用外，还可以通过其他的途径加以利用：经过提纯净化处理可以作为机动车燃料或者直接输送至天然气管道。本文主要对沼气的热电联产应用进行经济分析。当示范项目的规模扩大，可以扩建沼气提纯制取生物燃气部分。现有的热电联产机组可以继续保留，为项目提供热能需要。

5 经济分析

5.1 投资分析

车库式干法发酵示范项目（每年处理 3×10^4 t 源头分类收集的有机垃圾）总投资约4 253 900 €。投资估算详见表1。因示范项目是市政工程，因此投资成本里没有考虑用地成本。

5.2 成本分析

每年处理 3×10^4 t 源头分类收集的有机垃圾的干法发酵示范项目的年运行成本为853 866 €，折合成每吨有机垃圾的处理成本是28.46 €。成本分析里包括了每年6%的资本利息但没有包括有机垃圾收运成本。估算内容详见表2。

Table 1 Investment analysis of the pilot project

表1 示范项目的投资分析

投资	金额(€)
干法发酵和热电联产单元	2 100 000
地表和水泥作业	900 000
机械预处理和堆肥设备	500 000
运输车辆（铲车和物料筛分车）	250 000
参观、培训中心	250 000
其他费用（设计、审批、启动、培训和建设期资金利息等）	253 900
总计	4 253 900

5.3 收入分析

干法发酵示范项目（每年处理 3×10^4 t 源头分类收集的有机垃圾）的年收入约 75.86×10^4€。折合每吨收入25.29 €。成本分析中示范项目自需电能与热能已经从项目产出的能源中扣除，因此并网电量为净电量①。收入分析中，有机肥价格按20 €/t、并网电价按0.06 €/t、垃圾处理费按4 €/t进行估算。

①如果沼气发电厂的上网电价高于沼气发电厂从电网采购电力的价格，则可以选择将沼气厂的全部电力销售给电网，然后再从电网上采购工厂所需电能。

Table 2 Cost analysis of the pilot project

表 2 示范项目成本分析

成本分析	年成本(€)	折合每吨的成本(€/t)①
管理费	150 000	5.00
设备、设施维修费	53 500	1.78
人工工资	60 000	2.00
自来水及污水处理	320	0.01
柴油消耗	11 700	0.39
有机垃圾里剔除杂质的填埋费	10 000	0.33
项目保险(1%投资)	40 000	1.33
折旧(20 a)	212 695	7.17
环境教育和能力建设	55 000	1.83
资本利息 (总投资额的利息 6%,20 a)	255 234	8.51
成本总计	853 866	28.46

Table 3 Estimated incomes of the pilot project

表 3 示范项目收入分析

收入来源	总价	单价
有机肥	15 400 t/a	20.00 €/t
并网电量(净输出)	5 510 000 kWh/a	0.060 €/kWh
减排 CO_2 当量.	7 648 t/a	0.00
垃圾处理	30 000 t/a	4.00 €/t
收入		折合处理每吨垃圾的收入(€)
有机肥	308 000	10.27
并网电量	330 600	11.02
垃圾处理费	120 000	4.00
合计	758 600	25.29

综合车库式干法发酵示范项目(年处理 3×10^4 t 源头分类收集的有机垃圾)在表 1、表 2 和表 3 的投资、成本及收入分析,示范项目将处于亏本运行的状态。因为处理每吨有机垃圾的收入比处理成本低 3.18 €。但是这并不说明项目不可行。我们还需要进一步分析垃圾处理成本与有机肥和电价之间动态关系,从而找到示范项目实现盈亏平衡机会。我们知道各国政府及民众对气候保护、环境安全的要求都在日益提高,而中国城市垃圾的处理需求也在不断增加,垃圾填埋场的选址越来越远离城市,其征地成本、运输成本及运行费用也在增高。下面将主要分析垃圾处理费涨到什么水平时,有机垃圾干法发酵项目才能盈亏平衡。

5.4 垃圾处理费的征收水平

示范项目不但生产清洁能源,也产生有机肥促进改良土壤和提高农产品质量。这是不是意味着车库式干法发酵系统的实施会大大增加政府或居民需要支付的垃圾处理费?这里分两种情况讨论示范项目实现盈亏平衡时垃圾处理费的征收水平。第一种情况是电价不变,有机肥价格变化对垃圾处理费的影响。第二种情况是有机肥价格不变,而电价变化对垃圾处理费的影响。示范项目经济可行性分析的主要目的就是要明确在何种情况下需要征收垃圾处理费以及征收多少。

5.4.1 有机肥价格对垃圾处理费的影响

为了分析有机肥价格对垃圾处理费的影响,我们设定电力并网收入不变,有机肥产量保持在15 400 t/a,

①有机垃圾处置的吨成本 = 年成本/年处理量

有机肥价格在－4～60 €/t 之间波动，有机肥价格为负值表示有机肥产品没有卖掉，需要支付费用进行填埋处理。示范项目的有机肥收益随着有机肥价格的变化有很大变化。因此，为了实现示范项目盈亏平衡，垃圾处理费随着有机肥价格而波动，有机肥价格提高，征收的垃圾处理费降低，有机肥价格降低，则垃圾处理费要提高。图 2 展示了二者之间的量化对应关系，不同的有机肥价格，对应了一个需要征收垃圾处理费的值。当垃圾处理费为负值时，说明当有机肥价格达到某水平时，即使不征收垃圾处理费，示范项目也有盈余。

如果垃圾处理费维持目前 4 €/t 水平，那么有机肥价格最少要达到 25 €/t，示范项目才能达到盈亏平衡。当有机肥产品完全没有市场，而且还需要支付 4 €/t 来填埋处理时，垃圾处理费高达 17.48 €/t，示范项目才能达到盈亏平衡。当有机肥价格超过 30 €/t，垃圾处理费为零也能实现盈亏平衡。若有机肥价格达到市场预期 60 €/t 时，处理每吨源头分类收集的有机垃圾将获得 15.37 €/t 的税前收益。

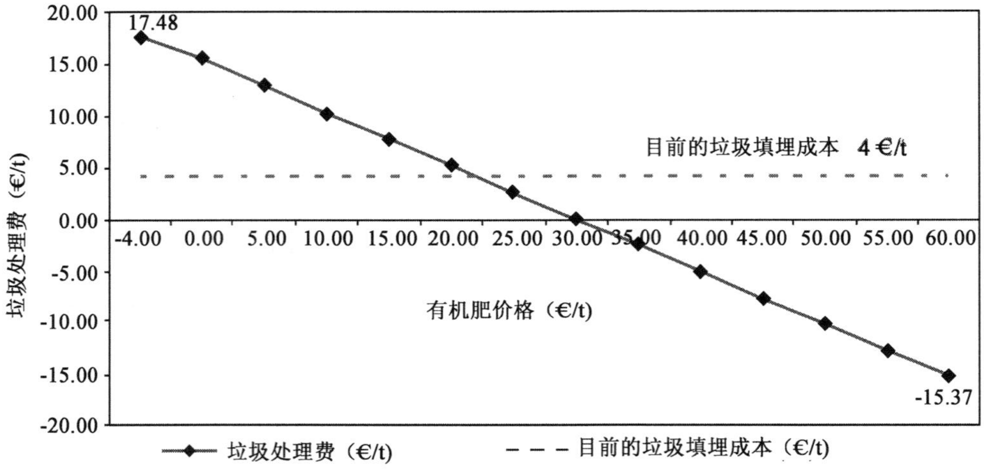

Figure 2 The impact of organic fertilizer price on waste disposal fee in the pilot project

图 2 示范项目有机肥价格对垃圾处理费的影响

5.4.2 上网电价对垃圾处理费的影响

2006 年中国政府出台了《可再生能源法》。对于沼气发电这类生物质能项目的发电补助是在当地脱硫标杆价电价的基础上每度电再加 0.25 元人民币。由于地方政府的经济水平及垃圾处理压力的不同，示范项目可能获得的上网电价可能有较大的地区差异。此外能源价格的不断上涨，也将推动上网电价。为了分析上网电价对垃圾处理费的影响，我们设定有机肥产量 15 400 t/a 和价格 20 €/t 保持不变，但上网电价变化范围为 0.06～0.10 €/kWh。示范项目并网电量(净输出)5 510 000 kWh/a。为了确保示范项目实行盈亏平衡，不同并网电价对应一个需要征收垃圾处理费的值，计算结果如图 3 所示。当垃圾处理费为负值时，说明当并网电价达到某个水平时，示范项目即使不征收垃圾处理费，项目也有盈余。

图 3 显示，当上网电价达到 0.067 €/kWh，有机肥价格 20 €/t，示范项目维持目前垃圾处理费水平(4 €/t)，就可以保持项目盈亏平衡。如果上网电价仍然是 0.06 €/kWh，那么为了维持示范项目盈亏平衡，垃圾处理费将提高至 5.16 €/t。当上网电价高于 0.088 €/kWh，即使垃圾处理费为零，示范项目也将能盈亏平衡。如果上网电价能实现 0.01 €/kWh，则示范项目处理每吨垃圾可产生 2.17 €的税前收入。

6 结论与建议

6.1 结论

本文基于德国车库式干法发酵工艺，分析中国建设一个示范项目的经济可行性。该示范项目年处理 3×10^4 t 源头单独的有机垃圾示范项目。项目投资 4 253 900 €，年运行成本 853 866 €，年运行成本折合为处理每吨有机垃圾的成本是 28.46 €。项目收益来自并网售电，有机肥销售和垃圾处理费。示范项目的并网电量为 5 510 000 kWh/a 的净输出。有机肥产量是 15 400 t/a。如果上网电价为 0.06 €/kWh 和有机肥

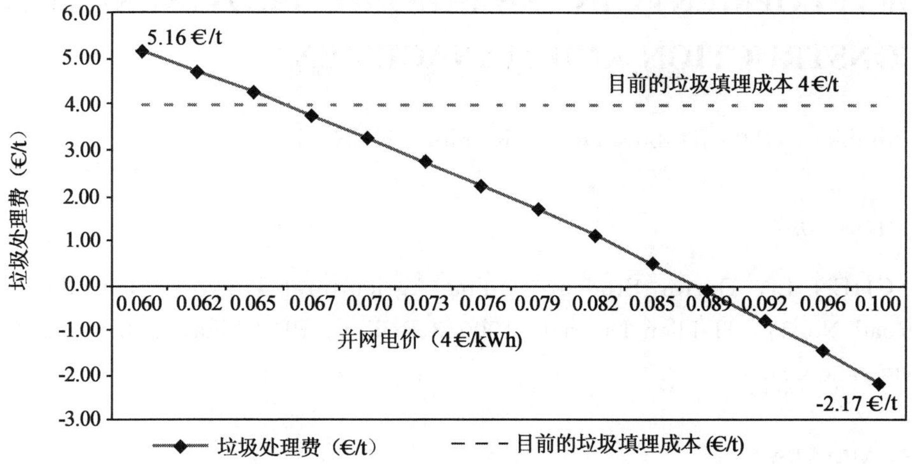

Figure 3 The impact of electricity price to grid on waste disposal fee in the pilot project

图 3 并网电价对垃圾处理费的影响

售价是 20 €/t 时，那么每吨有机垃圾的处理费需要是 5.16 €/t，这样才能维持示范项目的盈亏平衡。当并网电价、有机肥价格或垃圾处理高于上述水平时，示范项目将有盈余。例如，上网电价维持 0.06 €/kWh 水平，当有机肥价格达到 30 €/t，垃圾处理费为零也能实现盈亏平衡。有机肥价格为 60 €/t，处理每吨有机垃圾的税前收益约 15 €。

6.2 建议

中国城市生活垃圾的可持续管理是促进中国经济可持续增长的一个机会。而有机垃圾如何管理影响重大。本文对有机垃圾资源化进行经济分析。有机垃圾送入干法厌氧发酵厂，产生沼气并发电，产生的沼渣加工成高品质的有机肥。计算结果表明有机肥价格对项目的经济可行性有很大影响。例如，当示范项目的有机肥价格低于 10 €/t 时，征收的垃圾处置费需要高于 10.29 €/t，才能确保示范项目实现盈亏平衡。反之，如果示范项目的有机肥品质很好，能到达市场预期的销售价格 60 €/t 时，则示范项目处置每吨有机垃圾将获得 15.37 €的税前收入。有机肥价格的一个决定因素是有机肥品质。为了提高有机肥品质，就必须减少或避免有机垃圾在存储和收运环节中被污染，因此有机垃圾不宜与其他垃圾混合投放和混合储运。所以有机垃圾的源头单独投放及收运系统是有机垃圾资源化和产业化的一个重要前提。

项目及基金支持

感谢中德合作“中国城镇、农业和养殖业有机垃圾资源化”项目（德国教育与研究部 BMBF 0330847A 和中国科技部 MOST 2010DFA94350）以及四川省科技厅 2010Z20001 和四川大学青年基金 2009SCU11053，2010SCU11058 的支持。

参考文献

[1] 中华人民共和国国家统计局. 中国统计年鉴 2010. 北京：中国统计出版社，2010.

[2] Zhang D. Q., Tan S. K., Gersberg R. M. Municipal solid waste management in China: status, problems and challenges[J]. Environmental Management, 2010, 91(8): 1623-1633.

[3] Li Zhenshan. Municipal solid waste management in Beijing City[J]. Waste Management, 2009, 29: 2596-2599.

[4] Institut für Energetik und Umwelt GmbH, Bundesforschungsanstalt für Landwirtschaft, Kuratorium für Technik und Bauwesen in der Landwirtschaft e. V. Handreichung Biogasgewinnung und -nutzung [J]. Gülzow Fachagentur Nachwachsende Rohstoffe e. V. (FNR), 2009, 95.

BIOGAS DEVELOPMENT IN SHANDONG PROVINCE: INVESTMENT, DESIGN, CONSTRUCTION AND MANAGEMENT

XIE Jing, WU Shubiao, GUO Jianbin, DONG Renjie, CHEN Li

China Agricultural University

Contact: Mrs. CHEN Li, Doctor Student, China Agricultural University, College of Engineering, Qinghua East Road No. 17, Haidian District, 100083 Beijing, PR China, Tel.: +86-10-6273-7885, E-Mail: chenli329@yahoo.cn

EXECUTIVE SUMMARY

A case study of biogas plants in Shandong province was conducted in three cities in 2009. Construction information, digestion process parameters of the biogas plants, the production and use of biogas and biogas digester, plant operation and management, as well as the surrounding environment were investigated and assessed. The analysis results show that the government contributed to a very high degree to the overall project investment and construction costs. Generally, the single fermenter volume is mainly between 100 m^3 and 500 m^3, the actual amount of gas production is lower than the designed amount and is further generally lower during the winter. Most of the biogas plants were built on livestock and poultry farms and treat related wastes. Coal was used to heat the plants in winter which resulted in increased costs and environmental pollution. CSTR and USR are the main types of digesters. Biogas and its digesters capacity haven't been used sufficiently. All biogas plants are of poor technical operation and low economic benefits.

Key words: Biogas plant; Shandong; operation; economic benefits

山东省沼气工程发展调研报告:投资、设计、建设与运行

谢晶,吴树彪,郭建斌,董仁杰,陈理*

(中国农业大学能源工程与低碳技术研究室,北京 100083)

【摘要】调研小组于 2009 年底对山东省三个城市的大中型沼气工程的发展情况进行了抽样调研。此次调研内容包括沼气工程的建设基本概况;发酵过程相关工艺参数;“三沼”的生产和利用状况;工程运行管理及周边环境的调查。调查结果表明:工程投资与建设,政府在其中扮演了很重要的角色,部分工程政府起主导地位;单体容积大多为 100～500 m^3,实际日产气量小于设计产气量,冬季产气率低;沼气工程类型单一,大部分依托畜禽养殖厂建设,处理畜禽废弃物依然是建设沼气工程的主要目的;冬季加热方式多采用燃煤加热,增加了工程的运行成本而且同样的污染了环境;厌氧消化工艺以技术较成熟的 CSTR 和 USR 工艺为主;“三沼”没有充分利用。工程整体技术含量低、经济效益差。

【关键词】沼气工程;山东;运行;经济效益

* 通信作者:陈理,女,硕士研究生,副教授,中国农业大学工学工学院能源工程与低碳技术研究室,北京市海淀区清华东路 17 号 100083, chenli329@yahoo.cn, 010-62737885

1 引言

自20世纪农业部提出“生态家园富民计划”，并在全国范围内组织实施示范工程建设以来，以沼气为纽带的各种类型能源生态模式和工程技术得到了政府的大力支持。中央每年投入大量专项资金予以支持。截止到2009年年底，仅中央财政对农村沼气建设的投入已经高达1.9×10^{10}多元，使得21世纪前10年成为我国沼气发展最快的历史时期[1]。

尽管如此，沼气工程的建设仍处于政府推广扶持阶段，中国的沼气工程将如何健康独立的发展下去，仍是一个亟需解决的难题。目前，我国的沼气发展到什么阶段，是什么影响了中国沼气的发展，从哪些方面改变可以扭转现在沼气工程亏损的局面，带着这些疑问我们对山东省三市的沼气工程进行了抽样调研。2011年，山东省GDP总量位居全国第三，经济发展迅速，是中国重工业基地。山东省政府很重视可再生能源的发展，从2006年开始，省财政增加专项资金用于扶持农村沼气等可再生能源建设[2]，大力发展省内的可再生能源事业。综合以上优势选取山东省Z、Y、Q三个具代表性的城市对山东省沼气发展状况进行了抽样调研。

2 调研情况

2.1 沼气工程基本情况

三个城市均为全国的重要经济发展基地，在带来高GDP的同时也伴随着严重的环境污染。通过调研了解到这三个城市为治理环境污染，改善生态环境，政府都鼓励发展沼气工程，设置专项资金大力促进沼气工程的推广建设工作。

由表1可以看出在工程设计、建设及监管的各个环节都可以看到政府的身影，尤其在城市Y中政府则包揽了工程的设计施工和监督的50%以上的工作。

Table 1 Fundamental situation of biogas plant

表1 沼气工程基本情况

	所有单位		设计单位		施工单位		监督单位	
	企业	政府	企业	政府	企业	政府	企业	政府
城市Z(%)	75	25	81	19	81	19	34	56
城市Y(%)	90	10	35	65	35	65	20	80
城市Q(%)	100	0	92	8	92	8	0	100

结合图1和图2可以看出城市Y抽样调查的沼气工程的单体容积有70%在100～500 m^3，同时图5显示该市工程的厌氧消化工艺多是常规沼气池，此工艺所需建设资金较少，故65%的建设资金是自筹。城市Q和城市Z的沼气工程半数以上的工程是超大型，需要大量的建设资金，所以三方投资(国家、地方政府及企业共同投资)为主要投资模式。

2.2 沼气工程建设目的

沼气工程能源生态型与能源环保型最主要的区别是厌氧消化液的利用方式。前者周围有足够的农田来消纳厌氧发酵后的沼液、沼渣；后者是将厌氧消化液处理后达标排放或以回用为最终目标的处理工艺。后者的工艺处理单元效率高，工程规范化、管理及操作自动化水平高，沼气回收与污水达标、环境治理结合得较好，但工程投资较高，一般的养殖企业无法同时要求工程管理人员有较高技术水平；相对而言，能源生态型工程投资少、运行成本低、投资回收期较短、可获得沼气、工艺简单、管理和操作方便，其缺点也较明显：工艺处理单元的效率不高，处理后的污染物浓度仍很高。能源环保型的温室气体的减排效果也要优于能源生态型[3]。在中国以能源生态为目的建设的工程居多。

此次的调研也恰好反映了这一现象(见图3)，城市Q调研的13座沼气工程均是能源生态型，城市Z与城市Y能源环保型的工程的比例也较小，采用此种类型的均是大型的企业，所需资金投入较大。其中，由于

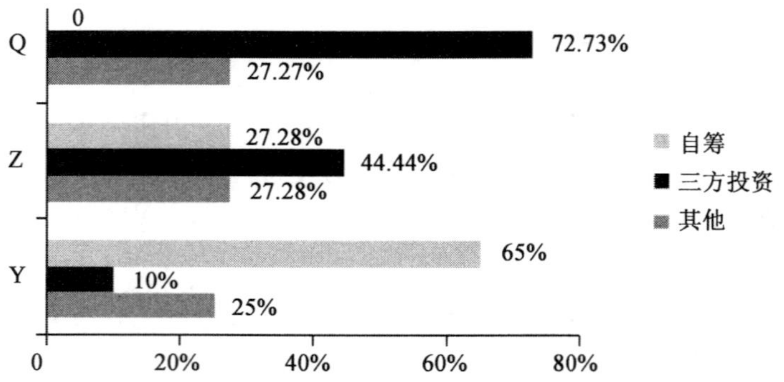

Figure 1 Investment proportion

图 1 投资比例图

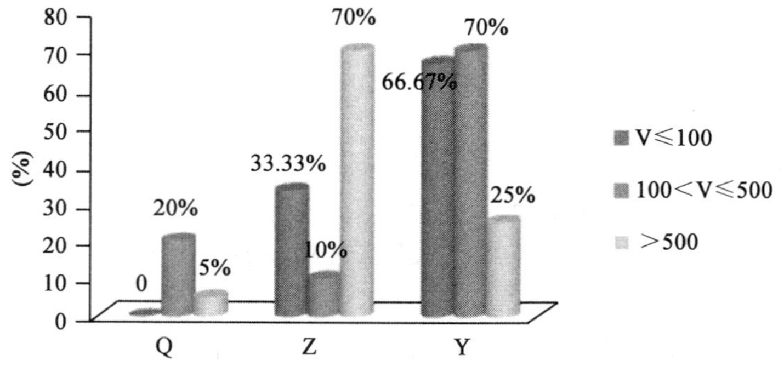

Figure 2 Monomer volume

图 2 单体容积图

工程的周边没有足够的农田消纳，抽样调查的能源生态型的沼气工程的沼渣沼液没有实现百分之百的利用，这也造成了另一种的环境污染和资源浪费。

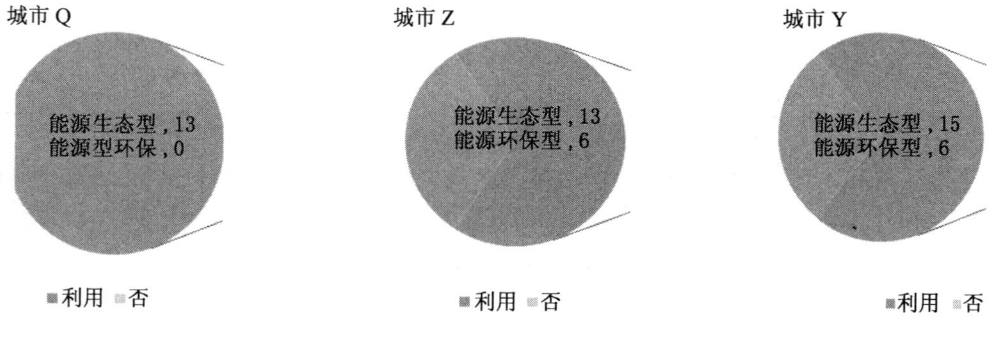

Figure 3 Construction objective of biogas plant

图 3 工程目的

2.3 工程原料

三市的发酵原料都是以畜禽废弃物为主，这也反映了我国沼气工程的特点：沼气工程多是依托畜禽养殖场建设，处理养殖场粪污依然是中国建设大中型沼气工程的主要目的。厌氧消化工艺主要采用的是 CSTR 和 USR，城市 Y 的情况比较特殊，由于城市 Y 发展沼气较早，现在使用中的沼气工程多是以前的户用沼气池的扩大和改装版，所以多是常规沼气池。详情见图 4 和图 5。

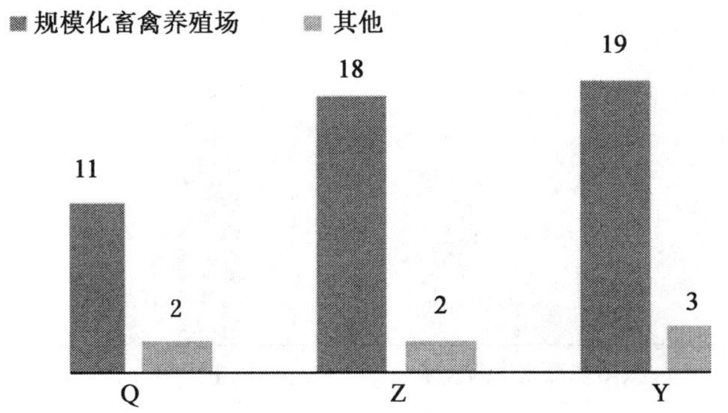

Figure 4　Type of biogas plant

图 4　沼气工程类型

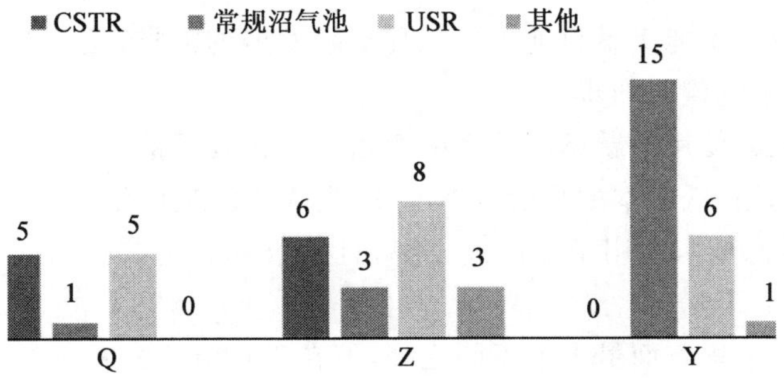

Figure 5　Anaerobic digestion technology

图 5　厌氧消化工艺

2.4　冬季保温方式

温度是影响沼气发酵产气率的关键因素之一，由于山东所处地理位置在冬季采取适当的保温措施才能使发酵工程常年稳定运行，保持恒定、高效的产气量[4]。沼气工程冬季加热会增加运行成本，故采用经济高效的加热升温方式是影响大中型沼气工程发展的关键。沼气池的加热方式主要包括电热膜加热、太阳能加热、化石能源热水锅炉加热、沼气锅炉加热、沼气发电余热利用和燃池式加热等 6 种[5]。当今国内外多采用热水循环加热法对发酵罐保温，目前实际应用较广的加热热水的方法主要有沼气锅炉加热、太阳能加热和沼气发电余热利用加热 3 种方式[6-7]。这三种加热方式无论从经济方面还是环保方面都优于化石能源热水锅炉加热。沼气锅炉加热、太阳能加热和沼气发电余热利用加热 3 种方式的投资比为 2∶11.2∶1，沼气发电余热利用加热方式的费用年值仅为沼气锅炉加热和太阳能加热方式的 60%和 12%[8]；沼气发电余热利用加热方式无论从经济效益、技术性能，还是适应性、运行持久性等方面都明显优于太阳能加热和沼气锅炉加热。因此，基于上述优点，沼气工程宜选择利用沼气发电余热加热的方式。

但根据各市的冬季加热方式对比图 6 显示，除城市近一半的沼气工程采用沼气发电余热保温外，其余的沼气工程多数仍采用燃煤加热。由于温度较低冬季沼气工程本身产气率低，加上仍需花费大量资金购买燃煤用于保温，使冬季工程的收益大幅下降，甚至多数工程在冬季是亏损运行。

2.5　“三沼”利用情况

相比于欧洲沼气用于发电并网，中国的沼气主要用于集中供气。其中的原因与国家的政策导向和经济发展水平密不可分。在德国国会对《可再生能源优先法》进行了修订，使小型农场沼气发电上网更有吸引力。除了上网电价优惠外，对装机低于 70 kW 的沼气工程还可获得 15 000 欧元的补助金以及低息贷款同时简化入网的繁琐手续[9]；相比之下中国虽然相继颁布了《农业法》、《节约能源法》、《可再生能源法》和《畜牧法》，其中都明确提到了要加强农村沼气建设，并将农村沼气建设列入《中国 21 世纪议程》、《国家国民经济和社会发展中长期规划》和《可再生能源中长期发展规划》。农业部相继制定了“十一五”、“十二五”全国农村沼气发展

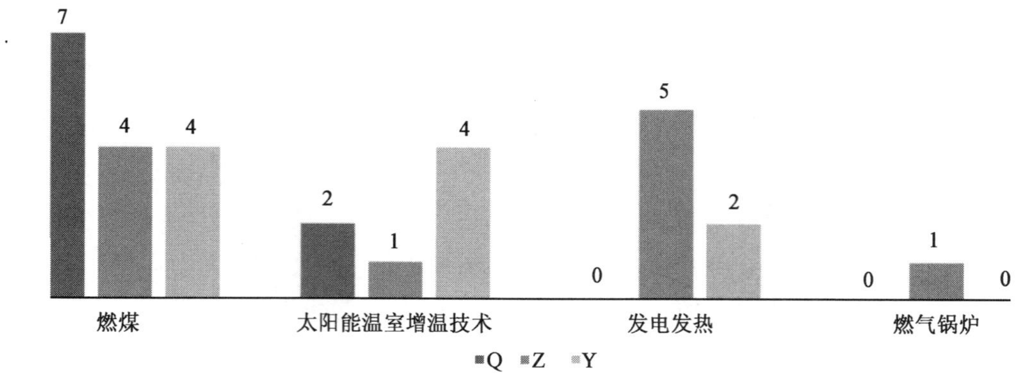

Figure 6 Winter Heating comparison in Winter

图 6 冬季加热方式对比

规划,但是这些政策过于粗泛,不能满足目前沼气工程实际发展的需求,尤其对发电入网这一项的规定不明确,繁琐的入网手续往往让企业望而却步。

沼气工程的厌氧发酵液的使用问题是制约大中型沼气工程发展的另一个重要因素,国内外沼液处理相继出现了低成本的资源性利用、低成本的自然生态净化、高成本的工厂化处理和高附加值的开发性处理等四种方式[10]。前二者属于"生态型",适用于全混式发酵,且要求具有足够消纳的生态单元;后二者属于"环保型",适用于干湿分离发酵。

中国与发达国家在沼气工程方面最大的不同之处,就在于对沼液的后处理:发达国家没有对于沼液进行深度处理,都是经过长期贮存后作为肥料在大田消纳;中国的沼气工程企业很少有足够的土地用来消纳沼液,但如得不到充分合理的处置,其高含量的有机物、氮、磷以及病源性微生物进入环境,将会造成二次污染和资源的浪费。在这方面山东省也没有更好的方法,三个城市的沼渣和沼液能够出售的只占工程的小部分,其余工程则是沼液和沼渣养殖场自用或免费给农户用。

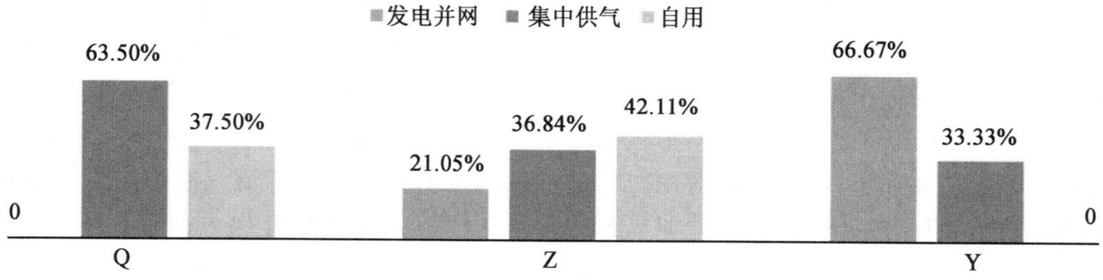

Figure 7 Use patterns of Biogas

图 7 沼气利用

3 问题分析

山东省的沼气工程反映了我国"三沼"利用的基本形式。首先是"自给自足式",目前国内的沼气工程多依附于养殖场,这种形式的养殖场大都规模不大,所以沼气产量也很少。产生的沼气足以满足养殖场内日常的保暖、用电及炊事;走进养殖场同时可以发现周边有一片田地,这片田地就用来消纳厌氧消化液。这种方式的理想形式是形成一个封闭的绿色循环,实现零污染零排放,带来的经济收益无非是节约的电费和购买燃煤的费用,但是面对庞大的前期建设资金,若没有政府的政策扶持。这种模式若想实现收支平衡并不容易。

其次是"供气式",这种模式是国内较常见的。沼气工程的规模大于上面的形式,产生的沼气除满足厂区自用外还可以铺设管道运送到周边的农户用于炊事,具体收费各个地区没有统一的标准。产生的沼渣沼液多数放在储液池中供农户随意取用。此模式铺设输气管道需要大量的资金，管道的具体线路需要政府的规划与协调，所以需要与政府密切合作，同时这种模式也有一个很大的限制因素就是沼气工程周边有密集的居民点，然而根据《沼气工程技术规范》和《畜禽养殖业污染防治技术规范》的规定，养殖场和沼气工程的选址要与居民点保持一定的安全距离这又与“供气式”周边有密集的居民点相矛盾。所以对于规模较小的工程点来说“供气式”的模式是不适用的。

第三种是“发电入网式”，这种模式是现在国际上推崇发展的模式同时也是今后沼气产业的发展的必然趋势。建设此种模式的工程需要大量的建设资金和专业技术人员提供技术支持和对工程的日常运行进行管理。所以国内“发电入网式”的工程很少，多数新建的超大型工程是这种模式，目前国内运行较好的是北京德清源和山东民和牧业，这两个企业也同时申请了 CDM 项目，通过碳交易又增加了一部分的收益。所以国内应建设该模式的工程，建议将几个行政村的畜禽废弃物集中处理建设超大型的工程，工程的日常运行、管理及维护交给专业的技术人员，申请 CDM 项目增加额外收益，测试当地土壤成分，添加一些必要的微量元素将沼渣沼液制成适应当地土壤肥效需求的有机肥。

4 山东沼气工程投资回收分析案例

本调研同时获取了山东某沼气工程的运营数据，以此为例，在不同的假设补充条件下对该沼气工程进行了投资回收期的分析，如图 8 所示，在良好管理的前提下，加大可以投入，扩大盈利面。沼气产业的经营是一条很好的盈利创收之路。

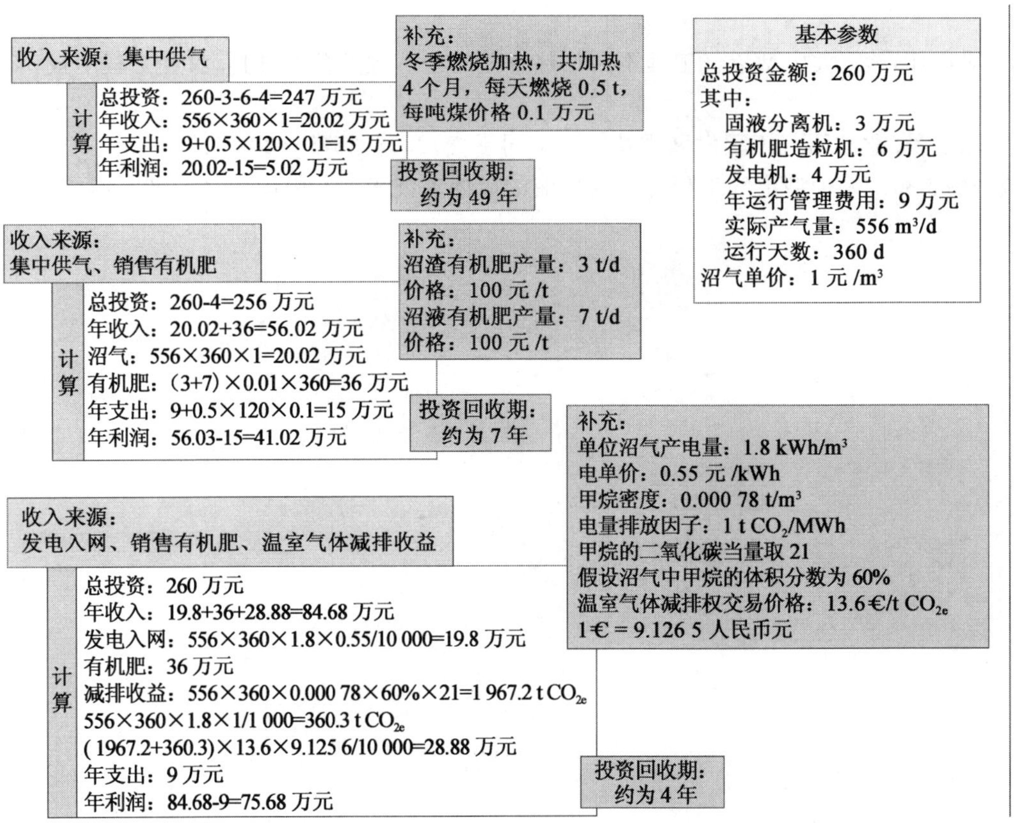

Figure 8 Payback period analysis of different biogas plant patterns

图 8 不同沼气工程模式回收期分析图

经过几十年的发展，虽然我国的沼气工程的各方面还存在着很多的不足，与国外的工程相比还有很大的差距，但这也正显示出我国沼气工程的还有很大地上升空间，随着相关新技术的不断研发，中国的沼气工程产业必将会迎来一个大发展时期。

项目及基金支持

“十一五”国家科技支撑计划“规模化沼气工程沼液、沼渣高值利用技术研究与工程示范”课题(2008BADC4B00);北京市教育委员会科学研究与科研基地建设项目“规模化养殖场粪污双向清洁处理与利用研究”。

参考文献

[1] 李景明，薛梅. 中国沼气产业发展的回顾与展望[J]. 可再生能源,2010,28(3).

[2] 季英德. 山东专项资金扶持可再生资源[J]. 中国环境报,2006,(1).

[3] 王坤，刘慧英，马宗虎,等. 规模猪场能源生态型与能源环保型粪污处理模式温室气体减排量评估[J]. 可再生能源,2009,24(7):55-65.

[4] 樊美婷，刘科，刘建禹. 高寒地区沼气工业化生产配套加热系统的初步设计[J]. 农机化研究,2008(3):105-107.

[5] 石惠娴，王韬，朱洪光，等. 地源热泵式沼气池加温系统[J]. 农业工程学报，2010，26(2):268-273.

[6] A. C. Yiannopoulos, I. D. Manariotisb, C. V. Chrysikopoulosb. Design and analysis of a solar reactor for anaerobic wastewater treatment [J]. Bioresource Technology, 2008(99):7742-7749.

[7] 孙静，郑茂余，吴飞. 严寒地区利用太阳能加热制沼气的试验研究[J]. 可再生能源，2008，26(1):46-49.

[8] 蒲小东，邓良伟，等.大中型沼气工程不同加热方式的经济效益分析.[J].农业工程学报，2010，26(7):281-284.

[9] 邓良伟，陈子爱.欧洲沼气工程发展现状[J].中国沼气,2007,25(5):23-31.

[10] 陈玉成，杨志敏. 大中型沼气工程厌氧发酵液的后处置技术[J].中国沼气,2009,28(1):14-20.